Energy Dissipation and Vibration Control: Modeling, Algorithm and Devices

Special Issue Editors

Gangbing Song
Steve C.S. Cai
Hong-Nan Li

MDPI • Basel • Beijing • Wuhan • Barcelona • Belgrade

MDPI

Special Issue Editors

Gangbing Song
University of Houston
USA

Steve C.S. Cai
Louisiana State University
USA

Hong-Nan Li
Dalian University of Technology
China

Editorial Office
MDPI AG
St. Alban-Anlage 66
Basel, Switzerland

This edition is a reprint of the Special Issue published online in the open access journal *Applied Sciences* (ISSN 2076-3417) from 2016–2017 (available at: http://www.mdpi.com/journal/applsci/special_issues/energy_dissipation).

For citation purposes, cite each article independently as indicated on the article page online and as indicated below:

Lastname, F.M.; Lastname, F.M. Article title. *Journal Name* **Year**, *Article number*, page range.

First Edition 2018

ISBN 978-3-03842-785-8 (Pbk)
ISBN 978-3-03842-786-5 (PDF)

Table of Contents

About the Special Issue Editors   v

Preface to "Energy Dissipation and Vibration Control: Modeling, Algorithm and Devices" . . vii

Gangbing Song, Steve C. S. Cai and Hong-Nan Li
Energy Dissipation and Vibration Control: Modeling, Algorithm, and Devices
doi: 10.3390/app7080801   1

Jae-Do Kang and Yasuhiro Mori
Evaluation of a Simplified Method to Estimate the Peak Inter-Story Drift Ratio of Steel
Frames with Hysteretic Dampers
doi: 10.3390/app7050449   3

Wei Lin, Gangbing Song and Shanghong Chen
PTMD Control on a Benchmark TV Tower under Earthquake and Wind Load Excitations
doi: 10.3390/app7040425 18

Weiqing Fu, Chunwei Zhang, Li Sun, Mohsen Askari, Bijan Samali, Kwok L. Chung and
Pezhman Sharafi
Experimental Investigation of a Base Isolation System Incorporating MR Dampers with the
High-Order Single Step Control Algorithm
doi: 10.3390/app7040344 41

F.Palacios-Quiñonero, J. Rubió-Massegú, J.M. Rossell and H.R. Karimi
Integrated Design of Hybrid Interstory-Interbuilding Multi-Actuation Schemes for
Vibration Controlof Adjacent Buildings under Seismic Excitations
doi: 10.3390/app7040323 56

Chi-Ying Lin and Hong-Wu Jheng
Active Vibration Suppression of a Motor-Driven Piezoelectric Smart Structure Using
Adaptive Fuzzy Sliding Mode Control and Repetitive Control
doi: 10.3390/app7030240 79

Bingbing He, Huajiang Ouyang, Xingmin Ren and Shangwen He
Dynamic Response of a Simplified Turbine Blade Model with Under-Platform Dry Friction
Dampers Considering Normal Load Variation
doi: 10.3390/app7030228 96

Zheng Lu, Yuling Yang, Xilin Lu and Chengqing Liu
Preliminary Study on the Damping Effect of a Lateral Damping Buffer under a Debris
Flow Load
doi: 10.3390/app7020201 120

Dahai Zhao, Yang Liu and Hongnan Li
Self-Tuning Fuzzy Control for Seismic Protection of Smart Base-Isolated Buildings Subjected
to Pulse-Type Near-Fault Earthquakes
doi: 10.3390/app7020185 133

Huanguo Chen, Jianyang Shen, Wenhua Chen, Chuanyu Wu, Chunshao Huang, Yongyu Yi and Jiacheng Qian
The Bivariate Empirical Mode Decomposition and Its Contribution to Grinding Chatter Detection
doi: 10.3390/app7020145 . . . . **154**

Zhe Zhang, Jinping Ou, Dongsheng Li and Shuaifang Zhang
Optimization Design of Coupling Beam Metal Damper in Shear Wall Structures
doi: 10.3390/app7020137 . . . . **171**

Huanguo Chen, Pei Chen, Wenhua Chen, Chuanyu Wu, Jianmin Li and Jianwei Wu
Wind Turbine Gearbox Fault Diagnosis Based on Improved EEMD and Hilbert Square Demodulation
doi: 10.3390/app7020128 . . . . **183**

Hui Zhang, Meng-ke Liu, Bao-chun Fan, Zhi-hua Chen, Jian Li and Ming-yue Gui
Investigations on the Effects of Vortex-Induced Vibration with Different Distributions of Lorentz Forces
doi: 10.3390/app7010061 . . . . **198**

Lianchun Wang, Jinhui Li, Danfeng Zhou and Jie Li
An Experimental Validated Control Strategy of Maglev Vehicle-Bridge Self-Excited Vibration
doi: 10.3390/app7010038 . . . . **218**

Demetris Demetriou and Nikolaos Nikitas
A Novel Hybrid Semi-Active Mass Damper Configuration for Structural Applications
doi: 10.3390/app6120397 . . . . **234**

About the Special Issue Editors

Gangbing Song [John and Rebeca Moores Professor] Dr. Song received his Ph.D. and MS degrees from the Department of Mechanical Engineering at Columbia University in the City of New York in 1995 and 1991, respectively. He received his B.S. degree in 1989 from Zhejiang University, China. Dr. Song is the founding Director of the Smart Materials and Structures Laboratory and a Professor of Mechanical Engineering, Civil and Environmental Engineering, and Electrical and Computer Engineering at the University of Houston (UH). He has expertise in smart materials and structures, structural vibration control, piezoceramics, ultrasonic transducers, structural health monitoring and damage detection. He has developed two new courses in smart materials and published more than 400 papers, including 200 peer-reviewed journal articles. Dr. Song is also an inventor or co-inventor of 11 US patents and 11 pending patents. Dr. Song is a member of ASCE, ASME, and IEEE.

Steve C.S. Cai, a PE since 1995, is a Professor at the Department of Civil and Environmental Engineering. He is serving as the coordinator of Structures Group and director of Bridge Innovative Research and Dynamics of Structures laboratory. He has been awarded the Edwin B. and Norma S. McNeil Distinguished Professorship since 2010. Dr. Cai was elected Fellow of ASCE in 2010. Dr. Cai received his Ph.D. degree in 1993 from the Department of Civil Engineering, University of Maryland, College Park, Maryland; M.S. degree in 1987 from the Department of Civil Engineering, Tsinghua University, Beijing, China; and B.S. degree in 1983 from the Department of Civil Engineering, Zhejiang University, Hangzhou, China. Dr. Cai began his employment in the Department of Civil and Environmental Engineering at Louisiana State University (LSU) as a tenure-track Assistant Professor in August, 2001, was appointed as a tenured associate professor in August, 2006, and was promoted to full professor in August, 2010. Prior to his arrival at LSU, Dr. Cai had one year of experience as a tenure-track Assistant Professor at Kansas State University (2000–2001); four years of experience as a structural researcher and development senior engineer at the Florida Department of Transportation (1996–2000); and three years of experience as a consulting engineer at Michael Baker Jr., Inc. (1993–1996). Since joining LSU in 2001, Dr. Cai has served as Principal Investigator for more than 50 federal, state government, and university funded projects. His research interests include bridge performance evaluation/instrumentation/testing, traditional and new material applications in infrastructures, performance and hazard mitigation of coastal structures under wave/wind actions, and long-span bridge aerodynamics. Dr. Cai has published over 390 technical papers in journals (over 200) and conference proceedings in these areas mainly related to bridges. He has graduated 16 Ph.D. students at LSU as the major professor. His students are well placed including a few on the faculty of universities in US and abroad such as University of Connecticut, Colorado State University, Hunan University, Nanyang Institute of Technology, etc. Dr. Cai is currently serving on a few national and international committees including former chair of Experimental Analysis and Instrumentation Committee, ASCE. He served and has been serving on many editorial boards including as Associate Editor of Journal of Bridge Engineering, and Journal of Engineering Mechanics. He has also served as an advisor for ASCE and other student organizations. Other major professional services include serving as Secretary and Treasurer of American Association for Wind Engineering, and serving on the Engineering Project Selection committee, East Baton Rouge Parish, representing LSU.

Hong-Nan Li is the dean of Faculty of Infrastructure Engineering, Dalian University of Technology (DUT), China. He received his Ph. D., M. S. and B. S. in 1990, 1987 and 1982, respectively. He is a Cheung Kong Scholars Program engaged professor, serves as Vice Chairman of China Panel, International Association for Structural Control and Monitoring; and Vice Chairman of Advanced Materials and Structures, ASCE Aerospace Division; Chairman of the Panel of the National Natural Science Foundation of China (NSFC). He is the editor-in-chief of Structural Monitoring and Maintenance, an international journal and associate editor-in-chief of ASCE Journal of Aerospace Engineering. His research interests are in structural control and monitoring, disaster prevention and reduction and earthquake engineering. He is the recipient of more than 30 research grants from governments and industrial companies, and his new techniques and methods have been applied to more than 50 practical engineering projects. He owns 38 patents and has published seven books and more than 400 peer-reviewed journal papers. He has won three national awards for Science and Technology and 10 provincial awards. He actively cooperates internationally with the United States, Japan, Australia, Korea, among other countries.

Preface to "Energy Dissipation and Vibration Control: Modeling, Algorithm and Devices"

"If you want to find the secrets of the universe, think in terms of energy, frequency and vibration."
—Nikola Tesla

From the sub-atomic world to the realm of orbital mechanics, energy, frequency, and vibration often play the role of the defining parameters. The importance of these parameters to the advancement of STEM, and by extension, civilization, cannot be understated. Volumes upon volumes of books, magazines, journals, and essays have been dedicated to the further understanding of these three elements of nature. The vast sea of literature on this topic begs the question of why yet another book such as the one you are reading now is needed.

Well, the vastness of the topic is precisely what drives us to assemble this book. Furthermore, not only is the topic too vast to be covered in a single, or even multiple volumes, but we feel that the depth of each sub-area of study merits its own attention. Every day, researchers around the world work tirelessly to discover new truths that can be gleaned from this field, and try to figure out how these truths can be used for our collective benefit. While we cannot hope to cover all that has been accomplished, we can, through our collective experiences and training, select several articles that we feel represent the state of the art in energy dissipation and vibration control. In particular, this book covers recent advances in vibration theory and the role of energy dissipation and vibration control in improving structural integrity in civil and energy infrastructure. The editors and authors hope the reader (likely a graduate student, researcher, postdoc, faculty member, or just someone interested in this field) will gain a better understanding of energy dissipation and vibration control in general and have a chance to see some of the latest activities that aim to improve robustness and quality across different strata of infrastructure.

Gangbing Song, Steve C.S. Cai and Hong-Nan Li
Special Issue Editors

applied sciences

MDPI

Editorial

Energy Dissipation and Vibration Control: Modeling, Algorithm, and Devices

Gangbing Song [1,2,*], Steve C. S. Cai [3] and Hong-Nan Li [2]

1 Department of Mechanical Engineering, University of Houston, Houston, TX 77004, USA
2 School of Civil Engineering, Dalian University of Technology, Dalian 116023, China; hnli@dlut.edu.cn
3 Department of Civil and Environmental Engineering, Louisiana State University, Baton Rouge, LA 70803, USA; cscai@lsu.edu
* Correspondence: GSong@UH.EDU; Tel.: +1-713-743-4525

Received: 31 July 2017; Accepted: 2 August 2017; Published: 7 August 2017

The topic of vibration control and energy dissipation is among the oldest and most relevant in the field of engineering. This area encompasses exciting frontier problems ranging from the control of vibrations of drill bits buried thousands of feet underground, to the isolation of towering skyscrapers against seismic excitation, and finally to the stabilization of solar panels attached to a spacecraft. Armed with an ever increasing arsenal of sensing and actuation technologies and theoretical progress, a vibrant community of researchers have gathered to tackle some of the world's toughest challenges in vibration control and energy suppression problems.

While this issue is not all-inclusive due to the extreme breadth and depth of the field, this issue does showcase some of the state-of-the-art in vibration control and energy dissipation. Specifically, topics in this issue cover: "Evaluation of a Simplified Method to Estimate the Peak Inter-Story Drift Ratio of Steel Frames with Hysteretic Dampers" [1], "PTMD Control on a Benchmark TV Tower under Earthquake and Wind Load Excitations" [2], "Experimental Investigation of a Base Isolation System Incorporating MR Dampers with the High-Order Single Step Control Algorithm" [3], "Integrated Design of Hybrid Interstory-Interbuilding Multi-Actuation Schemes for Vibration Control of Adjacent Buildings under Seismic Excitations" [4], "Active Vibration Suppression of a Motor-Driven Piezoelectric Smart Structure Using Adaptive Fuzzy Sliding Mode Control and Repetitive Control" [5], "Dynamic Response of a Simplified Turbine Blade Model with Under-Platform Dry Friction Dampers Considering Normal Load Variation" [6], "Preliminary Study on the Damping Effect of a Lateral Damping Buffer under a Debris Flow Load" [7], "Self-Tuning Fuzzy Control for Seismic Protection of Smart Base-Isolated Buildings Subjected to Pulse-Type Near-Fault Earthquakes" [8], "The Bivariate Empirical Mode Decomposition and Its Contribution to Grinding Chatter Detection" [9], "Optimization Design of Coupling Beam Metal Damper in Shear Wall Structures" [10], "Wind Turbine Gearbox Fault Diagnosis Based on Improved EEMD and Hilbert Square Demodulation" [11], "Investigations on the Effects of Vortex-Induced Vibration with Different Distributions of Lorentz Forces" [12], "An Experimental Validated Control Strategy of Maglev Vehicle-Bridge Self-Excited Vibration" [13], and "A Novel Hybrid Semi-Active Mass Damper Configuration for Structural Applications" [14].

This issue only provides a brief glimpse into the possibilities offered by research in vibration control and energy dissipation. As the current research builds upon the efforts of previous investigators, the results presented in this issue will now help to open the doors for increasingly advanced topics that are sure to offer benefits and push society forward.

Acknowledgments: This research was partially supported by the Major State Basic Development Program of China (973 Program, grant number 2015CB057704), Innovative research group project (grant number 51421064) and general project (grant number 51478080 and 51278084) of Natural Science Foundation of China. The authors would like to acknowledge for these financial supports.

Conflicts of Interest: The authors declare no conflict of interest.

References

1. Kang, J.; Mori, Y. Evaluation of a Simplified Method to Estimate the Peak Inter-Story Drift Ratio of Steel Frames with Hysteretic Dampers. *Appl. Sci.* **2017**, *7*, 449. [CrossRef]
2. Lin, W.; Song, G.; Chen, S. PTMD Control on a Benchmark TV Tower under Earthquake and Wind Load Excitations. *Appl. Sci.* **2017**, *7*, 425. [CrossRef]
3. Fu, W.; Zhang, C.; Sun, L.; Askari, M.; Samali, B.; Chung, K.; Sharafi, P. Experimental Investigation of a Base Isolation System Incorporating MR Dampers with the High-Order Single Step Control Algorithm. *Appl. Sci.* **2017**, *7*, 344. [CrossRef]
4. Palacios-Quiñonero, F.; Rubió-Massegú, J.; Rossell, J.; Karimi, H. Integrated Design of Hybrid Interstory-Interbuilding Multi-Actuation Schemes for Vibration Control of Adjacent Buildings under Seismic Excitations. *Appl. Sci.* **2017**, *7*, 323. [CrossRef]
5. Lin, C.; Jheng, H. Active Vibration Suppression of a Motor-Driven Piezoelectric Smart Structure Using Adaptive Fuzzy Sliding Mode Control and Repetitive Control. *Appl. Sci.* **2017**, *7*, 240. [CrossRef]
6. He, B.; Ouyang, H.; Ren, X.; He, S. Dynamic Response of a Simplified Turbine Blade Model with Under-Platform Dry Friction Dampers Considering Normal Load Variation. *Appl. Sci.* **2017**, *7*, 228. [CrossRef]
7. Lu, Z.; Yang, Y.; Lu, X.; Liu, C. Preliminary Study on the Damping Effect of a Lateral Damping Buffer under a Debris Flow Load. *Appl. Sci.* **2017**, *7*, 201. [CrossRef]
8. Zhao, D.; Liu, Y.; Li, H. Self-Tuning Fuzzy Control for Seismic Protection of Smart Base-Isolated Buildings Subjected to Pulse-Type Near-Fault Earthquakes. *Appl. Sci.* **2017**, *7*, 185. [CrossRef]
9. Chen, H.; Shen, J.; Chen, W.; Wu, C.; Huang, C.; Yi, Y.; Qian, J. The Bivariate Empirical Mode Decomposition and Its Contribution to Grinding Chatter Detection. *Appl. Sci.* **2017**, *7*, 145. [CrossRef]
10. Zhang, Z.; Ou, J.; Li, D.; Zhang, S. Optimization Design of Coupling Beam Metal Damper in Shear Wall Structures. *Appl. Sci.* **2017**, *7*, 137. [CrossRef]
11. Chen, H.; Chen, P.; Chen, W.; Wu, C.; Li, J.; Wu, J. Wind Turbine Gearbox Fault Diagnosis Based on Improved EEMD and Hilbert Square Demodulation. *Appl. Sci.* **2017**, *7*, 128. [CrossRef]
12. Zhang, H.; Liu, M.; Fan, B.; Chen, Z.; Li, J.; Gui, M. Investigations on the Effects of Vortex-Induced Vibration with Different Distributions of Lorentz Forces. *Appl. Sci.* **2017**, *7*, 61. [CrossRef]
13. Wang, L.; Li, J.; Zhou, D.; Li, J. An Experimental Validated Control Strategy of Maglev Vehicle-Bridge Self-Excited Vibration. *Appl. Sci.* **2017**, *7*, 38. [CrossRef]
14. Demetriou, D.; Nikitas, N. A Novel Hybrid Semi-Active Mass Damper Configuration for Structural Applications. *Appl. Sci.* **2016**, *6*, 397. [CrossRef]

applied
sciences

MDPI

Article

Evaluation of a Simplified Method to Estimate the Peak Inter-Story Drift Ratio of Steel Frames with Hysteretic Dampers

Jae-Do Kang [1],* and Yasuhiro Mori [2]

[1]   Earthquake Disaster Mitigation Research Division, National Research Institute for Earth Science and Disaster Resilience, Miki 673-0515, Japan
[2]   Graduate School of Environmental Studies, Nagoya University, Nagoya 464-8603, Japan; yasu@sharaku.nuac.nagoya-u.ac.jp
*   Correspondence: kang@bosai.go.jp; Tel.: +81-794-85-8211

Academic Editor: Gangbing Song
Received: 4 March 2017; Accepted: 20 April 2017; Published: 27 April 2017

Abstract: In this paper, a simplified method is proposed to estimate the peak inter-story drift ratios of steel frames with hysteretic dampers. The simplified method involved the following: (1) the inelastic spectral displacement is estimated using a single-degree-of-freedom (SDOF) system with multi-springs, which is equivalent to a steel frame with dampers and in which multi-springs represent the hysteretic behavior of dampers; (2) the first inelastic mode vector is estimated using a pattern of story drifts obtained from nonlinear static pushover analysis; and (3) the effects of modes higher than the first mode are estimated by using the jth modal period, jth mode vector, and jth modal damping ratio obtained from eigenvalue analysis. The accuracy of the simplified method is estimated using the results of nonlinear time history analysis (NTHA) on a series of three-story, six-story, and twelve-story steel moment resisting frames with steel hysteretic dampers. Based on the results of a comparison of the peak inter-story drift ratios estimated by the simplified method and that computed via NTHA using an elaborate analytical model, the accuracy of the simplified method is sufficient for evaluating seismic demands.

Keywords: simplified estimation method; steel hysteretic damper; peak inter-story drift ratio; inelastic mode vector; equivalent SDOF system

1. Introduction

Previous studies conducted over the past two decades have proposed energy dissipation systems that use displacement-dependent dampers and/or velocity-dependent dampers [1]. Currently, a major concern is associated with the need to design a structure with dampers; therefore, many studies have been conducted on this subject [2]. An energy-based design procedure based on achieving a balance between the mean energy dissipated per cycle by a structure and that dissipated by dampers was developed for the seismic retrofitting of existing buildings [3]. By using parameters such as a maximum damper ductility value and an elastic stiffness ratio between the bracing-hysteretic device system and the moment frame system, a design procedure was investigated for frames equipped with a hysteretic energy dissipation device [4,5]. By using a displacement-based design method, a design procedure was investigated for buildings with hysteretic dampers [6–8]. To determine the optimal damper volume related to seismic performance, extant studies have used a performance curve (called as performance spectra) for an elastic system [9,10] and an inelastic system [2,11–13] in the design process of buildings with dampers. However, these studies have not considered the turn from an elastic mode vector to an inelastic mode vector caused by the yielding of a main structure. This would lead to the loss of accuracy in a large drift range. On the one hand, nonlinear time history analysis (NTHA) using an elaborate analytical model is used to estimate the seismic performance of buildings with dampers. However, this is time-consuming and it is more convenient to perform a seismic performance assessment.

In performance-based seismic design (PBSD), it is essential for the risk associated with the performance level of a structure to be clear and transparent, such that decision makers can understand the expected seismic performance of the structures [14]. Specifically, with respect to PBSD, NTHA using hundreds (or thousands) of ground motion records is required to perform a probabilistic seismic performance assessment of buildings. In recent years, for the seismic performance and probabilistic collapse resistance assessment of buildings with energy dissipation systems, many studies that use incremental dynamic analysis [15] and perform an NTHA using an elaborate analytical model have been conducted [16–20]. However, NTHA using an elaborate analytical model also requires intensive computations to estimate seismic demands [21]. Therefore, a simplified estimation method of inelastic seismic demands is useful, and thus several estimation methods have been proposed by previous studies [22–26]. These methodologies mostly focus on estimating the seismic demands of general buildings and involve the use of equivalent single-degree-of-freedom (SDOF) systems converted from a building structure. In most equivalent SDOF systems, a single skeleton curve is adopted to represent the characteristics of a building structure [24–26]. In order to consider the behavior of dampers, an equivalent SDOF system that includes two springs have been proposed for reinforced concrete buildings with hysteretic dampers by using the results of nonlinear static pushover analysis (NSPA) [27]. However, this model requires interpolation [27], and therefore, the model may necessitate time and effort to approximate the behavior of the springs for the dampers.

In this paper, a simplified method is proposed to estimate the peak inter-story drift ratio of steel frames with hysteretic dampers, and the method is evaluated for use in structural performance assessment. In the method, an equivalent inelastic SDOF system of a steel frame with dampers is presented to estimate the inelastic spectral displacement. The method is also employed to estimate the first inelastic mode vector by using the pattern of story drifts obtained from NSPA. Additionally, in order to consider the effects of modes higher than the first mode, the modal elastic spectral displacement and the participation function are also estimated by using the jth modal period, jth mode vector, and jth modal damping ratio, as obtained from eigenvalue analysis (EVA). In order to estimate the accuracy of the simplified method, the simplified method is compared by using the results of NTHA using an elaborate analytical model on a series of three-story, six-story, and twelve-story steel moment resisting frames with dampers. Parametric analyses are performed for all the frames to confirm the effects of damper properties, such as stiffness and yield deformation, on the accuracy and stability of the simplified method.

2. Simplified Method to Estimate the Peak Inter-Story Drift Ratio of Steel Frames with Hysteretic Dampers

In this section, a simplified method is proposed to estimate the inter-story drift ratio of a steel frame with steel hysteretic dampers (called steel metallic dampers), such as buckling-restrained brace dampers and added damping and stiffness dampers. The basic concept employed was presented in a previous study [28]. The method is based on an inelastic modal predictor (IMP) [26], which is an extension of an elastic response spectrum method with a square-root-of-sum-of-squares rule to the inelastic response and is based on extant studies [22,23]. The simplified method also considers the change in the mode vector caused by the yielding of a main building. Additionally, the method accounts for additional steps in the IMP methodology to generate the equivalent SDOF system for estimating the inelastic spectral displacement and for determining the step number N necessary to derive the inelastic mode vector.

2.1. Estimation of Peak Inter-Story Drift Ratios

In this study, based on a previous study [26,28], the peak inter-story drift ratio θ_i^P of the ith story of steel frames with steel hysteretic dampers is evaluated as follows:

$$\theta_i^P = \sqrt{\left(PF_{1,i}^I \cdot Sd_1^I \right)^2 + \sum_{j=2}^{n} \left(PF_{j,i}^E \cdot Sd_j^E \right)^2} \tag{1}$$

where Sd_j^E is a jth modal elastic spectral displacement that is obtained from response spectrum analysis using a jth modal period and a jth modal damping ratio obtained from EVA. Additionally, $PF_{j,i}^E$ denotes the participation function of an inter-story drift ratio for an elastic jth mode that is defined as follows:

$$PF_{j,i}^E = \frac{\Gamma_j^E \cdot \left({}_s\phi_{j,i}^E - {}_s\phi_{j,i-1}^E \right)}{h_i} \tag{2}$$

where h_i and ${}_s\phi_{j,i}^E$ signify the height of the ith story and the elastic jth mode vector obtained from EVA of a steel frame with dampers, respectively. In this paper, the subscripts 's' and 'f' before the symbols refer to the entire system and bare frame that is a multi-story frame without dampers, respectively. Furthermore, Γ_j^E denotes the participation factor for the elastic jth mode that is defined as follows:

$$\Gamma_j^E = \frac{\sum\limits_{i=1}^{n} m_i \cdot {}_s\phi_{j,i}^E}{\sum\limits_{i=1}^{n} m_i \cdot \left({}_s\phi_{j,i}^E \right)^2} \tag{3}$$

where m_i denotes the mass at the ith story.

In Equation (1), Sd_1^I implies an inelastic spectral displacement, estimated using the equivalent inelastic SDOF system with multi-springs that is described in Section 2.2. Moreover, $PF_{1,i}^I$ denotes an inelastic participation function of the inter-story drift ratio evaluated via Equations (2) and (3) with ${}_s\phi_{j,i}^E$ replaced by ϕ_i^I, which is estimated using the pattern of story drifts obtained from NSPA at the Nth step, corresponding to the Sd_1^I value obtained from NTHA using the equivalent inelastic SDOF system with multi-springs. The procedure for estimating the inelastic spectral displacement Sd_1^I and inelastic mode vector ϕ_i^I is explained in Section 2.3.

2.2. Equivalent Inelastic SDOF System for Steel Frames with Hysteretic Dampers

This section describes the methodology for converting steel frames with steel hysteretic dampers into an equivalent inelastic SDOF system.

As shown in Figure 1a, the steel hysteretic dampers can be described using shear springs that depend on inter-story drifts. In order to describe an inelastic behavior, a hysteresis rule of a spring in the equivalent SDOF system can be defined by an approximated skeleton curve with respect to a base shear force versus roof drift curve obtained from NSPA, as shown in Figure 1b. When all of the structural components of a building behave in a similar manner, it is possible for an equivalent SDOF system with an assumed single skeleton curve to estimate the inelastic seismic performance of the entire system. However, if the hysteresis rules of each structural member, such as the dampers and frame members, are significantly different, then the assumed single skeleton curve leads to a loss of accuracy in estimating the seismic performance, because the assumed bilinear (or tri-linear) single skeleton curve may not be sufficient to consider the behavior of the dampers.

In the shear frame with dampers (hereafter referred to as the MDOF system), the story shear force is computed by accumulating the lateral force, which is resisted by the column elements and the dampers above the story level. Therefore, the base shear force, which is the shear force on the first story, is computed by accumulating two lateral forces, namely, the lateral force resisted by the column elements of each story and the lateral force resisted by the dampers of each story. In other words, the lateral force resisted by the dampers can also be divided into several forces, as shown in Figure 1d. Therefore, as shown in Figure 1e, the equivalent SDOF system of the MDOF system can be described using multi-springs with skeleton curves approximated by the base shear force relative to the roof drift curve, as shown in Figure 1d. Based on the fore-mentioned approach, the basic concept of an equivalent inelastic SDOF system with multi-springs was presented in a previous study [28]. Parts of the theoretical development of the method used in this paper are also based on a previous study [28]. The equivalent inelastic SDOF system with multi-springs comprises an effective mass $_s\overline{M_1}$ and an effective height $_s\overline{H_1}$ that are described in Section 2.2.1; an inelastic spring equivalent to the bare frame that is described Section 2.2.2; and inelastic springs equivalent to the dampers that are described Section 2.2.3.

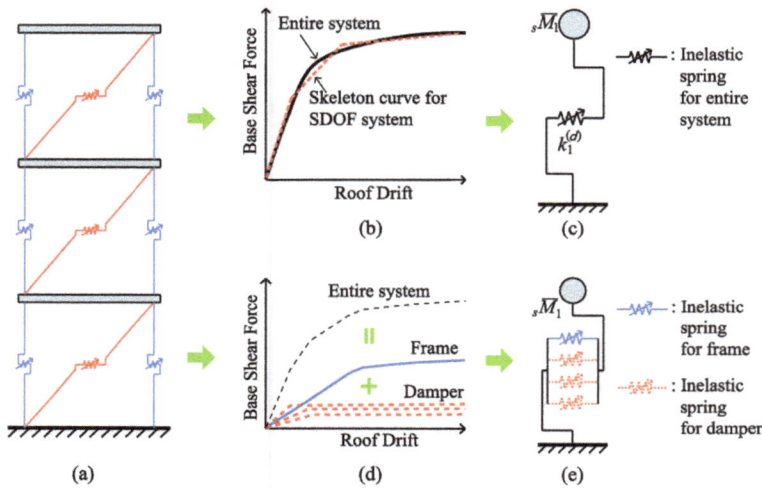

Figure 1. A multi-story frame with steel hysteretic dampers: (**a**) a shear frame with dampers; (**b**) a single skeleton curve; (**c**) an equivalent inelastic SDOF system with a single inelastic spring; (**d**) multi-skeleton curves; and (**e**) an equivalent inelastic SDOF system with inelastic multi-springs.

2.2.1. Mass and Height of Equivalent Inelastic SDOF System

With respect to the first mode, the effective mass $_s\overline{M_1}$ that corresponds to the mass of the equivalent SDOF system, and the effective height $_s\overline{H_1}$ that corresponds to the height of the equivalent SDOF system are expressed, respectively, as follows:

$$_s\overline{M_1} = \frac{\left(\sum\limits_{i=1}^{n} m_i \cdot_s \phi_{1,i}^E \right)^2}{\sum\limits_{i=1}^{n} m_i \cdot_s \phi_{1,i}^E} \tag{4}$$

$$_s\overline{H_1} = \frac{\sum\limits_{i=1}^{n} \left(m_i \cdot_s \phi_{1,i}^E \cdot H_i \right)}{\sum\limits_{i=1}^{n} m_i \cdot_s \phi_{1,i}^E} \tag{5}$$

where n and H_i are the number of stories and the height of the ith story above the base, respectively.

2.2.2. Inelastic Spring Equivalent to Steel Frame

With respect to the first mode, the period T_f of the bare frame can be obtained from EVA. Additionally, this period is also calculated as follows:

$$T_f = 2\pi \sqrt{\frac{{}_f\overline{M_1}}{k_f}} \tag{6}$$

where k_f denotes the effective elastic stiffness of the bare frame and ${}_f\overline{M_1}$ is the effective mass of the bare frame that is defined using Equation (4) with ${}_s\phi^E_{1,i}$ replaced by ${}_f\phi^E_{1,i}$, obtained from EVA of the bare frame.

Equation (6) is used to represent the elastic stiffness k_f of the bare frame, as follows:

$$k_f = \frac{4 \cdot \pi^2 \cdot {}_f\overline{M_1}}{\left(T_f\right)^2} \tag{7}$$

It is assumed that the period calculated by the effective elastic stiffness ${}_sk_f$ of the spring, which is represented as the only bare frame in the equivalent SDOF system with multi-springs, and the effective mass ${}_s\overline{M_1}$ is equal to the period of the equivalent SDOF system of the bare frame, as shown in Equation (6). This assumption is represented as follows:

$$T_f = 2\pi \sqrt{\frac{{}_s\overline{M_1}}{{}_sk_f}} = 2\pi \sqrt{\frac{{}_f\overline{M_1}}{k_f}} \tag{8}$$

Based on Equations (7) and (8), the effective elastic stiffness of the spring representing the bare frame in the equivalent SDOF system with multi-springs is defined as follows:

$$_sk_f = \frac{4 \cdot \pi^2 \cdot {}_s\overline{M_1}}{\left(T_f\right)^2} \tag{9}$$

In order to consider the post-elastic behavior of the bare frame, the stiffness reduction factors ($\gamma1$ and $\gamma2$) after the elastic behavior are calculated based on the same method used in a previous study [26], in which a tri-linear skeleton curve is assumed. Further details on the post-yield stiffness can be found in a previous study [26].

The first and second yield drifts on the skeleton curve for the bare frame can be defined as follows:

$$\delta = 1\frac{\theta_{roof,1} \cdot H_{roof}}{\Gamma^E_1 \cdot {}_s\phi^E_{1,roof}}, \; \delta2 = \frac{\theta_{roof,2} \cdot H_{roof}}{\Gamma^E_1 \cdot {}_s\phi^E_{1,roof}} \tag{10}$$

where $\theta_{roof,1}$ and $\theta_{roof,2}$ are the first and second yield drift ratios on the skeleton curve [26], respectively; and H_{roof} and ${}_s\phi^E_{1,roof}$ denote the height of the roof above the base and the first mode vector of the roof of the steel frame with dampers, respectively.

The stiffness reduction factors and yield drifts are used to calculate the yield forces $Q1$ and $Q2$, as follows:

$$Q1 = {}_sk_f \cdot \delta1, \; Q2 = {}_sk_f[\delta1 + \gamma1(\delta2 - \delta1)] \tag{11}$$

Subsequently, the skeleton curve of the inelastic spring that is equivalent to the bare frame is determined by Equations (10) and (11).

2.2.3. Inelastic Springs Equivalent to Dampers

This section describes the methodology used to estimate the skeleton curve of the inelastic springs that are equivalent to the dampers in the equivalent SDOF system with multi-springs.

The force versus drift curve of the damper at the *i*th story of the MDOF system and the *i*th inelastic spring of the equivalent SDOF system are shown in Figure 2a,b, respectively.

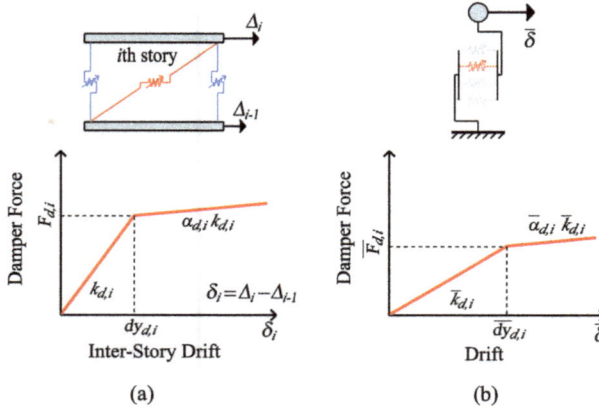

Figure 2. Skeleton curves of the dampers and inelastic springs: (**a**) damper force versus inter-story drift curve of the damper at the *i*th story in the MDOF system and (**b**) force versus drift curve of the *i*th spring in the SDOF system.

As shown in the top portion of Figure 2a, by considering only the first mode, a lateral drift Δ_i of the *i*th story of the MDOF system can be expressed using the drift $\bar{\delta}$ of the equivalent SDOF system, as shown in the top portion of Figure 2b. This can be expressed as follows:

$$\Delta_i = {}_s\phi_{1,i}^E \cdot \Gamma_1^E \cdot \bar{\delta} \tag{12}$$

where ${}_s\phi_{1,i}^E$ signifies the first elastic mode vector of the *i*th story as obtained from EVA. Thus, the inter-story drift δ_i at the *i*th story of the MDOF system can be expressed as follows:

$$\delta_i = \Delta_i - \Delta_{i-1} = \left({}_s\phi_{1,i}^E - {}_s\phi_{1,i-1}^E\right) \cdot \Gamma_1^E \cdot \bar{\delta} \tag{13}$$

It is assumed that the energy $E_{d,i}$ dissipated by the damper at the *i*th story of the MDOF system is equivalent to the energy $\bar{E}_{d,i}$ dissipated by the *i*th inelastic spring of the equivalent SDOF system.

Based on this assumption and Equation (13), in the range of $\delta_i < dy_{d,i}$ that corresponds to the yield deformation of the damper at the *i*th story of the MDOF system, the stiffness $\bar{k}_{d,i}$ of the *i*th inelastic spring of the equivalent SDOF system can be expressed as follows:

$$\bar{k}_{d,i} = k_{d,i} \left({}_s\phi_{1,i}^E - {}_s\phi_{1,i-1}^E\right)^2 \left(\Gamma_1^E\right)^2 \tag{14}$$

where $k_{d,i}$ denotes an elastic stiffness of the damper at the *i*th story.

Based on a previous assumption that indicates $E_{d,i} = \bar{E}_{d,i}$ and Equations (13) and (14), when $\delta_i = dy_{d,i}$, the yield deformation $\overline{dy}_{d,i}$ of the *i*th inelastic spring of the equivalent SDOF system can be expressed as follows:

$$\overline{dy}_{d,i} = \frac{dy_{d,i}}{\left({}_s\phi_{1,i}^E - {}_s\phi_{1,i-1}^E\right)\Gamma_1^E} \tag{15}$$

Based on a previous assumption $E_{d,i} = \overline{E}_{d,i}$ and Equations (13)–(15), in the range of $\delta_i > dy_{d,i}$, a post-elastic stiffness ratio $\overline{\alpha}_{d,i}$ of the ith inelastic spring of the equivalent SDOF system can be determined as follows:

$$\overline{\alpha}_{d,i} = \alpha_{d,i} \tag{16}$$

where $\alpha_{d,i}$ signifies the post-elastic stiffness ratios of the damper at the ith story.

2.3. Procedure for Estimating Inelastic Spectral Displacement and Inelastic Mode Vector

For estimating the peak inter-story drift ratio of the steel frame with steel hysteretic dampers by using Equation (1), the inelastic spectral displacement Sd^I and the inelastic mode vector ϕ_i^I can be estimated by the following steps (as shown in Figure 3):

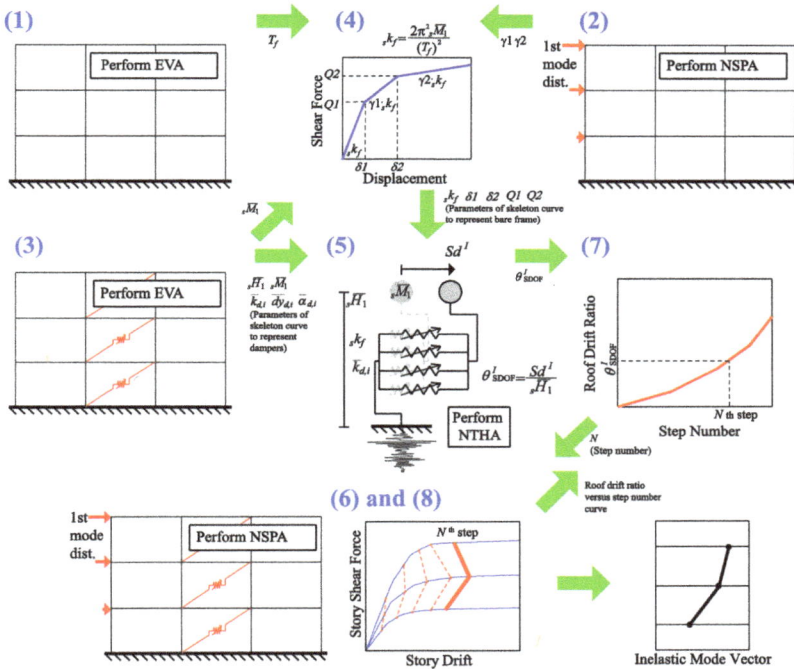

Figure 3. Procedure for estimating the inelastic spectral displacement Sd^I and the inelastic mode vector ϕ_i^I.

(1) Define the period T_f for the first mode of the bare frame, as obtained from EVA of the bare frame.

(2) By using the NSPA results of the bare frame, in which a lateral load pattern is based on the first mode vector, define the stiffness reduction factor to consider the post-elastic behavior of the bare frame.

(3) Define the mode vector $_s\phi_{1,i}^E$ by performing EVA of the steel frame with dampers. Subsequently, by using Equations (4), (5), and (14)–(16) with $_s\phi_{1,i}^E$, determine the effective mass $_s\overline{M_1}$, the effective height $_s\overline{H_1}$, and the skeleton curves of the inelastic springs equivalent to the dampers.

(4) By using Equations (9)–(11) with $_s\phi_{1,i}^E$ obtained from Step (3) and the stiffness reduction factor obtained from Step (2), determine the skeleton curve of the inelastic spring equivalent to the bare frame.

(5) Generate the equivalent inelastic SDOF system with multi-springs using the effective mass $_s\overline{M_1}$, effective height $_s\overline{H_1}$, and inelastic springs with the skeleton curves, as obtained from Steps (3) and (4). Subsequently, perform NTHA using the equivalent inelastic SDOF system with multi-springs to evaluate the peak drift ratio, as follows:

$$\theta^I_{SDOF} = \frac{Sd^I}{_s\overline{H_1}} \qquad (17)$$

where Sd^I denotes the inelastic spectral displacement of the steel frame with dampers for the first mode that is obtained from NTHA using an equivalent inelastic SDOF system with multi-springs.

(6) Obtain the shear force versus the drift curve for each story by performing NSPA of the steel frame with dampers, in which the lateral load pattern is based on the first mode vector.

(7) Define the step number N at which the response corresponds to θ^I_{SDOF} in the roof drift ratio versus the step number curve obtained from the NSPA result in Step (6).

(8) Determine the first inelastic mode vector ϕ^I_i using the pattern of story drifts of the shear force versus the drift curve defined in Step (6) at the Nth step defined in Step (7).

3. Numeral Examples

3.1. Building Models

In this paper, a numerical example is considered, as shown in Figure 4, with three two-dimensional steel moment resisting frames (SMRFs) to estimate the accuracy of the simplified method. The SMRFs are designed to satisfy the current seismic requirements in Japan in terms of both strength and drift.

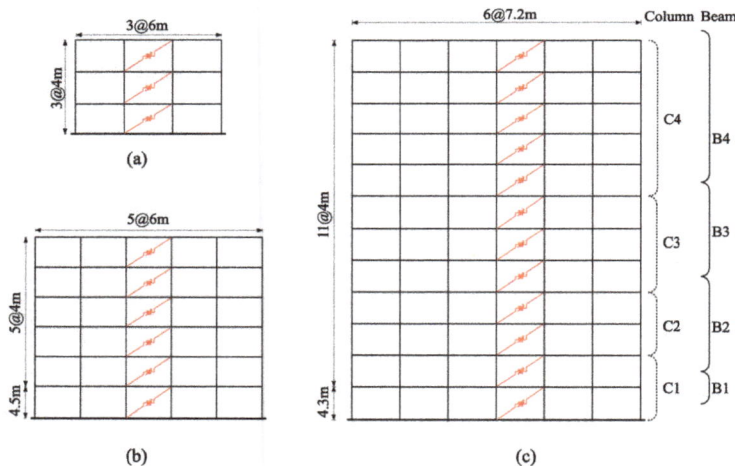

Figure 4. Analysis model: (**a**) a three-story steel moment resisting frame with dampers; (**b**) a six-story steel moment resisting frame with dampers; and (**c**) a twelve-story steel moment resisting frame with dampers.

Figure 4a shows the SMRF with three bays of 6.0 m and story heights of 4.0 m. Steel sections of H-350×350×12×19 and H-450×200×9×14 are used for all the columns and all the beams, respectively. The yield strength of the steel material and Young's modulus are assumed to be 235 N/mm² and 205,000 N/mm², respectively. The mass of each story is 70 tons. According to EVA, the fundamental natural period corresponds to 0.758 s. Additionally, NTHA is performed using the nonlinear dynamic analysis program SNAP ver.7 [29]. In SNAP ver.7, the beams and columns are modeled using the beam element with elasto-plastic characteristics. The nonlinear behaviors of the beams and columns are represented using the concentrated plasticity concept with rotational springs. The rotational behavior of the plastic regions follows a trilinear hysteretic response based on ST3, which is the steel hysteretic material model. The plastic moments at the ends of the columns and beams assume that Mp = 592 kN·m for the columns and Mp = 387 kN·m for the beams. The beam and column elements have three degrees of freedom in each node. Tangent stiffness-proportional damping is considered with a 2% damping ratio for the first mode.

Figure 4b shows the SMRF with five bays and six stories. Steel sections of H-400×400×18×28 and H-500×200×10×16 are used for all the columns and all the beams, respectively. The yield strength of the steel material is assumed to be 325 N/mm². The mass of each story is 130 tons. The plastic moments at the ends of the columns and beams assume that Mp = 1634 kN·m for the columns and Mp = 692 kN·m for the beams. According to EVA, the fundamental natural period is 1.23 s.

Figure 4c shows the SMRF with six bays and twelve stories. Table 1 also lists the plastic moments at the ends of the columns and beams, and the steel sections. The yield strength of steel is assumed to be 325 N/mm². The mass of each story is 205 tons. Based on EVA, the fundamental natural period corresponds to 2.17 s.

Table 1. Sections and plastic moments of the columns and beams of a twelve-story steel moment resisting frame.

Element	Label	Section	Mp (kN·m)
Columns	C4	□-550×22	2991.7
	C3	□-550×25	3361.7
	C2	□-550×28	3723.0
	C1	□-550×32	4191.2
Beams	B4	H-550×250×12×22	1204.3
	B3	H-550×250×12×25	1322.3
	B2	H-550×250×12×28	1439.0
	B1	H-550×300×14×28	1718.4

3.2. Properties of the Hysteretic Dampers

The properties of the steel hysteretic dampers are assumed to be proportional to those of the bare frame. The stiffness ratio κ [10] that corresponds to the ratio of the damper stiffness to the story stiffness of the frame is used to determine the stiffness of the damper at the *i*th story, as follows:

$$k_{d,i} = \kappa \cdot k_{f,i} \tag{18}$$

where $k_{f,i}$ is the lateral story stiffness of the bare frame that is determined by using the NSPA results. The lateral stiffness ratio α $(= k_f/k_{total}$ where $k_{total} = k_f + k_d)$ of reference [5,13] is equivalent to $1/(\kappa + 1)$.

The yield drift ratio υ [27] that corresponds to the ratio of the yield deformation of the damper to the story yield drift of the bare frame is used to determine the yield deformation of the damper at the *i*th story, as follows:

$$dy_{d,i} = \upsilon \cdot dy_{f,i} \tag{19}$$

where $dy_{f,i}$ is the yield story drift of the bare frame that is determined using the NSPA results.

In this paper, the stiffness ratio, the drift ratio, and the post-elastic stiffness ratio of the dampers correspond to two values (1.5 and 3), two values (0.4 and 0.6), and one value (0.03), respectively.

3.3. Earthquake Ground Motion Records

In conjunction with the frame models described in the previous section, 73 ground motions selected from the PEER Strong Motion Database are used to investigate the simplified method. A detailed list of the earthquakes can be found in a previous study [22].

3.4. Results and Discussion

In this paper, with respect to the simplified method, the first-mode response; first-mode and second-mode responses; and first-mode, second-mode, and third-mode responses are considered for a three-story frame, a six-story frame, and a twelve-story frame with dampers, respectively. In Equation (1), the second and third modal elastic spectral displacements and the participation functions are estimated by using the *j*th modal period, *j*th mode vector, and *j*th modal damping ratio, respectively, as obtained from EVA of the steel frame with dampers.

To confirm the accuracy of the method used to generate an equivalent inelastic SDOF system, the periods obtained from the EVA of the elaborate analytical model are compared with the periods obtained from the EVA of the equivalent SDOF system. According to the comparison results, which are not shown in this paper, the periods of the equivalent SDOF system with multi-springs are equivalent to the periods of an elaborate analytical model.

The accuracy of the simplified method is expressed by the following: (i) its bias denoted as a that is calculated by the median of the ratio of the peak inter-story drift ratio θ_i^P, which is estimated by the simplified method, with respect to the corresponding peak inter-story drift ratio θ_i, which is computed via NTHA using the elaborate analytical model; and (ii) its dispersion denoted as σ that is calculated by the standard deviation of the natural logarithms of θ_i^P / θ_i. The parameters bias and the dispersion are equivalently obtained by performing a one-parameter log-log linear least-squares regression of θ_i on θ_i^P. A bias exceeding unity implies overestimation, while a bias less than unity implies an underestimation of the average by the simplified method. In the paper, the accuracy of the simplified method is confirmed by using 876 results of NTHA using the elaborate analytical model.

The structural demand parameter for evaluation is denoted by the maximum peak inter-story drift ratio θ_{max}, which corresponds to a maximum response over time and a peak with respect to the height of the structure, because θ_{max} correlates well with the structural and nonstructural damage in the structure [30]. Figures 5–7 illustrate the regressions of θ_{max} computed via NTHA using the elaborate analytical model on the θ_{max}^P estimated by the simplified method for a three-story frame, a six-story frame, and a twelve-story frame with dampers, respectively, that are subject to all the earthquake records. The horizontal axis represents the maximum peak inter-story drift ratio that is computed via NTHA using the elaborate analytical model, and the vertical axis corresponds to that estimated by the simplified method. The figures also present the values of bias (a) and dispersion (σ), for which the solid lines and the dotted lines denote the regression lines and the lines of $\theta_i^P / \theta_i = 1$, respectively. Based on Figures 5–7, the maximum peak inter-story drift ratios estimated by the simplified method agree fairly well with those computed via NTHA using the elaborate analytical model, irrespective of the difference in the damper parameters with respect to the stiffness ratio κ and the drift ratio υ. The biases and dispersions of the simplified method are in the range of 0.995–1.01 and 0.041–0.053, respectively, for the three-story frame, as shown in Figure 5; 0.992–1.012 and 0.051–0.075, respectively, for the six-story frame, as shown in Figure 6; and 1.005–1.033 and 0.059–0.075, respectively, for the twelve-story frame, as shown in Figure 7. The ranges of the bias and dispersion of the simplified method increase as the height of the structure increases. However, the value does not affect the estimation of the accuracy of the simplified method because the increase in the value is very small.

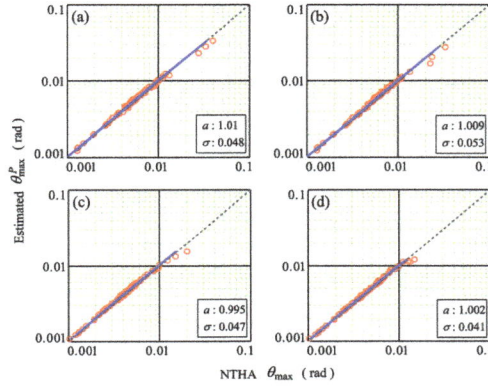

Figure 5. Regressions of θ_{max} on θ_{max}^P for a three-story frame with dampers: (**a**) $\kappa = 1.5$ and $\upsilon = 0.4$; (**b**) $\kappa = 1.5$ and $\upsilon = 0.6$; (**c**) $\kappa = 3$ and $\upsilon = 0.4$; and (**d**) $\kappa = 3$ and $\upsilon = 0.6$.

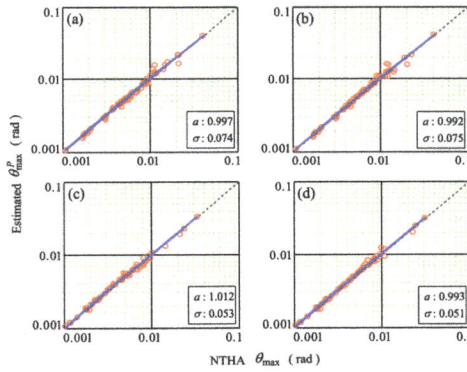

Figure 6. Regressions of θ_{max} on θ_{max}^P for a six-story frame with dampers: (**a**) $\kappa = 1.5$ and $\upsilon = 0.4$; (**b**) $\kappa = 1.5$ and $\upsilon = 0.6$; (**c**) $\kappa = 3$ and $\upsilon = 0.4$; and (**d**) $\kappa = 3$ and $\upsilon = 0.6$.

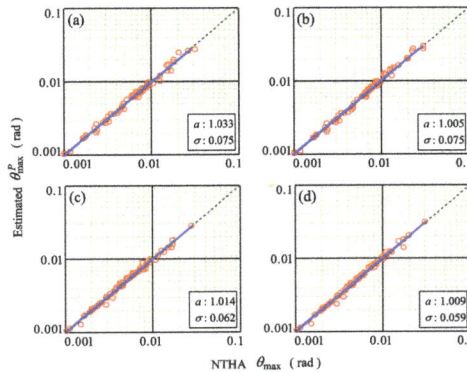

Figure 7. Regressions of θ_{max} on θ_{max}^P for a twelve-story frame with dampers: (**a**) $\kappa = 1.5$ and $\upsilon = 0.4$; (**b**) $\kappa = 1.5$ and $\upsilon = 0.6$; (**c**) $\kappa = 3$ and $\upsilon = 0.4$; and (**d**) $\kappa = 3$ and $\upsilon = 0.6$.

Figure 8 summarizes the bias (*a*) and the dispersion (σ) of the simplified method for: (a) a three-story frame; (b) a six-story frame; and (c) a twelve-story frame with dampers that are subjected to all the earthquake records. The biases and dispersions of θ_i^P estimated by the simplified method are in the range of 0.972–1.151 and 0.025–0.179, respectively, for all the frames. In the case where $\kappa = 1.5$, as shown in Figure 8c, the biases of θ_i^P estimated by the simplified method exceed 1.1 for the upper stories (i.e., 9–12 stories) of the twelve-story frame with dampers. However, the biases of θ_i^P for the other stories of all the frames are less than 1.1. Additionally, the dispersions of θ_i^P estimated by the simplified method are less than 0.02 for all the stories of all the frames. Therefore, the accuracy of the simplified method is sufficient for evaluating the seismic demands, irrespective of the difference in the damper parameters.

In this paper, the assumption $E_{d,i} = \overline{E}_{d,i}$ is used to convert the steel frame with dampers into an equivalent SDOF system. To confirm this assumption, Figure 9 shows a one-to-one comparison between the total energy dissipated by the dampers in the elaborate analytical model and the total energy dissipated by the inelastic springs in the equivalent SDOF system. The vertical axis represents the total energy dissipated by the springs of the equivalent SDOF system, and the horizontal axis signifies the total energy dissipated by the dampers of the elaborate analytical model. As shown in Figure 9a, the total energies dissipated by the springs agree well with those dissipated by the dampers. In contrast, the total energy dissipated by the springs fluctuates slightly when compared with the total energy dissipated by the dampers, as shown in Figure 9b,c. However, the total energy dissipated by the springs agrees well with the total energy dissipated by the dampers, irrespective of the difference in the magnitude of an earthquake. Therefore, the accuracy of the equivalent SDOF system is sufficient to evaluate the seismic demands, irrespective of the difference in the damper parameters with respect to the stiffness ratio κ and the drift ratio υ.

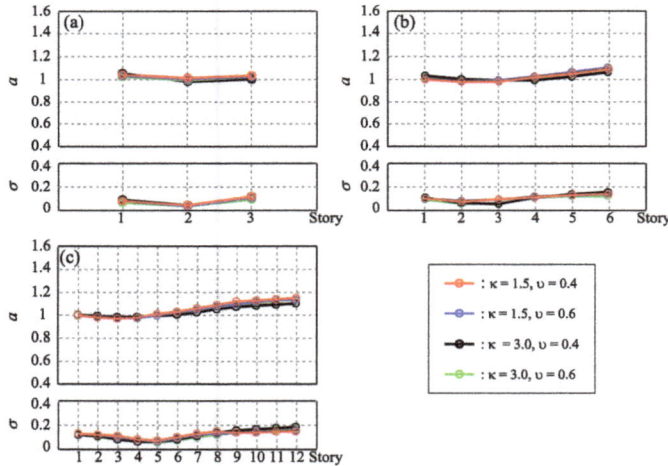

Figure 8. Bias (*a*) and dispersion (σ) of the simplified method: (**a**) for a three-story frame with dampers; (**b**) for a six-story frame with dampers; and (**c**) for a twelve-story frame with dampers.

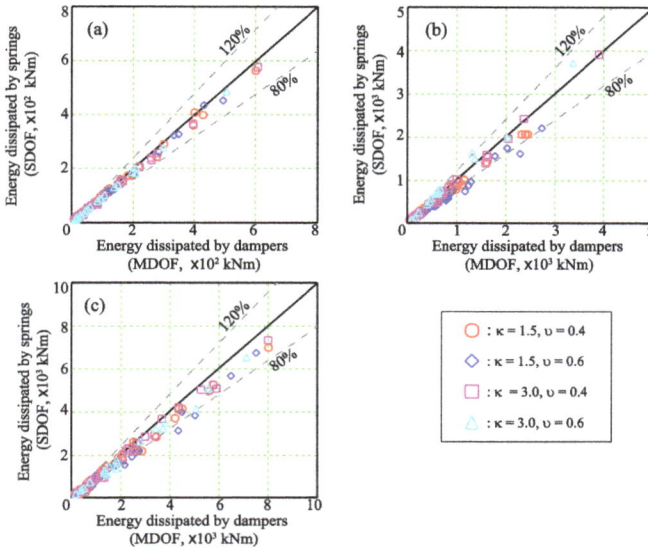

Figure 9. One-to-one comparison between energy dissipated by dampers in an elaborate analytical model and energy dissipated by springs in an equivalent inelastic SDOF system: (**a**) for a three-story frame with dampers; (**b**) for a six-story frame with dampers; and (**c**) for a twelve-story frame with dampers.

4. Conclusions

In this paper, a simplified method is proposed to estimate the peak inter-story drift ratio of steel multi-story frames with steel hysteretic dampers. In this method, with respect to the first mode, the inelastic spectral displacement of a steel frame with dampers is estimated by using an equivalent inelastic SDOF system that includes multi-springs in order to consider the hysteretic behavior of dampers, and the inelastic mode vector is estimated by using a pattern of story drifts that is obtained from NSPA of a steel frame with dampers with a lateral load pattern based on the first mode vector. Additionally, in order to consider the effects of modes higher than the first mode, the second and third modal elastic spectral displacement and the participation functions are also estimated by using the jth modal period, jth mode vector, and jth modal damping ratio, respectively, as obtained from EVA of the steel frame with dampers. In order to estimate the accuracy of the simplified method, the simplified method is compared by using the results of NTHA that use an elaborate analytical model on a series of three-story, six-story, and twelve-story SMRFs with dampers. In order to confirm the effects of the damper properties, such as stiffness and yield deformation, on the accuracy and stability of the simplified method, parametric analyses are performed for all the frames that are subjected to all the earthquake records. The results indicate that the accuracy of the simplified method is sufficient to evaluate the seismic demands, irrespective of the difference in damper parameters. Therefore, this method reduces the computational time and effort required to perform a seismic performance assessment.

Further, studies on different structural buildings such as a reinforced concrete structures and studies on different layouts of dampers are needed to gain more insight into the efficiency of the proposed method.

Acknowledgments: This work was supported by Grant-in-Aid for JSPS Fellows (No. 26.7166).

Author Contributions: Jae-Do Kang designed this work, conducted all the analytical work, and wrote the manuscript. Yasuhiro Mori supervised and reviewed this work.

Conflicts of Interest: The authors declare no conflict of interest.

References

1. Symans, M.; Charney, F.; Whittaker, A.; Constantinou, M.; Kircher, C.; Johnson, M.; McNamara, R. Energy dissipation systems for seismic applications: Current practice and recent developments. *J. Struct. Eng.* **2008**, *134*, 3–21. [CrossRef]
2. Pu, W.; Liu, C.; Zhang, H.; Kasai, K. Seismic control design for slip hysteretic timber structures based on tuning the equivalent stiffness. *Eng. Struct.* **2016**, *128*, 199–214. [CrossRef]
3. Benavent-Climent, A. An energy-based method for seismic retrofit of existing frames using hysteretic dampers. *Soil Dyn. Earthq. Eng.* **2011**, *31*, 1385–1396. [CrossRef]
4. Vargas, R.; Bruneau, M. Analytical response and design of buildings with metallic structural fuses. I. *J. Struct. Eng.* **2009**, *135*, 386–393. [CrossRef]
5. Tena-Colunga, A.; Nangullasmú-Hernández, H. Assessment of seismic design parameters of moment resisting RC braced frames with metallic fuses. *Eng. Struct.* **2015**, *95*, 138–153. [CrossRef]
6. Lin, Y.-Y.; Tsai, M.; Hwang, J.; Chang, K. Direct displacement-based design for building with passive energy dissipation systems. *Eng. Struct.* **2003**, *25*, 25–37. [CrossRef]
7. Kim, J.; Seo, Y. Seismic design of low-rise steel frames with buckling-restrained braces. *Eng. Struct.* **2004**, *26*, 543–551. [CrossRef]
8. Martínez, C.A.; Curadelli, O.; Compagnoni, M.E. Optimal placement of nonlinear hysteretic dampers on planar structures under seismic excitation. *Eng. Struct.* **2014**, *65*, 89–98. [CrossRef]
9. Kasai, K.; Fu, Y.; Watanabe, A. Passive control systems for seismic damage mitigation. *J. Struct. Eng.* **1998**, *124*, 501–512. [CrossRef]
10. Japan Society of Seismic Isolation. *Design and Construction Manual for Passively Controlled Buildings*, 2nd ed.; Japan Society of Seismic Isolation (JSSI): Tokyo, Japan, 2005.
11. Kasai, K.; Pu, W. Passive control design method for MDOF slip-hysteretic structure added with visco-elastic damper. *J. Struct. Constr. Eng.* **2010**, *75*, 781–790. [CrossRef]
12. Kasai, K.; Ogawa, R.; Pu, W.; Kiyokawa, T. Passive control design method for elasto-plastic frame added with visco-elastic dampers. *J. Struct. Constr. Eng.* **2010**, *75*, 1625–1633. [CrossRef]
13. Guo, J.W.W.; Christopoulos, C. Performance spectra based method for the seismic design of structures equipped with passive supplemental damping systems. *Earthq. Eng. Struct. Dyn.* **2013**, *42*, 935–952. [CrossRef]
14. Xue, Q.; Chen, C.C. Performance-based seismic design of structures: A direct displacement-based approach. *Eng. Struct.* **2003**, *25*, 1803–1813. [CrossRef]
15. Vamvatsikos, D.; Cornell, C.A. Incremental dynamic analysis. *Earthq. Eng. Struct. Dyn.* **2002**, *31*, 491–514. [CrossRef]
16. Seo, C.Y.; Karavasilis, T.L.; Ricles, J.M.; Sause, R. Seismic performance and probabilistic collapse resistance assessment of steel moment resisting frames with fluid viscous dampers. *Earthq. Eng. Struct. Dyn.* **2014**, *43*, 2135–2154. [CrossRef]
17. Silwal, B.; Ozbulut, O.E.; Michael, R.J. Seismic collapse evaluation of steel moment resisting frames with superelastic viscous damper. *J. Constr. Steel Res.* **2016**, *126*, 26–36. [CrossRef]
18. Karavasilis, T.L. Assessment of capacity design of columns in steel moment resisting frames with viscous dampers. *Soil Dyn. Earthq. Eng.* **2016**, *88*, 215–222. [CrossRef]
19. Dimopoulos, A.I.; Tzimas, A.S.; Karavasilis, T.L.; Vamvatsikos, D. Probabilistic economic seismic loss estimation in steel buildings using post-tensioned moment-resisting frames and viscous dampers. *Earthq. Eng. Struct. Dyn.* **2016**, *45*, 1725–1741. [CrossRef]
20. Kim, J.; Shin, H. Seismic loss assessment of a structure retrofitted with slit-friction hybrid dampers. *Eng. Struct.* **2017**, *130*, 336–350. [CrossRef]

21. Luco, N.; Mori, Y.; Funahashi, Y.; Cornell, C.A.; Nakashima, M. Evaluation of predictors of non-linear seismic demands using 'fishbone'models of SMRF buildings. *Earthq. Eng. Struct. Dyn.* **2003**, *32*, 2267–2288. [CrossRef]

22. Luco, N. Probabilistic Demand Analysis, SMRF Connection Fractures, and Near-Source Effects. Ph.D. Thesis, Stanford University, Stanford, CA, USA, 2002.

23. Luco, N.; Cornell, C.A. Structure-specific scalar intensity measures for near-source and ordinary earthquake ground motions. *Earthq. Spectra* **2007**, *23*, 357–392. [CrossRef]

24. Kuramoto, H.; Teshigawara, M.; Okuzono, T.; Koshika, N.; Takayama, M.; Hori, T. Predicting the earthquake response of buildings using equivalent single degree of freedom system. In Proceedings of the 12th World Conference on Earthquake Engineering, Auckland, New Zealand, 30 January–4 February 2000; No. 1039.

25. Chopra, A.K.; Goel, R.K. A modal pushover analysis procedure for estimating seismic demands for buildings. *Earthq. Eng. Struct. Dyn.* **2002**, *31*, 561–582. [CrossRef]

26. Mori, Y.; Yamanaka, T.; Luco, N.; Cornell, C.A. A static predictor of seismic demand on frames based on a post-elastic deflected shape. *Earthq. Eng. Struct. Dyn.* **2006**, *35*, 1295–1318. [CrossRef]

27. Oviedo, J.A.; Midorikawa, M.; Asari, T. An equivalent SDOF system model for estimating the response of R/C building structures with proportional hysteretic dampers subjected to earthquake motions. *Earthq. Eng. Struct. Dyn.* **2011**, *40*, 571–589. [CrossRef]

28. Kang, J.-D.; Mori, Y. Simplified method for estimating inelastic seismic demand of multistory frame with displacement-dependent passive dampers. In Proceedings of the 5th Asia Conference on Earthquake Engineering, Taipei, Taiwan, 16–18 October 2014; No. 149.

29. *SNAP Technical Manual*, 7th ed.; Kozo System, Inc.: Tokyo, Japan, 2015.

30. Tothong, P.; Cornell, C.A. Structural performance assessment under near-source pulse-like ground motions using advanced ground motion intensity measures. *Earthq. Eng. Struct. Dyn.* **2008**, *37*, 1013–1037. [CrossRef]

*applied
sciences*

MDPI

Article

PTMD Control on a Benchmark TV Tower under Earthquake and Wind Load Excitations

Wei Lin [1,*], Gangbing Song [2] and Shanghong Chen [1]

[1] School of Civil Engineering, Fuzhou University, Fuzhou 350116, China; chenshanghong@fzu.edu.cn
[2] School of Mechanical Engineering, University of Houston, Houston, TX 77204, USA; GSong@Central.UH.EDU
* Correspondence: cewlin@fzu.edu.cn; Tel.: +86-181-0606-0906

Academic Editors: Dimitrios G. Aggelis and Gino Iannace
Received: 16 February 2017; Accepted: 12 April 2017; Published: 22 April 2017

Abstract: A pounding tuned mass damper (PTMD) is introduced by making use of the energy dissipated during impact. In the proposed PTMD, a viscoelastic layer is attached to an impact limitation collar so that energy can be further consumed and transferred to heat energy. An improved numerical model to simulate pounding force is proposed and verified through experimentation. The accuracy of the proposed model was validated against a traditional Hertz-based pounding model. A comparison showed that the improved model tends to have a better prediction of the peak pounding force. A simulation was then carried out by taking the benchmark Canton Tower, which is a super-tall structure, as the host structure. The dynamic responses of uncontrolled, TMD-controlled and PTMD controlled system were simulated under wind and earthquake excitations. Unlike traditional TMDs, which are sensitive to input excitations and the mass ratio, the proposed PTMD maintains a stable level of control efficiency when the structure is excited by different earthquake records and different intensities. Particularly, more improvement can be observed when an extreme earthquake is considered. The proposed PTMD was able to achieve similar, or even better, control effectiveness with a lower mass ratio. These results demonstrate the superior adaptability of the PTMD and its applicability for protection of a building against seismic activity. A parametric study was then performed to investigate the influence of the mass ratio and the gap value on the control efficiency. A comparison of results show that better control results will be guaranteed by optimization of the gap value.

Keywords: vibration control; pounding tuned mass (PTMD) damper; energy dissipation; super-high structure; earthquake excitation

---

1. Introduction

Along with the construction of higher and more complex civil structures, there have been concerns about their safety under natural disasters, such as typhoons and earthquakes. It is nearly impossible for the designers to fully predict the excitations during the service stage; if the designers try to enhance the capacity of the structure against all possibilities, the loadings will largely increase the cost of the construction. Structural control is considered to be an effective approach to adapt the host structures to different kinds of harmful excitations and enhance their safety and serviceability [1–3]. Passive control approaches, such as tuned-mass-dampers (TMD), viscous dampers, viscoelastic dampers, and base isolation techniques have advantages of low cost and easy implementation [4]. Particularly, TMD is considered to be one of the most effective passive devices and has many applications, such as the CN tower in Toronto and the Taipei 101 building [5]. It possesses the merit of easy maintenance and can be implemented even on existing structures. The TMD absorbs inertial force from the host structure to reduce its motion, with its effectiveness determined by its dynamic characteristics, stroke, and the amount of added mass it employs. The potential of the TMD has been verified in reducing vibrations under a variety of excitations [6]. A TMD consist of a mass, a damping mechanism, and a restoring mechanism with certain stiffness connected to the host structure. During operation, the TMD is tuned to a certain frequency and a certain amount of the vibration energy will be transferred to the movement of the mass of the device. However, studies have pointed out that passive devices are hard to adjust once they are installed and their robustness is questionable when the host structure is subjected to different excitations. Since TMD can only reduce vibration components whose frequencies are close to the tuned frequencies, it is believed that the most significant limitation of the TMD is its narrow effective bandwidth and the high sensitivity to even a small change in the tuning [7–10]. Thus, it is considered to be unsuitable for vibration control under broadband earthquake inputs [11].

As a result, various schemes have been proposed to improve the TMD for better robustness and reliability [12]. For example, a nonlinear hysteretic damper has been supplemented for enhanced energy dissipation [13–17]. Chung et al. [18] demonstrated the effectiveness of such a TMD by implementation in the Taipei 101 building. These enhanced TMDs were found to be more effective than the conventional ones in a certain range of excitation frequencies. Another common alternative is to incorporate TMDs with active and semi-active devices [19,20]. Active mass dampers (AMD) have been developed and widely used in the civil engineering field [21–23]. Active tuning of TMDs can be achieved by using active springs [24,25], and some scholars tried to use shape-memory alloy (SMA) to achieve on-line adaptation [26,27]. Nagarajaiah and his coworkers developed semi-active tuned mass dampers (STMD) using semi-active variable stiffness systems [28,29], the STMD is able to be tuned to the desired frequencies instantaneously. Some researchers incorporated magnetorheological (MR) dampers into TMD to enhance adaptability [30]. The MR damper generates extra dissipation energy in the TMD and emulates positive and negative stiffness in order to adjust the stiffness of the TMD [31]. Cai et al. [32] conducted an experiment on a cable installed with TMD-MR and great control results were obtained. Eason et al. [33] found that multiple semi-active tuned mass dampers can give significant reduction on both the steady-state and transient responses.

Although the active and semi-active TMDs possess better performance, they are also liable to control algorithms and must be equipped with sensors and external power sources, which are not guaranteed to function normally under severe events. The complicated setup of the system and time delay caused by computational times of the control system will also affect the control results. The actuator used in the control system may also have its own dynamics and interactions with other structural components, all of which need to be considered. They also have higher maintenance costs, thus limiting their applicability. Therefore, this paper attempts to seek possible application of a pounding tuned mass damper (PTMD) in the reduction of vibration of high-rise structures induced by typhoons and earthquakes. An improved numerical model to predict the pounding force from the impact between the steel mass and the viscoelastic layer will be proposed and validated through experimental data. By taking a super tall benchmark TV tower as the host structure, a comprehensive numerical study will be carried out to verify the control efficiency of the PTMD under wind load and earthquake excitations, and possibly surpass in performance of traditional TMD. A parameter study will be carried out to analyze the influence of parameters on the control results, while seeking better performance of the control system.

2. Schematic Model of a Pounding Tuned Mass Damper (PTMD)

Based on the fact that a large amount of energy can be dissipated during impact, a pounding tuned mass damper (PTMD) is developed by setting up a motion limitation around the mass of a tuned mass damper (TMD) device [34]. The basic schematic model of a PTMD device is shown in Figure 1. As shown in the figure, a viscoelastic (VE) layer is attached to the limitation collar. Unlike some limitation collars found in traditional TMD devices, the limitation can not only function to prevent extreme vibration of the mass, but it can also further dissipate vibration energy through the deformation of VE layer during impact. When a certain gap is set between the mass and the limitation, impact will happen between the mass and the VE layer as soon as their relative displacement reaches the gap value. As a result, during operation, this device can absorb vibration energy of the host structure and transfer it to the kinetic energy and potential energy of the attached mass, as well as some heat energy from the impact. The amount of kinetic energy is largely dependent on the weight and the moving speed of the mass; the potential energy is affected by the stiffness of the connection; the amount of impact energy will be dependent on the relative velocity and deformation of the VE layer. Therefore, by changing the mass, stiffness, and gap of the TMD will definitely vary the control efficiency under different cases.

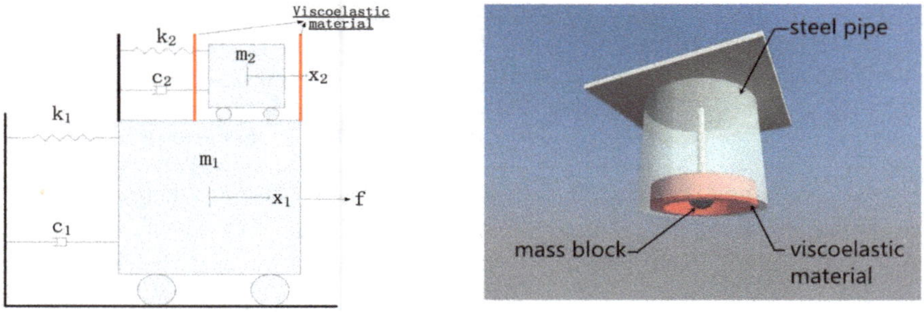

(a) Schematic model of a PTMD   (b) PTMD Prototype

Figure 1. Schematic model and prototype of the pounding tuned mass damper (PTMD).

3. Modeling of Structure Installed with PTMD

Assuming the host structure has n degree of freedoms (DOFs), a controlled structure coupled with a PTMD has the equation of motion as:

$$M\ddot{X} + C\dot{X} + KX = EF + DP \tag{1}$$

where M, C, K are the mass, damping, and stiffness matrices of the coupled system, which has $n + 1$ DOFs; F is the excitation force and P is the control force acting between the added mass and the host structure; matrices E and D are the index matrix indicating the DOFs of the excitation and the control forces, respectively.

By reordering the DOFs of the controlled system, making n represent the DOF of the host structure connected with PTMD, and r to represent the DOF of the added mass, the coupled equation is rewritten as:

$$
\begin{bmatrix} m_1 & & & \\ & \ddots & & \\ & & m_n & \\ & & & m_r \end{bmatrix}
\begin{bmatrix} \ddot{x}_1 \\ \vdots \\ \ddot{x}_n \\ \ddot{x}_r \end{bmatrix}
+
\begin{bmatrix} c_1 & \cdots & c_{1n} & 0 \\ \vdots & \ddots & \vdots & \vdots \\ c_{n1} & \cdots & c_n + c_r & -c_r \\ 0 & \cdots & -c_r & c_r \end{bmatrix}
\begin{bmatrix} \dot{x}_1 \\ \vdots \\ \dot{x}_n \\ \dot{x}_r \end{bmatrix}
+
\begin{bmatrix} k_1 & \cdots & k_{1n} & 0 \\ \vdots & \ddots & \vdots & \vdots \\ k_{n1} & \cdots & k_n + k_r & -k_r \\ 0 & \cdots & -k_r & k_r \end{bmatrix}
\begin{bmatrix} x_1 \\ \vdots \\ x_n \\ x_r \end{bmatrix}
=
\begin{bmatrix} f_1 \\ \vdots \\ f_n \\ f_r \end{bmatrix}
+
\begin{bmatrix} 0 \\ \vdots \\ p \\ -p \end{bmatrix}
\tag{2}
$$

where m_r represents the mass of the PTMD, c_r and k_r are, respectively, the connecting damping and stiffness of the PTMD, and p is the pounding force between the attached mass and the VE layer. If decoupling the system, one is able to obtain the internal forces, which is the control force exerted on the host structure. Extracting the last line in Equation (2), the equation of motion referring to the dynamic response of added mass is expressed as:

$$m_r\ddot{x}_r - c_r\dot{x}_n + c_r\dot{x}_r - k_r x_n + k_r x_r = f_r - p \tag{3}$$

By moving those terms related to x_n to the right side of the equation yields the governing equation of the added mass as:

$$m_r \ddot{x}_r + c_r \dot{x}_r + k_r x_r = k_r x_n + c_r \dot{x}_n + f_r - p \tag{4}$$

and $k_r x_n + c_r \dot{x}_n$ in above equation can be considered as the external force applied to the PTMD mass.

Further extracting those equations corresponding to the original system from Equation (2), the governing equation of the system is rewritten as:

$$\left[\begin{array}{cc} \mathbf{M}_n & 0 \end{array}\right]\left[\begin{array}{c} \ddot{X}_n \\ \ddot{x}_r \end{array}\right] + \left(\left[\begin{array}{cc} \mathbf{C}_n & 0 \end{array}\right] + \left[\begin{array}{cccc} 0 & \cdots & 0 & 0 \\ \vdots & \ddots & \vdots & \vdots \\ 0 & \cdots & c_r & -c_r \end{array}\right]\right)\left[\begin{array}{c} \dot{X}_n \\ \dot{x}_r \end{array}\right] + \left(\left[\begin{array}{cc} \mathbf{K}_n & 0 \end{array}\right] + \left[\begin{array}{cccc} 0 & \cdots & 0 & 0 \\ \vdots & \ddots & \vdots & \vdots \\ 0 & \cdots & k_r & -k_r \end{array}\right]\right)\left[\begin{array}{c} X_n \\ x_r \end{array}\right] = \mathbf{F} + \left[\begin{array}{c} 0 \\ p \end{array}\right] \tag{5}$$

where K_n, M_n, and C_n, respectively, represent the stiffness, mass, and damping matrices of the host structure. Reconstruction of the above equation yields:

$$M_n \ddot{X}_n + C_n \dot{X}_n + K_n X_n = F + C_d \left[\begin{array}{c} \dot{X}_n \\ \dot{x}_r \end{array}\right] + K_d \left[\begin{array}{c} X_n \\ x_r \end{array}\right] + \left[\begin{array}{c} 0 \\ p \end{array}\right] \tag{6}$$

where:

$$C_d = \left[\begin{array}{cccc} 0 & \cdots & 0 & 0 \\ \vdots & \ddots & \vdots & \vdots \\ 0 & \cdots & c_r & -c_r \end{array}\right], K_d = \left[\begin{array}{cccc} 0 & \cdots & 0 & 0 \\ \vdots & \ddots & \vdots & \vdots \\ 0 & \cdots & k_r & -k_r \end{array}\right] \tag{7}$$

By comparing with uncontrolled equation of motion, the control force F_{PTMD} acting on the host structure can be decoupled and extracted as:

$$F_{PTMD} = C_d \left[\begin{array}{c} \dot{X}_n \\ \dot{x}_r \end{array}\right] + K_d \left[\begin{array}{c} X_n \\ x_r \end{array}\right] + \left[\begin{array}{c} 0 \\ p \end{array}\right] \tag{8}$$

From the above equation one can see that, if the impact does not happen, $p = 0$ in Equation (8). In this case the PTMD is performing as a traditional TMD.

4. Improved Pounding Model

4.1. Improved Hertz Contact-Based Pounding Model

One of the most popular contact force models used in the impact events is proposed by Lankarani and Nikravesh and further improved for higher-impact velocities [35,36], this model gives an expression for the hysteresis damping factor relating the kinetic energy loss due to internal damping, and it has been utilized in many domains [37–39]. More recently, Gonthier et al. proposed a three-dimensional contact force model that can be used to simulate fully elastic to completely plastic impacts [40]. Another Hertz contact theory-based contact force model recently published was developed by Flores et al. [41]. This model uses a hysteresis damping parameter that accommodates the loss of energy during the contact process. Most of these pounding models have their respective suitable domains of applications, and most of them can only be accurate when the impact objects are considered to be rigid, the contact area is small, and the impact velocity is not too large.

For the particular pounding cases in the proposed PTMD, the pounding happens between a rigid mass and a VE layer with high nonlinearity. Viscoelastic material is featured by the large deformation ability and energy-dissipating capacity and, thus, energy can be further dissipated by the damping effect. A numerical pounding force model is required to decide the interaction forces between the host structure and the added mass. A non-linear viscoelastic model based on the Hertz contact law in conjunction with a damper that is active only during the approach period of the impact has been used to analyze the structural pounding in the previous study [42]. The model can be denoted as:

$$
F = \begin{cases}
\beta(u_1 - u_2 - g_p)^{3/2} + c(\dot{u}_1 - \dot{u}_2) & u_1 - u_2 - g_p > 0 \text{ and } \dot{u}_1 - \dot{u}_2 > 0 \\
\beta(u_1 - u_2 - g_p)^{3/2} & u_1 - u_2 - g_p > 0 \text{ and } \dot{u}_1 - \dot{u}_2 < 0 \\
0 & u_1 - u_2 - g_p < 0
\end{cases}
\tag{9}
$$

where u_1 and u_2 are the displacements of the device and pounding layers, and g_p is the distance between the device and pounding layers. $u_1 - u_2 - g_p$ is the relative pounding displacement and $\dot{u}_1 - \dot{u}_2$ is the pounding velocity. β is the pounding stiffness coefficient that mainly depends on material properties and the geometry of colliding bodies, and c is the impact damping which, at any instant of time, can be obtained from the formula:

$$
c = 2\xi \sqrt{\beta \sqrt{u_1 - u_2 - g_p} \frac{m_1 m_2}{m_1 + m_2}}
\tag{10}
$$

$$
\xi = \frac{9\sqrt{5}}{2} \frac{1 - e^2}{e(e(9\pi - 16) + 16)}
\tag{11}
$$

where m_1 and m_2 are the masses of the two colliding bodies, and ξ is the impact damping ratio correlated with the coefficient of restitution e, which is defined as the relation between the post-impact (final) relative velocity, $\dot{u}_1^f - \dot{u}_2^f$ and the prior-impact (initial) relative velocity, $\dot{u}_1^0 - \dot{u}_2^0$, of two colliding bodies:

$$
e = \frac{\left| \dot{u}_1^f - \dot{u}_2^f \right|}{\dot{u}_1^0 - \dot{u}_2^0}
\tag{12}
$$

The coefficient of restitution e can also be determined by calculating the ratio of the rebound height h^f and the original height h^0:

$$
e = \sqrt{\frac{h^f}{h^0}}
\tag{13}
$$

The case when $e = 1$ denotes a fully elastic collision, whereas $e = 0$ represents a perfectly plastic impact.

However, the above Hertz contact-based model expressed by Equations (9)–(13) is more accurate when describing the pounding force between two layers with higher stiffness. In the previous application, this model tends to underestimate the pounding forces. After a careful examination of the deformation features of the viscoelastic layer during impact, a modified pounding model is proposed as follows: In the proposed modified model, it is assumed that during the approaching period, the impact damping, c, is not only affected by the stiffness, but also influenced by the different extent of compressed areas which can be considered as a function of the approaching velocity, denoted as:

$$
\bar{c}(t) = 2\bar{\xi}_1 \sqrt{\bar{\beta} \sqrt{\delta(t)} \frac{m_1 m_2}{m_1 + m_2}} + \bar{\xi}_2 \dot{\delta}(t)^{s_2}
\tag{14}
$$

in which $\bar{\xi}_1$ is the damping ratio correlated with the coefficient of restitution e, which is expressed by Equation (12), and $\bar{\xi}_2$ is the damping ratio correlated with approaching velocity.

$$
\begin{aligned}
F(t) &= \bar{\beta}\delta^{s_1}(t) + \bar{c}(t)\dot{\delta}(t) \quad \left(\dot{\delta}(t) > 0 \right) \\
F(t) &= \bar{\beta}\delta^{s_1}(t) \quad \left(\dot{\delta}(t) < 0 \right)
\end{aligned}
\tag{15}
$$

$\bar{\beta}, \bar{\xi}_2, s_1, s_2$ are parameters to be decided in this model. During application, these parameters are relevant to the characteristic of the viscoelastic material.

4.2. Experimental Validation

A small-scale experiment is carried out for validation of the improved numerical pounding force model [34]. Figure 2 shows the setup of the experiment. A mass is driven by an attached motor and impact happens between the mass and the surface of a semi-ring to which seven layers of 3M VHB4936 tape adhere. A force sensor is installed beneath the semi-ring to measure the pounding force, and a Keyence Lbl1 laser sensor (Keyence, Osaka, Japan) is installed to capture the displacement of the mass. The data is collected by dSPACE 1104 (dSPACE, Shanghai, China) and the sampling frequency is set as 1000 Hz. The measured pounding force and displacement are shown in Figure 3.

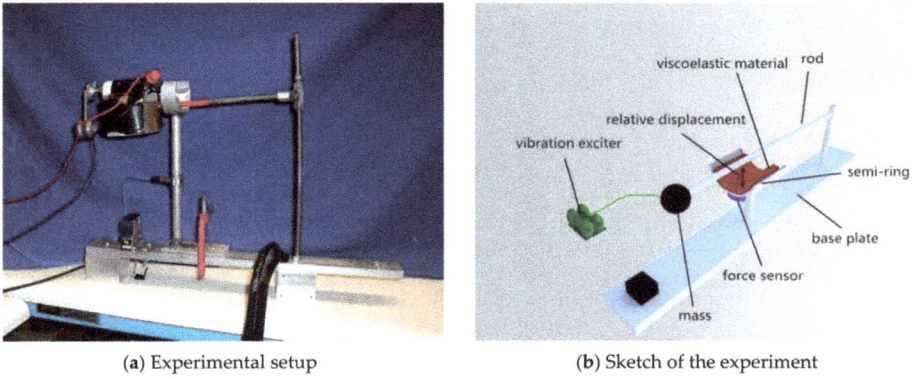

(**a**) Experimental setup

(**b**) Sketch of the experiment

Figure 2. Experimental setup for the validation experiment. Upon excitation of the motor, the horizontal rod holding the mass will impact repeatedly on the viscoelastic layer.

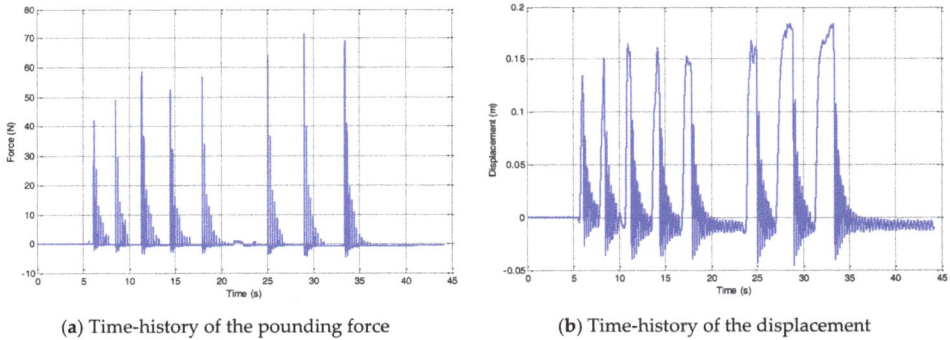

(**a**) Time-history of the pounding force

(**b**) Time-history of the displacement

Figure 3. Pounding force and displacement recorded during the experiment.

4.3. Parameter Tuning

For the selected material, the parameters for both the traditional Hertz-based model, as well as the five parameters for the modified pounding model, are optimized. The minimization problem is defined and the following residual vector is to be minimized:

$$J = \mathbf{R}^T \mathbf{W} \mathbf{R} \tag{16}$$

$$R = \left\{ \begin{array}{cc} R^p & R^s \end{array} \right\}^T \tag{17}$$

Here W is a weighting factor applied to different residual items, R^p represents the differences of the peak pounding force between the experimental and simulation data, and R^s represents the similarity of the pounding force vector. R^p and R^s are defined as:

$$R^p(i) = \frac{|\max(F^E) - \max(F^S)|}{\max(F^E)} \tag{18}$$

$$R^S(i) = 1 - \frac{\{\phi_i^E\}^T \{\phi_i^S\}}{\left(\{\phi_i^E\}^T \{\phi_i^E\}\right)\left(\{\phi_i^S\}^T \{\phi_i^S\}\right)} \tag{19}$$

in which $\{\phi\}$ is the pounding force vector, the superscripts 'E' and 'S' represent the items associated with the experimental and the simulated data, respectively, and 'T' denotes vector (matrix) transpose.

A Hertz contact law-based pounding model is adopted in the previous study by Zhang et al. [34]. In the previous model, only the pounding stiffness β needs to be estimated. In this modified model, a trust-region based optimization method with global minimization is adopted to decide the parameters β_1, ξ_2, s_1, s_2. Ten sets of pounding data were chosen for model tuning and the optimized parameters are listed in Table 1. The tuned model is then used to predict the pounding force from the recorded displacement. The comparison of the simulated and experimental pounding force is shown in Figure 4. An obvious improvement is observed from these figures. By compensating the influence of the changing impact damping, the proposed model can now better predict the peak value of the pounding force during impact.

Table 1. Optimized parameters of the modified pounding model.

Parameters	β_1	ξ_2	s_1	s_2
value	10,560	0.8	1.3	1.1

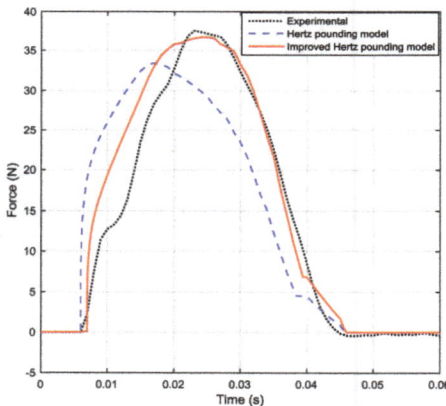

(a) Pounding Case 1   (b) Pounding Case 2

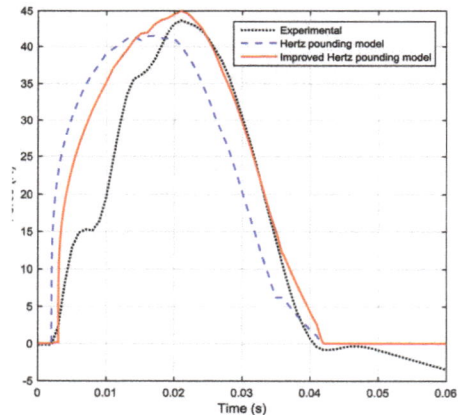

Figure 4. *Cont.*

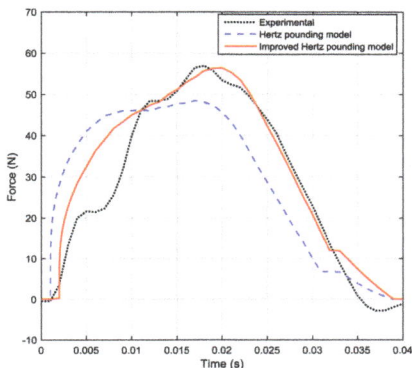

(c) Pounding Case 3

Figure 4. Comparison of the improved and traditional pounding force models.

5. SHM System and Benchmark Model of the Canton Tower

In order to study the feasibility of the PTMD on vibration control of civil structures under wind load and earthquake excitations, a benchmark model of the Canton Tower is selected as the host structure. The Canton Tower is located in Guangzhou, China. It has a total height of 600 m, includes a 454 m high main tower and a 146 m high antenna mast. The composite main tower has a reinforced concrete tube covered by a steel lattice. A sophisticated SHM system consisting of 16 types of more than 700 sensors has been installed in the tower. The SHM system has continuously acquired response data and loading data, such as temperature and wind speed, since 2010. As of today, the SHM system has successfully recorded the dynamic responses of the structure under several severe events, such as earthquakes and typhoons. Sets of ambient data have been adopted to verify the correctness of the full-order and reduced-order models in the previous study [43]. Particular wind speed data and wind rose diagrams drawn based on the monitor on top of the main tower during typhoons are shown in Figure 5. In the following simulation of this paper, two sets of the wind load data will be extracted from selected typhoons to verify the control effectiveness of the proposed PTMD device.

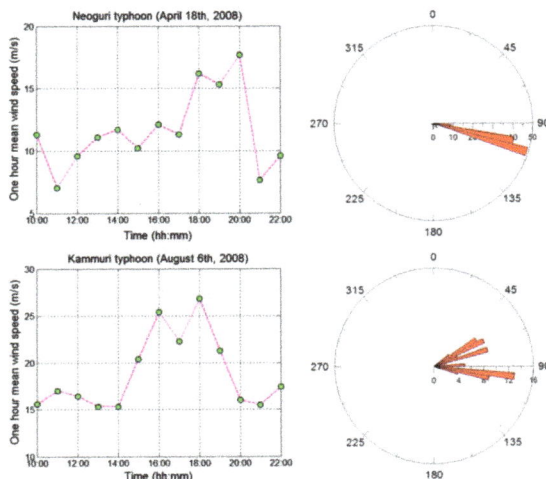

Figure 5. Wind speed and wind rose diagrams obtained during typhoons.

The benchmark problem for this tower originates from the structural health monitoring and damage detection problem of super-high-rise structures. Both a full-order FE model and a corresponding reduced-order model are established by the first author of this paper. Since the purpose of this paper will focus mainly on the global dynamic responses of the structure, the reduced order FE model is adopted in this study. The reduced-order model has been validated in the previous study [43] by field test data to have great precision and can describe the dynamic characteristic of the structure well. In the reduced-order model, the whole structure is modeled as 37 beam elements (27 for the main tower and 10 for the mast) and 38 nodes. Considering five DOFs for each node, which are the two horizontal translational and three rotational DOFs, a total number of 185 DOFs are included. The damping ratio of the structure was assumed to be 0.004 [44,45]. The layout of the model [46] is shown in Figure 6. Directions x and y represent the direction along the long axis and the short axis, respectively.

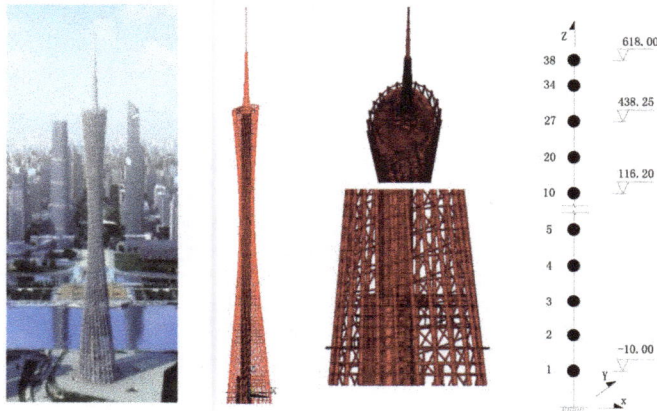

Figure 6. Benchmark model of Canton Tower.

6. Vibration Control with PTMD

6.1. Design of the PTMD

In this study, the PTMD is assumed to be effective in both directions, which means the impact may happen in any horizontal direction. Thus, pounding force can always be decomposed to the components along the x and y directions in every impact. The PTMD is first designed as an optimal TMD whose stiffness is first tuned to match a certain natural frequency of the host structure. The first 10 natural frequencies obtained from modal analysis results are listed in Table 2. The first and the second modes of the structure appear to be bending modes along the long and short axes. The PTMD here is assumed to be tuned as 0.11 Hz, the same as the first natural frequency of the structure. The thickness of the viscoelastic layer was set to 0.6 m during the simulation. As shown in Figure 1, the impact will occur as soon as the mass contacts the surface of the limitation device; and by assuming the viscoelastic material can be fully compressed, the maximum allowance for the deformation of the viscoelastic layer will also be 0.6 m.

Table 2. Natural frequencies of the reduced-order model.

Mode	1	2	3	4	5
Frequency (Hz)	0.110	0.159	0.347	0.368	0.399
Mode	6	7	8	9	10
Frequency (Hz)	0.460	0.485	0.738	0.902	0.997

Usually, for this kind of high-rise structure, the optimal position to install a TMD device is on top of the structure [47]. However, considering the mast is made of lightweight steel and has a relatively small section, the local safety cannot be guaranteed if too much weight is connected. Additionally, it will be difficult to install at a certain height. Therefore, in this study, the top of the main tower is selected to be the optimal position of the PTMD. The corresponding nodal number is 27 in the reduced order model. For TMD and PTMD controls, the control effectiveness is how effective the PTMD is in controlling vibration. Theoretically, for a TMD device, the weight of the mass has to reach a certain mass ratio for obvious control effectiveness. However, for real complex civil structures, it is difficult to reach the ideal weight due to the carrying capacity of local elements around the connection. Examples of the world's largest TMD systems include the Taipei 101 building with a weight of 660 t, the Shinjuku Nomura building at 1400 t (two units), and the largest at the Shinjuku Mitsui building at 1800 t (six units) [48]. The TMD is only designed and installed for wind-induced vibration control purposes. In this study, after trial calculation, the weight of the mass used in the PTMD device is chosen to be varied from 100 t to 500 t which, when compared to the first order modal mass, 2.17×10^7 kg, the mass ratio is between 0.005 and 0.023. The stiffness of the connection and the optimal damping ratio is calculated as [49]:

$$k = m_0(2\pi f_1)^2 \tag{20}$$

$$\zeta_{opt} = \sqrt{\frac{3\mu_t}{8 \cdot (1 + \mu_t)}} \tag{21}$$

where f_1 is the desired frequency to be tuned, usually set to be the first natural frequency of the structure, m_0 is the weight of the mass attached to the structure, and $\mu_t = m_0/m_1$ denotes the mass ratio of the mass to the generalized mass of the tuned mode.

The aforementioned pounding model is adopted for the simulation of the pounding force. The impact is considered to happen as soon as the relative displacement of the mass and the connected node is less than the selected gap. For the proposed PTMD device, the preset gap value significantly affects the control effectiveness. If the gap is too large, pounding can hardly happen or too small a pounding force is generated due to the small relative velocity between the mass and the limitation. However, if the preset gap value is too small, the movement of the mass will be confined to a relatively small amplitude, and the device will not be able to consume more pounding energy since the relative velocity will also be small in this case. As a result, it is believed the selection of the gap value will be influenced by different control purposes, the frequency components, and the intensity of the excitation. The influences of the gap value will be discussed later in the numerical simulation.

In the following simulation, the control effectiveness η was defined to as a metric to evaluate the control performance of the damper. The control effectiveness is expressed as:

$$\eta = \frac{R_u - R_c}{R_u} \times 100\% \tag{22}$$

where R_u and R_c refer to the uncontrolled and controlled responses, respectively.

6.2. Wind Load Excitation Cases

Time domain analyses are employed to calculate the responses of the structure under typical typhoons and stochastic wind loads. In typhoon cases, the time history of the wind speed is selected from the record of an anemometer installed on the Canton Tower. Typical recorded wind speeds and wind angle data are shown in Figure 7. It is assumed that wind load only acts on each node along the long and short axes. Information about the shape coefficient is used to generate the wind load from the wind speed, and is collected from the wind tunnel experiments done by Tongji University, China. The approximate windward area of each section is derived from the full-order FE model of the structure. After generating, the wind load will be decomposed into two directions, along the long-axis and short-axis, respectively. The composition force along the two directions will then be calculated and applied to the corresponding nodes of the reduced order model. For simplification, it is assumed that the segment between two adjacent nodes share the same wind load value. The time-history displacement and acceleration responses of the uncontrolled, TMD-controlled, and PTMD-controlled systems are shown in Figure 8. A summary of the peak and RMS responses on the mast top and the tower top of the structure are listed in Tables 3 and 4. In these cases, the weight of the mass is fixed as 300 t for comparison. It should be noted that, as shown in Figure 8b, even though the wind excitation can be quite severe, the PTMD did not contribute much vibration mitigation. The reason for the lack of vibration suppression can be seen in Figure 8c, where one can see that the pounding force was insufficient despite the wind input.

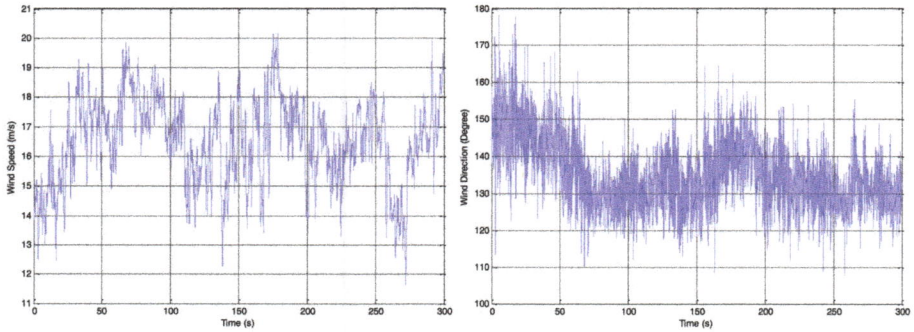

Figure 7. Time history record of wind speed and direction on top of the main tower (Typhoon Nanmado, 2011).

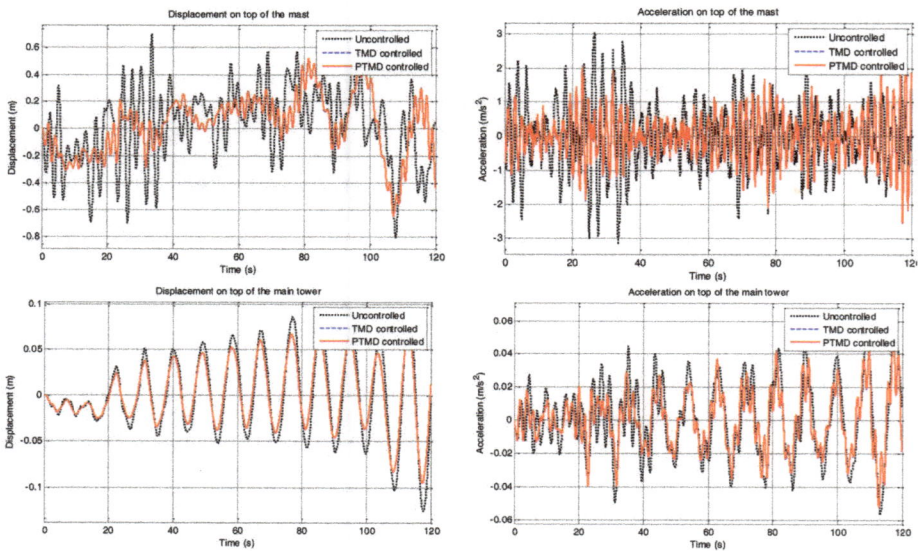

(**a**) Responses along *x*-axis

Figure 8. *Cont.*

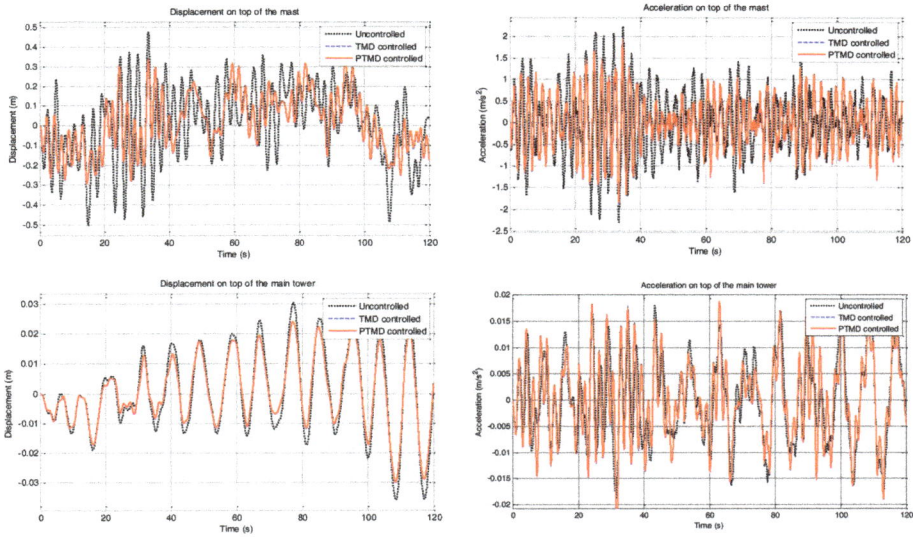

(b) Responses along *y*-axis

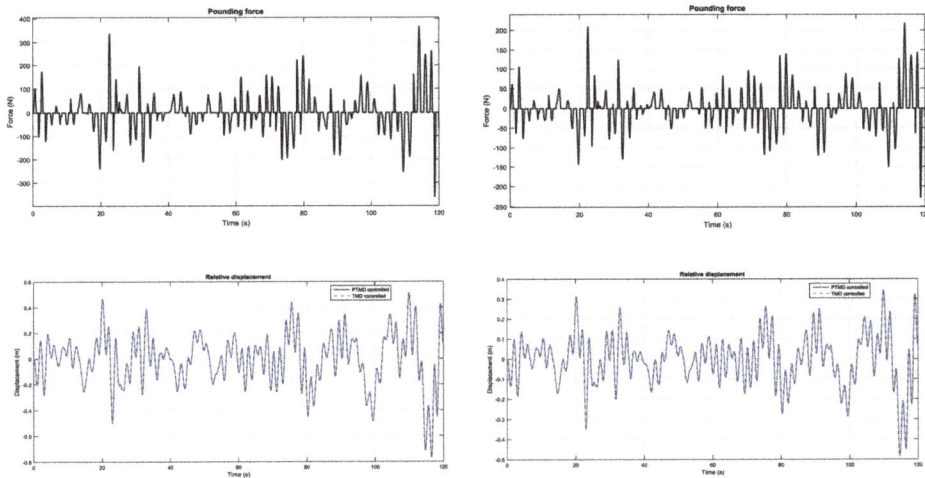

(c) Pounding forces and relative displacements along *x*- and *y*-axis (gap: 0.05 m)

Figure 8. Time-history responses under Typhoon Haima excitation (2011).

Under typhoon excitation, it is shown that the proposed PTMD shares similar control effectiveness; only slight improvement is noticed on the peak and the RMS values. It can be observed from the tables that both control approaches can reduce the peak and RMS responses. On the mast top, the peak displacement responses are reduced by 19.14% and 34.75% subjected to Typhoon Haima and Typhoon Nanmado excitations, respectively, and the RMS displacement responses are reduced by 21.50% and 33.23%. Meanwhile, the peak displacement response along the *y*-direction on top of the mast are reduced by 24.24% and 32.86% under these typhoon excitations. When it comes to the acceleration responses, 18.46% and 28.15% control effectiveness values are obtained for peak responses on top of the mast, and 30.84% and 32.53% for the reduction of the RMS responses. As listed in the table, the

PTMD is able to reduce the peak displacement responses on top of the main tower by 24.23% and 20.72%, but it is also important to notice that the peak acceleration responses along the short axis on the tower top are slightly increased when the TMD or PTMD is installed.

Table 3. Comparison of peak and RMS responses under Typhoon Haima excitation.

Responses		Uncontrolled	TMD Controlled	PTMD Controlled
Mast top (*long axis*)	Peak displacement (m)	0.802	0.637	0.636
	RMS displacement (m)	0.278	0.238	0.238
	Peak acceleration (m/s²)	3.138	2.375	2.346
	RMS acceleration (m/s²)	0.987	0.788	0.781
Tower top (*long axis*)	Peak displacement (m)	0.126	0.092	0.092
	RMS displacement (m)	0.044	0.034	0.034
	Peak acceleration (m/s²)	0.065	0.053	0.053
	RMS acceleration (m/s²)	0.023	0.018	0.018
Mast top (*short axis*)	Peak displacement (m)	0.502	0.397	0.396
	RMS displacement (m)	0.190	0.150	0.150
	Peak acceleration (m/s²)	2.288	1.558	1.543
	RMS acceleration (m/s²)	0.758	0.526	0.522
Tower top (*short axis*)	Peak displacement (m)	0.036	0.032	0.032
	RMS displacement (m)	0.014	0.012	0.012
	Peak acceleration (m/s²)	0.019	0.022	0.022
	RMS acceleration (m/s²)	0.008	0.008	0.008

Table 4. Comparison of peak and RMS responses under Typhoon Nanmado excitation.

Responses		Uncontrolled	TMD Controlled	PTMD Controlled
Mast top (*long axis*)	Peak displacement (m)	1.128	0.742	0.736
	RMS displacement (m)	0.376	0.251	0.251
	Peak acceleration (m/s²)	5.445	3.913	3.857
	RMS acceleration (m/s²)	1.713	1.155	1.145
Tower top (*long axis*)	Peak displacement (m)	0.081	0.064	0.064
	RMS displacement (m)	0.028	0.022	0.022
	Peak acceleration (m/s²)	0.070	0.065	0.065
	RMS acceleration (m/s²)	0.023	0.018	0.018
Mast top (*short axis*)	Peak displacement (m)	0.726	0.491	0.487
	RMS displacement (m)	0.271	0.164	0.164
	Peak acceleration (m/s²)	3.558	2.662	2.623
	RMS acceleration (m/s²)	1.289	0.773	0.766
Tower top (*short axis*)	Peak displacement (m)	0.039	0.033	0.033
	RMS displacement (m)	0.012	0.010	0.010
	Peak acceleration (m/s²)	0.031	0.037	0.037
	RMS acceleration (m/s²)	0.011	0.012	0.012

Stochastic wind load is generated from the Davenport spectrum to further verify the effectiveness of the devices. The AR method is utilized here for stochastic wind load generation. The time-history of the generated wind speed and the responses along the long-axis on the tower top and mast top are shown Figures 9 and 10 and Table 5. The PTMD still observed similar control efficiency, not much better than that of the TMD, but the control efficiency is better than the typhoon excitation cases. Both of them are capable of reducing the peak and RMS responses by 57.79% and 49.65%, respectively.

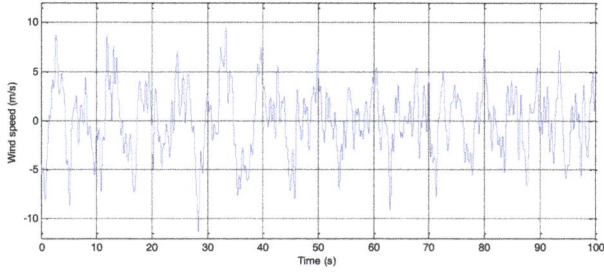

Figure 9. Generated wind speed on top of Canton Tower.

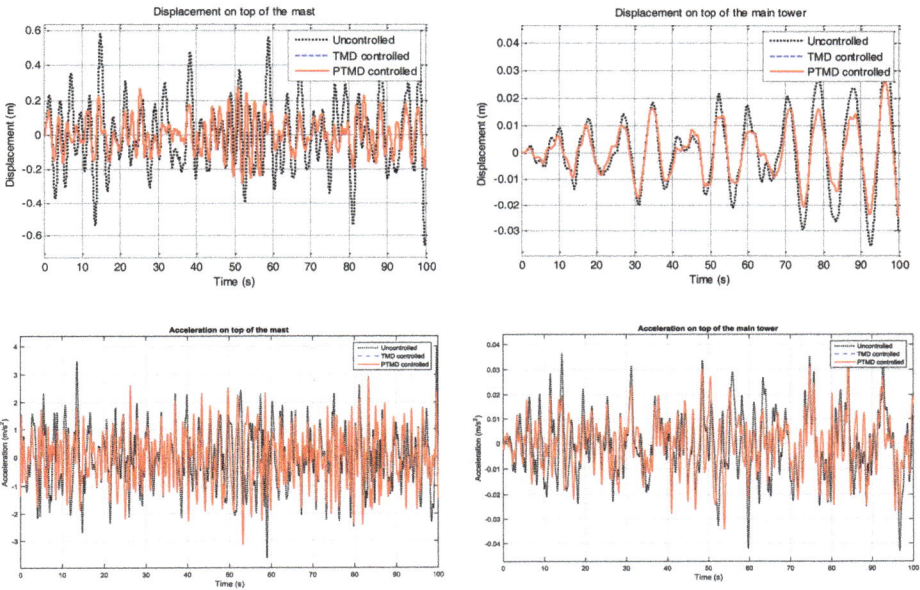

(**a**) Responses along long axis

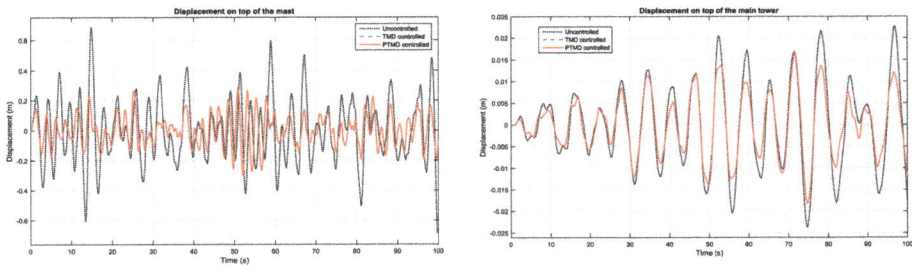

Figure 10. *Cont.*

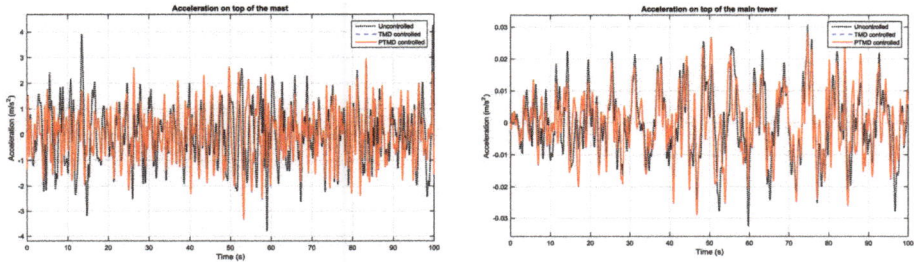

(**b**) Responses along short axis

Figure 10. Control results under a stochastic wind load.

Table 5. Comparison of peak and RMS responses under stochastic wind excitation.

Responses		Uncontrolled	TMD Controlled	PTMD Controlled
Mast top (*long axis*)	Peak displacement (m)	0.662	0.280	0.276
	RMS displacement (m)	0.203	0.102	0.101
	Peak acceleration (m/s^2)	3.976	3.170	3.139
	RMS acceleration (m/s^2)	1.068	0.935	0.929
Tower top (*long axis*)	Peak displacement (m)	0.042	0.026	0.026
	RMS displacement (m)	0.014	0.009	0.009
	Peak acceleration (m/s^2)	0.043	0.037	0.037
	RMS acceleration (m/s^2)	0.013	0.011	0.011
Mast top (*short axis*)	Peak displacement (m)	0.691	0.305	0.301
	RMS displacement (m)	0.213	0.105	0.104
	Peak acceleration (m/s^2)	4.253	3.359	3.323
	RMS acceleration (m/s^2)	1.130	0.955	0.949
Tower top (*short axis*)	Peak displacement (m)	0.031	0.020	0.020
	RMS displacement (m)	0.011	0.008	0.008
	Peak acceleration (m/s^2)	0.029	0.027	0.027
	RMS acceleration (m/s^2)	0.009	0.008	0.008

6.3. Optimization Control Effectiveness under Earthquake Excitation

The feasibility of the proposed PTMD on enhancing earthquake safety of the host structure is verified by subjecting the structure to real earthquake records. The El Centro earthquake (NS, 1980) and Tianjin earthquake records (NS, 1976) are adopted to excite the structure. A simulation is performed by tuning the peak acceleration of the excitations along the long-axis from 0.1 *g* to 0.3 *g*. The earthquake records are chosen based on the sites and the suggestions from the local design codes. Earthquake components along the long and short axes are considered simultaneously, and the ratio of peak acceleration is set to be 1:0.85 according to the Chinese Aseismic Design Code.

Unlike wind load excitation cases, earthquakes can generate more severe impacts between the mass and the host structure. Thus, a certain amount of energy can be consumed through the impact behavior. As mentioned before, the original gap value between the mass and the limitation should be selected carefully to guarantee beyond a certain impact level. Here the maximum gap value is decided based on the maximum relative displacement of the mass and the structure derived from TMD-controlled simulation results. Time history of uncontrolled, TMD-controlled, and PTMD-controlled displacement responses on the tower top are compared in Figure 11 when a 300 t mass is connected and the gap is set as 0.25 m. The corresponding pounding force is also shown in the figures. It is clearly observed from these results that the PTMD can produce much better performance on the reduction of earthquake-induced displacement. When the Tianjin earthquake strikes, the TMD failed to reduce the peak responses. On the contrary, the peak and RMS displacement responses are decreased by 60.10% and 64.82%, respectively, when the PTMD is installed. Data from Figures 11 and 12 suggests that higher acceleration is needed to enable significant displacement reduction. This was observed especially when the Tianjin earthquake record was used. On the other hand, the acceleration on the lower floor is not severe, as shown in Figure 13. For now this is a major drawback for PTMD control, and this problem may be improved by the use of multiple, smaller PTMDs to make the displacement over the whole structure smaller.

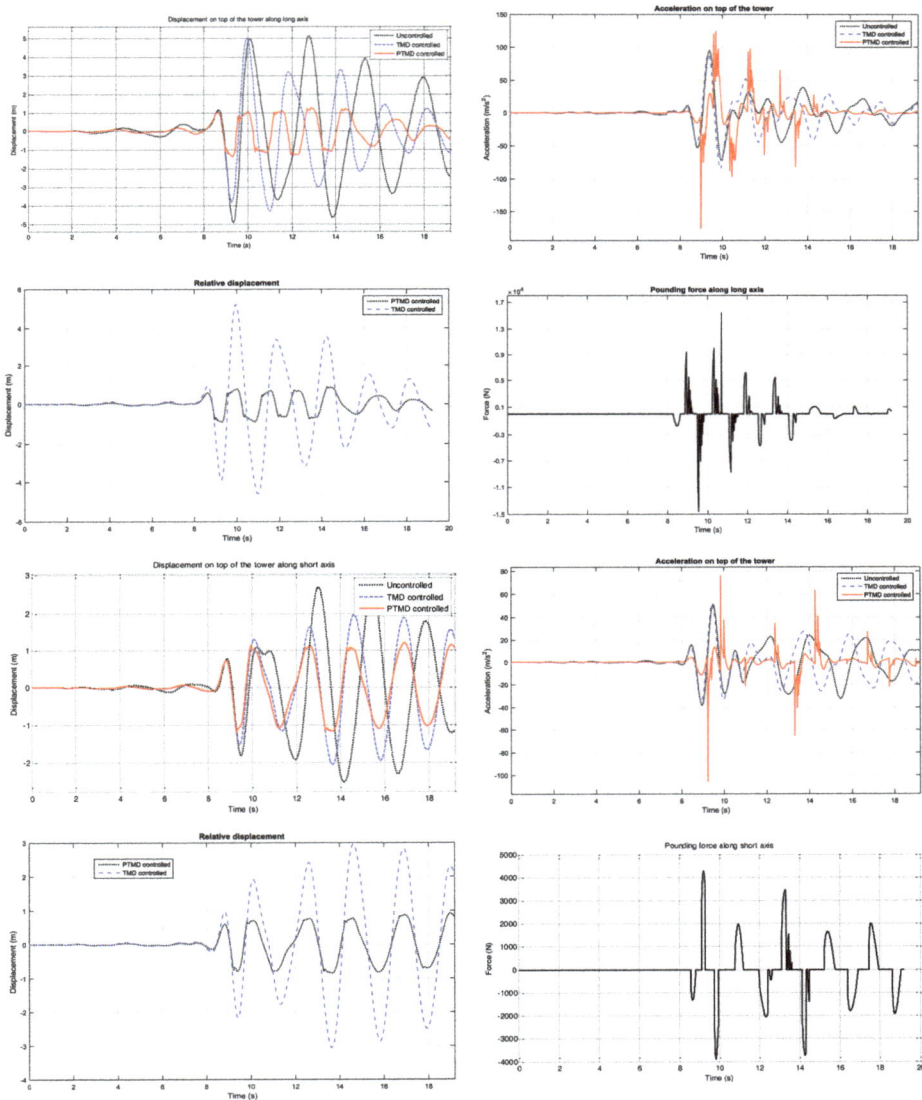

Figure 11. Comparison of the seismic responses (Tianjin earthquake record, 0.3 *g*).

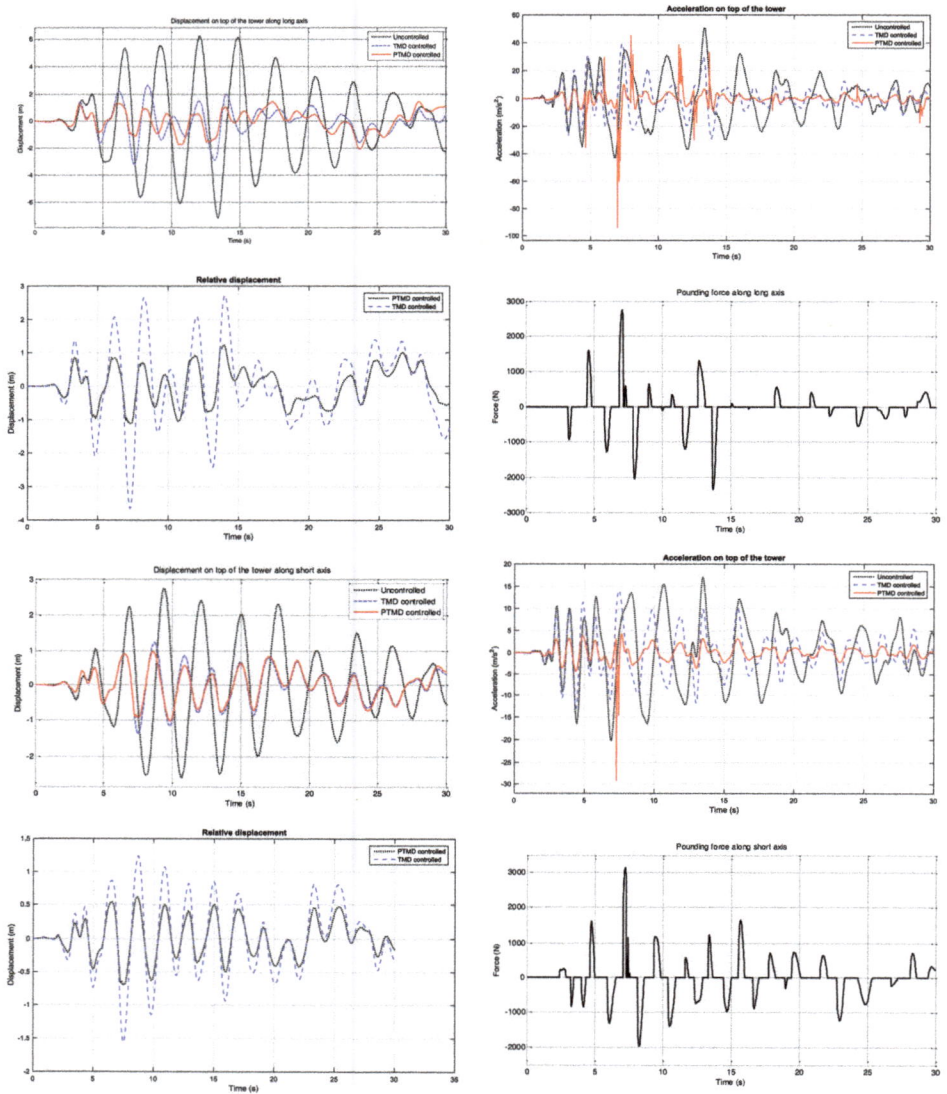

Figure 12. Comparison of the seismic responses. (El Centro earthquake record, 0.3 *g*).

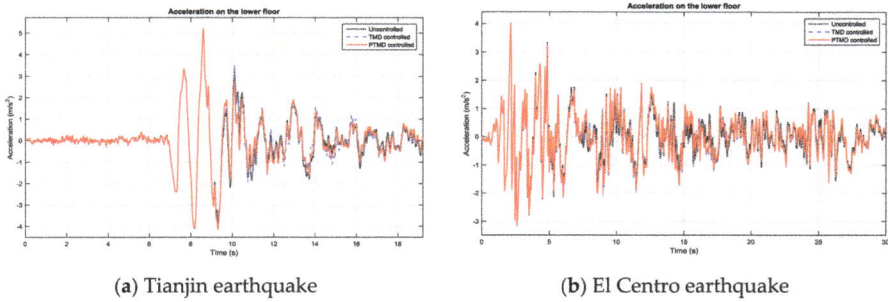

(**a**) Tianjin earthquake   (**b**) El Centro earthquake

Figure 13. Acceleration responses on the lower floor.

6.4. Control Performance with Improved Hertz Pounding Model

The comparison of time history displacement responses when using the improved pounding model and the Hertz-based pounding model is shown in Figure 14. As mentioned above, the traditional Hertz-based pounding model tends to underestimate the peak pounding forces especially when the relative velocities between the two impact bodies are large. The comparison results show that when the structure is subjected to the Tianjin earthquake, the improved Hertz model gives better control performances (approximate 32%) by fully considering the nonlinear damping coefficient during impact, and the dissipated energy during the impact behavior can be described more precisely. When the El Centro earthquake strikes, although the peak pounding force is larger, only slight improvement can be obtained when using the improved Hertz model. As similar comparison results between the control effectiveness of the TMD and PTMD controls is shown in Figure 12, the El Centro excitation did not generate a particularly severe impact compared to the Tianjin record. The dissipated energy through the impact behavior is relatively small and, in this case, the PTMD acts more like a TMD.

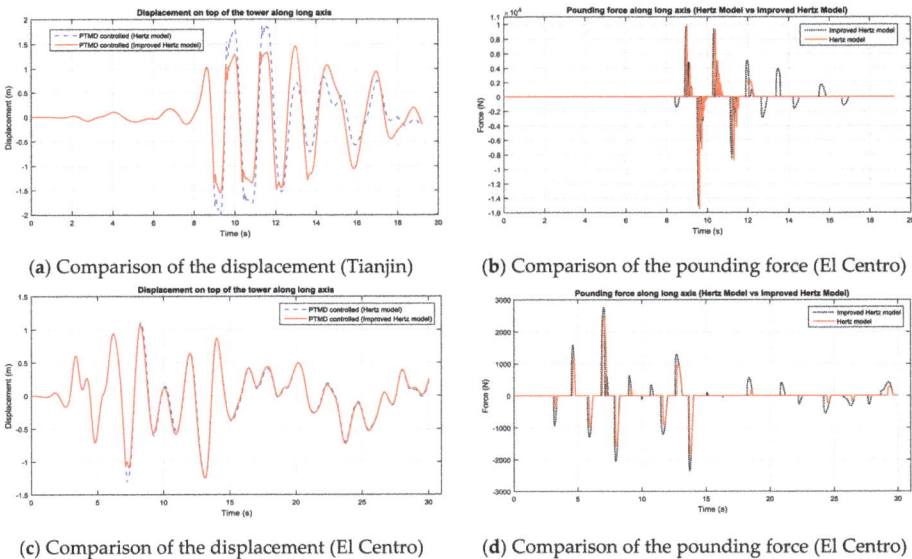

(**a**) Comparison of the displacement (Tianjin)   (**b**) Comparison of the pounding force (El Centro)

(**c**) Comparison of the displacement (El Centro)   (**d**) Comparison of the pounding force (El Centro)

Figure 14. Comparison of displacement responses and pounding force using different pounding force model.

6.5. Parameter Studies

In order to investigate the influence of the gap value on the control effectiveness, the peak and RMS displacement responses along the x direction are listed in Table 6 and Figure 15, with the gap value selected from 0.005 m to 3 m. The maximum gap is chosen with reference to the maximum relative displacement between the mass and the connected point during TMD cases' simulations. As listed in the table, when the gap becomes larger, the control effectiveness is approaching the TMD-controlled results, which means pounding can hardly happen or not enough impact energy is consumed. Along with the decreasing of the gap value, the control results improved until the gap reaches a certain value and starts to degenerate. This phenomenon arises from the fact that even though the gap value is small enough to constantly induce impact, the relative velocity before pounding is not large enough to dissipate more energy and yield deformation of the VE layer. In addition, the movement of the mass will also be confined by the small gaps and, thus, the mass will possess less kinetic and potential energy and reduce the control effectiveness.

Table 6. Comparison of control efficiency with the variation of gap value (mass: 200 t).

Excitation			Uncontrolled	TMD	PTMD Gap (m)								
					0.005	0.01	0.05	0.1	0.25	0.5	1	2	3
TJ	0.1 g	Peak	2.203	2.209	1.655	1.341	1.029	1.087	1.233	1.436	1.895	2.208	2.209
		RMS	0.883	0.768	0.434	0.377	0.336	0.360	0.423	0.540	0.680	0.768	0.768
	0.2 g	Peak	3.672	3.682	1.543	1.442	1.210	1.181	1.313	1.545	2.005	2.920	3.677
		RMS	1.472	1.281	0.531	0.487	0.464	0.470	0.475	0.501	0.673	1.055	1.279
	0.3 g	Peak	5.141	5.155	2.098	1.878	1.564	1.304	1.422	1.614	2.062	2.932	3.868
		RMS	2.060	1.792	0.534	0.506	0.493	0.507	0.566	0.655	0.681	1.030	1.436
EC	0.1 g	Peak	3.068	1.787	1.128	1.042	1.136	1.217	1.317	1.337	1.781	1.787	1.787
		RMS	1.252	0.499	0.420	0.422	0.437	0.457	0.483	0.494	0.498	0.499	0.499
	0.2 g	Peak	5.114	2.979	1.334	1.329	1.303	1.301	1.363	1.464	2.140	2.662	2.977
		RMS	2.087	0.831	0.555	0.555	0.550	0.558	0.592	0.657	0.785	0.828	0.831
	0.3 g	Peak	7.160	4.170	1.670	1.773	1.794	1.718	1.792	1.820	2.013	3.064	3.507
		RMS	2.922	1.164	0.697	0.698	0.743	0.721	0.752	0.773	0.894	1.130	1.155

By fixing the gap value to 0.05 m, the influence of the mass ratio can be examined (Table 7). As seen in the table, both the TMD and PTMD systems tend to perform better with a larger mass ratio. However, the performance of the TMD varies significantly with the changing mass value and the excitations. Results show that the TMD can perform well under the El Centro excitations—both the peak and the RMS responses can have a reduction over 40%. In the Tianjin earthquake excitation cases, even though the mass of the TMD increases from 200 t to 500 t, the control effectiveness of the peak responses can only increase from −0.28% to 12.65%. On the contrary, promising results in which less variation are observed in the PTMD-controlled cases with the changing value of the mass ratio. The control results on the peak and the RMS varies little and remains above 50% in every cases. These results also indicate that the PTMD is able to achieve better control results with a lower mass ratio, which will result in easier installation and lower cost of the whole controlled system. In addition, the trend of the PTMD control results presented with respect to the variation of the intensity of the earthquake input clearly shows better improvement on the control efficiency under extreme excitations.

(a) Control ratio of peak responses (Tianjin)

(b) Control ratio of peak responses (El Centro)

(c) Control ratio of RMS responses (Tianjin)

(d) Control ratio of RMS responses (El Centro)

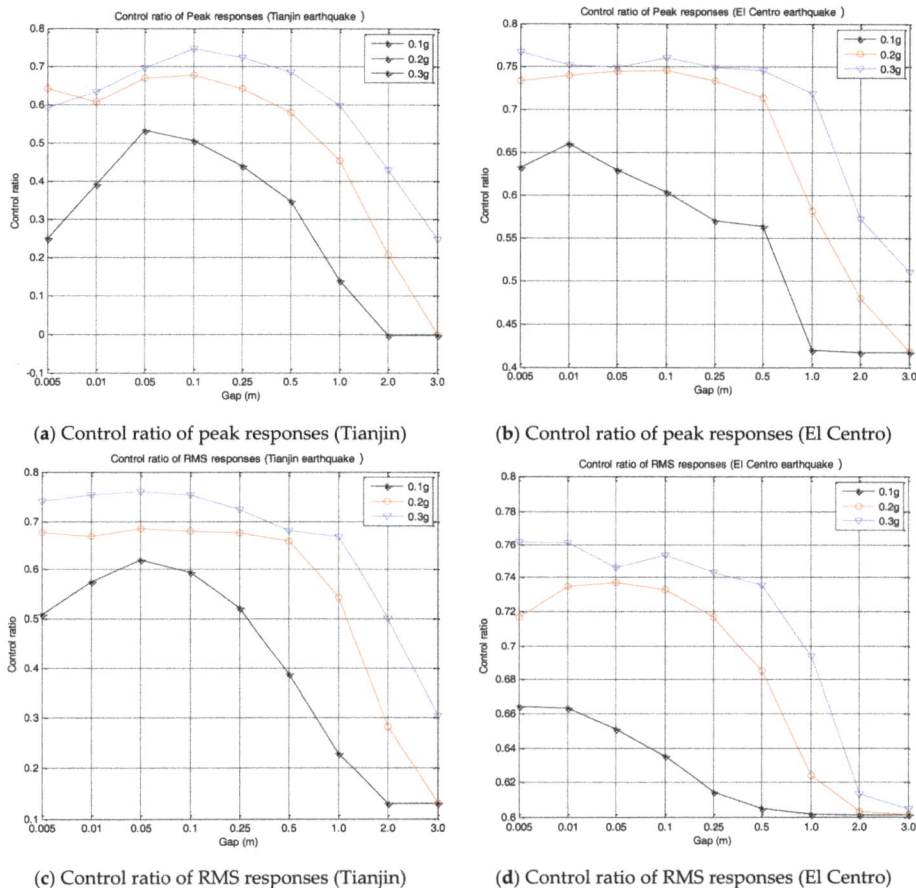

Figure 15. Control effectiveness with different gap value (mass: 200 t).

Table 7. Comparison of control efficiency with the variation of the mass (gap: 0.05 m).

Excitation		Terms	Uncontrolled	TMD			PTMD		
				200 t	300 t	500 t	200 t	300 t	500 t
Tianjin	0.1 g	Peak (m)	2.203	2.209	2.150	1.924	1.029	1.024	1.047
		RMS (m)	0.883	0.768	0.660	0.508	0.336	0.329	0.335
	0.2 g	Peak (m)	3.672	3.682	3.584	3.207	1.210	1.098	1.103
		RMS (m)	1.472	1.281	1.100	0.847	0.464	0.466	0.469
	0.3 g	Peak (m)	5.141	5.155	5.016	4.490	1.564	1.161	1.094
		RMS (m)	2.060	1.792	1.540	1.185	0.493	0.507	0.517
El Centro	0.1 g	Peak (m)	3.068	1.787	1.383	0.924	1.136	1.246	0.908
		RMS (m)	1.252	0.499	0.422	0.330	0.437	0.410	0.329
	0.2 g	Peak (m)	5.114	2.979	2.305	1.541	1.303	1.379	1.402
		RMS (m)	2.087	0.831	0.703	0.551	0.550	0.561	0.515
	0.3 g	Peak (m)	7.160	4.170	3.227	2.157	1.794	2.025	1.618
		RMS (m)	2.922	1.164	0.984	0.771	0.743	0.760	0.642

From these parameter study results, the basic design guidelines of a PTMD can be concluded thusly: (i) the mass ratio can be selected from 0.005 to 0.1; (ii) tune the frequency of the PTMD according to the first natural frequency of the host structure; (iii) calculate the TMD-controlled cases with moderate/designed earthquake input levels; (iv) the initial gap value between the mass and the viscoelastic limitation is selected as 1/3 of the maximum relative displacement of the mass and the connection point in the TMD control case; and (v) more earthquake inputs with different intensities are adopted for trial calculations to further adjust and optimize the gap value according to the control effectiveness.

7. Conclusions

Making use of the energy dissipation during impact, an application of a PTMD on controlling the responses of high-rise structures under wind load and earthquake inputs are presented in this paper. An improved pounding force model was proposed. After optimizing the parameters, the comparison results between the simulated and experimental data demonstrates that the proposed model appears to be more precise over the traditional Hertz-based pounding model in describing the pounding force between the mass and the viscoelastic layer. By taking the benchmark Canton Tower with a height of 600 m as the host structure, TMD- and PTMD-controlled cases were simulated and compared under wind and earthquake excitations. Particularly, the recorded typhoon data collected from the structural health monitoring system on the tower was adopted to excite the structure. The proposed PTMD will perform as a traditional TMD under wind load excitations. The generated impact will not be strong enough to dissipate energy even when trying to adjust the gap to force pounding. Under earthquake excitations, a TMD can only be effective under certain excitations and the control effectiveness is very sensitive to the mass ratio. On the contrary, the PTMD appears to have better performance over the traditional TMD. In addition, the robustness of the device is demonstrated by the stability of the control results under different earthquake inputs. The superiority of the device over traditional TMD is also verified by the fact that it is able to achieve similar, or even better, control effectiveness with less weight, and the improvement is even better when the intensity of the earthquake input increases. Simulation results indicate that the performance of the PTMD has a lot to do with the selected gap value and, thus, it should be carefully chosen with respect to the optimal value.

Acknowledgments: The authors are grateful for financial support by the National Natural Science Foundation of China (No. 51578159 and 51678158), the Program for New Century Excellent Talents at Fujian Province University (2016, No. 83016017), the Co-operative Project of Fujian Province Colleges and Universities (No. 2016H6011), and the Co-operative Leading Project of Fujian Province Colleges and Universities (No. 2017H0016).

Author Contributions: Wei Lin and Gangbing Song did the modelling work, built the PTMD simulation program, and wrote the paper. Shanghong Chen analyzed the simulation data.

Conflicts of Interest: The authors declare no conflict of interest.

References

1. Zhang, Z.; Ou, J.P. Optimization Design of Coupling Beam Metal Damper in Shear Wall Structures. *Appl. Sci.* **2017**, *7*, 7020137. [CrossRef]
2. Weber, F.; Distl, H.; Fischer, S.; Braun, C. MR Damper Controlled Vibration Absorber for Enhanced Mitigation of Harmonic Vibrations. *Appl. Sci.* **2016**, *5*, 5040027. [CrossRef]
3. Jansen, L.M.; Dyke, S.J. Semiactive control strategies for MR dampers: Comparative study. *J. Eng. Mech.* **2000**, *126*, 795–803. [CrossRef]
4. Soong, T.T.; Spencer, B.F., Jr. Supplemental energy dissipation: State-of-the-art and state-of-the-practice. *Eng. Struct.* **2002**, *24*, 243–259. [CrossRef]
5. Lee, C.L.; Chen, Y.T.; Chung, L.L.; Wang, Y.P. Optimal design theories and applications of tuned mass dampers. *Eng. Struct.* **2006**, *28*, 43–53. [CrossRef]
6. Mishra, S.K.; Gur, S.; Chakraborty, S. An improved tuned mass damper (SMA-TMD) assisted by a shape memory alloy spring. *Smart Mater. Struct.* **2013**, *22*, 095016. [CrossRef]

7. Nagarajaiah, S. Adaptive passive, semiactive, smart tuned mass dampers: Identification and control using empirical mode decomposition, hilbert transform, and short-term fourier transform. *Struct. Control Health Monit.* **2009**, *16*, 800–841. [CrossRef]

8. Casado, C.M.; Poncela, A.V.; Lorenzana, A. Adaptive tuned mass damper for the construction of concrete piers. *Struct. Eng. Int.* **2007**, *17*, 252–255. [CrossRef]

9. Occhiuzzi, A.; Spizzuoco, M.; Ricciardelli, F. Loading models and response control of footbridges excited by running pedestrians. *Struct. Control Health Monit.* **2008**, *15*, 349–368. [CrossRef]

10. Weber, B.; Feltrin, G. Assessment of long-term behavior of tuned mass dampers by system identification. *Eng. Struct.* **2010**, *32*, 3670–3682. [CrossRef]

11. Hoang, N.; Fujino, Y.; Warnitchai, P. Optimal tuned mass damper for seismic applications and practical design formulas. *Eng. Struct.* **2008**, *30*, 707–715. [CrossRef]

12. Demetriou, D.; Nikitas, N. A Novel Hybrid Semi-Active Mass Damper Configuration for Structural Applications. *Appl. Sci.* **2016**, *6*, 6120397. [CrossRef]

13. Inaudi, J.A.; Kelly, J.M. Mass damper using friction-dissipating devices. *J. Eng. Mech.* **1995**, *121*, 142–149. [CrossRef]

14. Rüdinger, F. Optimal vibration absorber with nonlinear viscous power law damping and white noise excitation. *J. Eng. Mech.* **2006**, *132*, 46–53. [CrossRef]

15. Rüdinger, F. Tuned mass damper with nonlinear viscous damping. *J. Sound Vib.* **2007**, *300*, 932–948. [CrossRef]

16. Alexander, N.A.; Schilder, F. Exploring the performance of a nonlinear tuned mass damper. *J. Sound Vib.* **2009**, *319*, 445–462. [CrossRef]

17. Kaloop, M.R.; Hu, J.W.; Bigdeli, Y. Identification of the Response of a Controlled Building Structure Subjected to Seismic Load by Using Nonlinear System Models. *Appl. Sci.* **2016**, *6*, 6100301. [CrossRef]

18. Chung, L.L.; Wu, L.Y.; Huang, H.H.; Chang, C.H.; Lien, K.H. Optimal design theories of tuned mass dampers with nonlinear viscous damping. *Earthq. Eng. Eng. Vib.* **2009**, *8*, 547–560. [CrossRef]

19. Ricciardelli, F.; Occhiuzzi, A.; Clemente, P. Semi-active tuned mass damper control strategy for wind-excited structures. *J. Wind Eng. Ind. Aerodyn.* **2000**, *88*, 57–74. [CrossRef]

20. Pinkaew, T.; Fujino, Y. Effectiveness of semi-active tuned mass dampers under harmonic excitation. *Eng. Struct.* **2001**, *23*, 850–856. [CrossRef]

21. Ikeda, Y. Active and semi-active vibration control of buildings in Japan—Practical applications and verification. *Struct. Control Health Monit.* **2009**, *16*, 703–723. [CrossRef]

22. Spencer, B.F., Jr.; Nagarajaiah, S. State of the art of structural control. *J. Struct. Eng. (ASCE)* **2003**, *129*, 845–856. [CrossRef]

23. Li, L.; Song, G.; Ou, J. Hybrid active mass damper (AMD) vibration suppression of nonlinear high-rise structure using fuzzy logic control algorithm under earthquake excitations. *Struct. Control Health Monit.* **2011**, *18*, 698–709. [CrossRef]

24. Gsell, D.; Feltrin, G.; Motavalli, M. Adaptive tuned mass damper based on pre-stressable leaf-springs. *J. Intell. Mater. Syst. Struct.* **2007**, *18*, 845–851. [CrossRef]

25. Rizos, D.; Feltrin, G.; Motavalli, M. Structural identification of a prototype pre-stressable leaf-spring based adaptive tuned mass damper: Nonlinear characterization and classification. *Mech. Syst. Signal Proc.* **2011**, *25*, 205–221. [CrossRef]

26. Williams, K.A.; Chiu, G.C.; Bernhard, R.J. Dynamic modelling of a shape memory alloy adaptive tuned vibration absorber. *J. Sound Vib.* **2005**, *280*, 211–234. [CrossRef]

27. Sun, W.Q.; Li, Q.B. TMD semi-active control with shape memory alloy. *J. Harbin Inst. Technol.* **2009**, *41*, 164–168. [CrossRef]

28. Varadarajan, N.; Nagarajaiah, S. Wind response control of building with variable stiffness tuned mass damper using empirical mode decomposition/Hilbert transform. *J. Eng. Mech.* **2004**, *130*, 451–458. [CrossRef]

29. Nagarajaiah, S.; Varadarajan, N. Short time Fourier transform algorithm for wind response control of buildings with variable stiffness TMD. *Eng. Struct.* **2005**, *27*, 431–441. [CrossRef]

30. Weber, F.; Maślanka, M. Frequency and damping adaptation of a TMD with controlled MR damper. *Smart Mater. Struct.* **2012**, *21*, 055011. [CrossRef]

31. Weber, F.; Boston, C.; Maślanka, M. An adaptive tuned mass damper based on the emulation of positive and negative stiffness with an MR damper. *Smart Mater. Struct.* **2011**, *20*, 015012. [CrossRef]

32. Cai, C.S.; Wu, W.J.; Araujo, M. Cable vibration control with a TMD-MR damper system: Experimental exploration. *J. Struct. Eng.* **2007**, *133*, 629–637. [CrossRef]

33. Eason, R.P.; Sun, C.; Dick, A.J.; Nagarajaiah, S. Attenuation of a linear oscillator using a nonlinear and a semi-active tuned mass damper in series. *J. Sound Vib.* **2013**, *332*, 154–166. [CrossRef]

34. Zhang, P.; Song, G.; Li, H.N.; Lin, Y.X. Seismic Control of Power Transmission Tower Using Pounding TMD. *J. Eng. Mech.* **2012**, *139*, 1395–1406. [CrossRef]

35. Lankarani, H.M.; Nikravesh, P.E. A contact force model with hysteresis damping for impact analysis of multibody systems. *J. Mech. Des.* **1990**, *112*, 369–376. [CrossRef]

36. Lankarani, H.M.; Nikravesh, P.E. Continuous contact force models for impact analysis in multibody systems. *Nonlinear Dyn.* **1994**, *5*, 193–207.

37. Wasfy, T.M.; Noor, A.K. Computational strategies for flexible multibody systems. *Appl. Mech. Rev.* **2003**, *56*, 553–613. [CrossRef]

38. Schwab, A.L.; Meijaard, J.P.; Meijers, P. A comparison of revolute joint clearance models in the dynamic analysis of rigid and elastic mechanical systems. *Mech. Mach. Theory* **2002**, *37*, 895–913. [CrossRef]

39. Pereira, C.M.; Ambrósio, J.A.; Ramalho, A.L. A methodology for the generation of planar models for multibody chain drives. *Multibody Syst. Dyn.* **2010**, *24*, 303–324. [CrossRef]

40. Gonthier, Y.; Mcphee, J.; Lange, C.; Piedbceuf, J.C. A Regularized Contact Model with Asymmetric Damping and Dwell-Time Dependent Friction. *Multibody Syst. Dyn.* **2004**, *11*, 209–233. [CrossRef]

41. Flores, P.; Ambrósio, J.; Claro, J.P.; Lankarani, H.M. Dynamic behaviour of planar rigid multi-body systems including revolute joints with clearance. *Proc. Inst. Mech. Eng. Part K J. Multi-Body Dyn.* **2007**, *221*, 161–174. [CrossRef]

42. Flores, P.; Machado, M.; Silva, M.T.; Martins, J.M. On the continuous contact force models for soft materials in multibody dynamics. *Multibody Syst. Dyn.* **2011**, *25*, 357–375. [CrossRef]

43. Ni, Y.Q.; Xia, Y.; Lin, W.; Chen, W.H.; Ko, K.M. SHM benchmark for high-rise structures: A reduced-order finite element model and field measurement data. *Smart Mater. Struct.* **2012**, *10*, 411–426. [CrossRef]

44. Chen, W.H.; Lu, Z.R.; Lin, W.; Chen, S.H.; Ni, Y.Q.; Xia, Y.; Liao, W.Y. Theoretical and experimental modal analysis of the Guangzhou New TV Tower. *Eng. Struct.* **2011**, *33*, 3628–3646. [CrossRef]

45. Li, Q.S.; Fang, J.Q.; Jeary, A.P.; Wong, C.K.; Liu, D.K. Evaluation of wind effects on a supertall building based on full-scale measurements. *Earthq. Eng. Struct. Dyn.* **2000**, *29*, 1845–1862. [CrossRef]

46. Lin, W.; Lin, Y.; Song, G.; Li, J. Multiple Pounding Tuned Mass Damper (MPTMD) control on benchmark tower subjected to earthquake excitations. *Earthq. Struct.* **2016**, *11*, 1123–1141.

47. Zhang, H.Y.; Zhang, L.J. Tuned Mass Damper System of High-rise Intake Towers Optimized by Improved Harmony Search Algorithm. *Eng. Struct.* **2017**, *138*, 270–282. [CrossRef]

48. Yaguchi, T.; Kurino, H.; Kano, N.; Nakai, T.; Fukuda, R. Development of large tuned mass damper with stroke control system for seismic upgrading of existing high-rise building. *Int. J. High Rise Build.* **2016**, *5*, 167–176. [CrossRef]

49. Rana, R.; Soong, T.T. Parametric study and simplified design of tuned mass dampers. *Eng. Struct.* **1998**, *20*, 193–204. [CrossRef]

Article

Experimental Investigation of a Base Isolation System Incorporating MR Dampers with the High-Order Single Step Control Algorithm

Weiqing Fu [1,2], Chunwei Zhang [1,3,*], Li Sun [4], Mohsen Askari [3], Bijan Samali [3], Kwok L. Chung [1] and Pezhman Sharafi [3]

[1] School of Civil Engineering, Qingdao University of Technology, Qingdao 266033, China; fuweiqing@qut.edu.cn (W.F.); klchung@qut.edu.cn (K.L.C.)

[2] School of Civil Engineering, Heilongjiang University, Harbin 150080, China

[3] Centre for Infrastructure Engineering, Western Sydney University, Penrith NSW 2751, Australia; m.askari@westernsydney.edu.au (M.A.); b.samali@westernsydney.edu.au (B.S.); p.sharafi@westernsydney.edu.au (P.S.)

[4] School of Civil Engineering, Shenyang Jianzhu University, Shenyang 110168, China; sunli@sjzu.edu.cn

[*] Correspondence: zhangchunwei@qut.edu.cn; Tel.: +86-532-8507-1693

Academic Editor: César M. A. Vasques
Received: 19 December 2016; Accepted: 25 March 2017; Published: 30 March 2017

Abstract: The conventional isolation structure with rubber bearings exhibits large deformation characteristics when subjected to infrequent earthquakes, which may lead to failure of the isolation layer. Although passive dampers can be used to reduce the layer displacement, the layer deformation and superstructure acceleration responses will increase in cases of fortification earthquakes or frequently occurring earthquakes. In addition to secondary damages and loss of life, such excessive displacement results in damages to the facilities in the structure. In order to overcome these shortcomings, this paper presents a structural vibration control system where the base isolation system is composed of rubber bearings with magnetorheological (MR) damper and are regulated using the innovative control strategy. The high-order single-step algorithm with continuity and switch control strategies are applied to the control system. Shaking table test results under various earthquake conditions indicate that the proposed isolation method, compared with passive isolation technique, can effectively suppress earthquake responses for acceleration of superstructure and deformation within the isolation layer. As a result, this structural control method exhibits excellent performance, such as fast computation, generic real-time control, acceleration reduction and high seismic energy dissipation etc. The relative merits of the continuity and switch control strategies are also compared and discussed.

Keywords: structural vibration control; base isolation control system; magnetorheological (MR) damper; shaking table test; high order single step integration algorithm

1. Introduction

The conventional base isolation method with rubber bearing is fundamentally a damping technology. It has the advantages of obvious damping effect, safety, reliability and low cost, and hence has been widely used in a number of low-rise buildings [1–3]. However, the isolation structure with the rubber bearings will normally have a large isolation layer deformation during rare earthquakes. In order to control the displacement of the isolation layer, more isolation bearings are generally embedded into the isolation layer system. After adding some stiffness and damping for the isolation layer, the acceleration of the superstructure and the displacement of the layer may be enlarged, sometimes even in low-level earthquakes [4,5]. This excessive acceleration will cause equipment damage in certain buildings, such as hospitals, communication centers, government buildings, etc. Moreover, articles falling in the building may cause secondary damages [6]. The need for security protection for the overall structure to mitigate such damages has led to the development of novel isolation technologies [7–9]. The mitigation of secondary damages is one of the design objectives in the present study.

Magnetorheological (MR) damper is a variable damping controller with excellent performance, which has been demonstrated by many research cases either experimentally or practically. It has the characteristics of a damping force in wide range, rapid response and requires little energy input. It has been widely used in hybrid or semi-active controlled structures [10–13]. In broader applications, MRD has been applied in protecting civil infrastructure systems against severe earthquake and wind loading [14], in semi-active seat suspension systems and in payload launch vibration isolation of a spacecraft [15,16]. Bharti et al. proposed a coupled building control scheme interconnecting the inline floors of two closely spaced adjacent buildings with semi-active Magnetorheological (MR) dampers [17]. It was noted that the control scheme is quite effective in response mitigation of both the buildings under a wide range of ground motions [18]. The efficacy of this smart system in reducing structural responses for a wide range of loading conditions was demonstrated in a series of experiments [19]. Amongst various control devices, seismic base isolation has proved to be a time tested method and semi-active MR dampers have also emerged as a very attractive proposition for a control device [20]. Maddaloni found that when the semi-active control is suitably designed and implemented, the seismic performance of the structure can be significantly improved [21]. A series of large scale experimental tests was conducted on a mass equipped with a seismic base isolation system that consists of high-damping rubber bearings and an MR damper [22,23]. Researchers have been exploring various approaches of seismic hazard mitigation of closely spaced adjacent buildings, by way of employing various control devices [24].

This study will focus on the base isolation control system using MR damper and rubber bearings with the high-order single-step algorithm (HSA) [25]. HSA is known to be well-suited for real-time control applications due to its excellent capability in high-speed computation and fast convergence. The new structural vibration control involves the use of two control strategies, namely, continuity and switch control. In this paper, both the numerical simulation and experimental test were subjected to seismic waves with different characteristics and intensities. The research work establishes a theoretical and experimental basis for the novel isolation system applications.

2. Magnetorheological Smart Isolation System

The base isolation system is composed of MR dampers, ordinary rubber bearings, real-time data acquisition, signal processing, control determination and actuation parts, etc., as shown in Figure 1. The rubber bearings are capable of providing restoring forces. The MR dampers can provide variable and controllable damping forces to the model structure by controlling the excitation voltages to the damper coils. The damping force of the damper in every moment is obtained by optimal control algorithm calculation and control rate adjustment. The control calculation uses real-time feedback data including displacement, velocity and acceleration response of the structure (Figure 1). The adjustment of the control approach for the control force should consider whether the damper is able to provide a damping control force in principle, although in rare cases the opposite situation could apply. This ensures that the damping force provided by the isolation device to the isolation layer is optimal and can be realized at every single temporal moment.

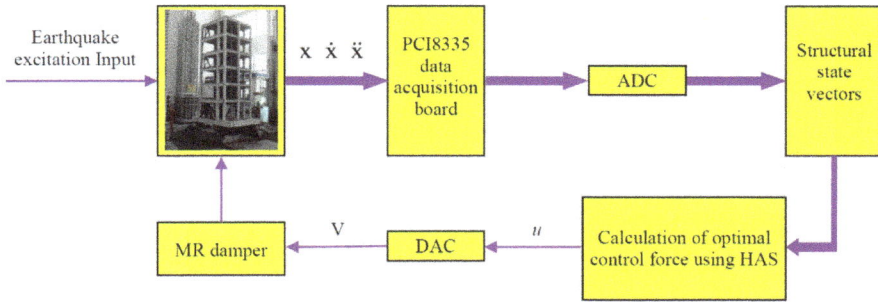

Figure 1. Flow chart for the configuration of control system.

2.1. Test Model

The test model is a six storey frame structure with three frames and two spans. The prototype structure is designed using PKPM software. PKPM software is developed by the China Academy of Building Research (Beijing, China), and can be used for the design and calculation of the frame structure. The geometric ratio is 1:5. In the model, the cross section area of column is 90 mm × 90 mm, the cross section area of the main beam is 120 mm × 60 mm, and cross section size of secondary beam is 80 mm × 60 mm. The thickness of each floor is 30 mm. Figure 2 shows the photo of the model structure. Figure 3 shows the structure's model dimension.

Figure 2. Photo of the experimental structure model.

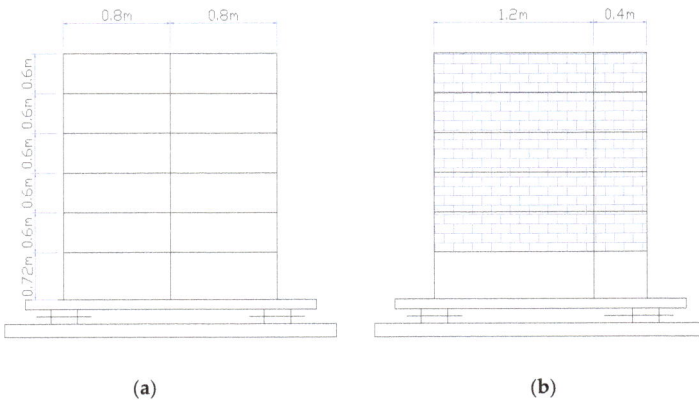

Figure 3. Structure's model dimension. (**a**) Secondary shock direction; (**b**) The main shock direction.

In the model structure, the beam column is made of C30 concrete material, and the slab is made of C20 concrete material, which is the same as the prototype structure. The concrete cube compressive strength standard value is equal to 30 in terms of the scale of concrete material; its strength grade is expressed as C30. C20 also has the same definition. The reinforcement including diameter, length, and number of roots of the model structure is obtained by the conversion of the actual structure according to the geometric ratio of 1:5.

Table 1 shows a similar coefficient of the model. According to the similarity coefficient, the model was made and the whole experiment was finished.

Table 1. The similar coefficient of the model.

Physical Quantity	Symbol	Dimension	Similar Coefficient
length	S_l	L	1/5
elastic modulus	S_E	$ML^{-1}T^{-2}$	1
stiffness	S_k	MT^{-2}	1/5
acceleration	S_a	L/T^2	1
time	S_T	T	$1/\sqrt{5}$
velocity	S_v	L/T	$1/\sqrt{5}$
displacement	S_x	L	1/5
mass	S_m	M	1/25

The two bottom plates in the model structure had a weight of about 3.75×10^3 kg while the empty frame model had a weight of about 2.25×10^3 kg. Since the actual structure weight was 281.55×10^3 kg, the structure model needed to have an additional weight of approximately 4 tons according to the relationship of similarity ratio configuration. The additional weight was assigned evenly into six storeys, thus each storey weighed about 0.7×10^3 kg. In order to adjust the dynamics of the model structure close to the actual structure, the concrete block wall was built from the second to the sixth storey of the model, while the bottom layer retained a larger space.

2.2. Control Algorithm and Strategy

The HSA method is a structural dynamic time history analysis method, which has the characteristics of high accuracy and simple calculation consumption. In real-time control scenarios, the issue of time delay is often encountered. When the system uses only one step in time delay, the control force produced by the HSA method will have minimum time delay. When the delay time is equal to the multiple (e.g., N) integral steps, the method can estimate the state vector of the $(N-1)$th step, then generate the optimal control force of the Nth time step for the structure by state vector calculation.

The formula for the structure dynamic response can be expressed by the following HSA method:

$$\left. \begin{array}{l} x_{n+1} = \mathbf{G_{11}}x_n + \mathbf{G_{12}}\dot{x}_n - \mathbf{Q_{12}}\ddot{x}_n + \mathbf{R_{12}}(1\ddot{x}_{g,n+1} - \mathbf{M}^{-1}\mathbf{BU}_{n+1}) \\ \dot{x}_{n+1} = \mathbf{G_{12}}x_n + \mathbf{G_{22}}\dot{x}_n - \mathbf{Q_{22}}\ddot{x}_n + \mathbf{R_{22}}(1\ddot{x}_{g,n+1} - \mathbf{M}^{-1}\mathbf{BU}_{n+1}) \\ \ddot{x}_{n+1} = -1\ddot{x}_{g,n+1} + \mathbf{M}^{-1}(\mathbf{BU}_{n+1} - \mathbf{C}\dot{x}_{n+1} - \mathbf{K}x_{n+1}) \end{array} \right\} \tag{1}$$

The performance index function is defined as follows:

$$\mathbf{J} = \frac{1}{2}x^T W_1 \mathbf{K}x + \frac{1}{2}\dot{x}^T W_2 \mathbf{M}\dot{x} + \frac{1}{2}(\mathbf{BU})^T \mathbf{K}^{-1}\mathbf{BU} \tag{2}$$

where W_1 and W_2 are the weight parameters of adjustment control effect for the displacement and velocity response of the isolation layer, where I is the identity matrix. The specific parameter values can be determined by simulation calculation. Detailed calculation procedure is given in the literature [26]. By setting $\delta J_{n+1} = 0$, yields

$$\mathbf{BU}_{n+1} = \mathbf{D_1}x_n + \mathbf{D_2}\dot{x}_n - \mathbf{D_3}\ddot{x}_n + \mathbf{D_4}I\ddot{x}_{g,n+1} \tag{3}$$

In order to realize the closed-loop control, by giving up $\mathbf{DI}\ddot{x}_{g,n+1}$, so each $\delta\mathbf{J}_{n+1}$ is no longer zero. This is only an approximate optimization:

$$\mathbf{BU}_{n+1} = \mathbf{D}_1 x_n + \mathbf{D}_2 \dot{x}_n - \mathbf{D}_3 \ddot{x}_n \tag{4}$$

where $\mathbf{D}_1$, $\mathbf{D}_2$ and $\mathbf{D}_3$ are all constant matrices, and $\mathbf{U}$ is the optimal control force vector for each single time step. The constant matrix $\mathbf{B}$ indicates the location or configuration of the dampers within the structure [16].

$$u(i) = \begin{cases} u_{\max} & \text{if} \quad f_{opt} \times \dot{x}_b < 0 \\ u_{\min} & \text{if} \quad f_{opt} \times \dot{x}_b \geq 0 \end{cases} \tag{5}$$

For an MR control system with variable damping, the control strategy is usually divided into two types. One is similar to the switch control mode. The voltage regulation for the MR damper is either maximum or minimum in temporal series. The other is the continuity control mode, which considers the characteristics with continuously adjustable damping force of the MR damper. The control strategy is used to regulate the optimal control force in real-time [27–29].

$$u(i) = \begin{cases} u_{\max} & \text{if } f_{opt} \times \dot{x}_b < 0 \quad \text{and} \quad |f_{opt}| > |u_{\max}| \\ f_{opt} & \text{if } f_{opt} \times \dot{x}_b < 0 \quad \text{and} \quad |u_{\max}| \geq |f_{opt}| > |u_{\min}| \\ u_{\min} & \text{if } f_{opt} \times \dot{x}_b < 0 \quad \text{and} \quad |f_{opt}| \leq |u_{\min}| \\ u_{\min} & \text{if } f_{opt} \times \dot{x}_b > 0 \end{cases} \tag{6}$$

where $u_{\max}$ and $u_{\min}$ are the maximum and minimum damping forces which can be provided by the MR damper subject to velocity variations; $\dot{x}_b$ is the relative velocity of the isolation layer, and f_{opt} is the calculated optimal control force, whereas u is the adjusted damping force realized by the damper, which can be achieved by controlling the output current regulator.

2.3. Isolation Bearing

For this experiment, four ordinary rubber sandwich isolation pads were fabricated. They were placed on the four corners of the model structure. The isolation bearing diameter used in the experiment is 100 mm and the total thickness of the rubber layer is 22.5 mm. Other basic parameters for the rubber bearings have been shown in Table 2.

Table 2. The basic parameters of designed rubber bearing.

Total Height	87.5 mm	Thickness of Protective Layer	5 mm
External diameter	110 mm	Middle hole diameter	18 mm
Height (excluding connection plate)	63.5 mm	Rubber layer thickness	1.5 mm
Effective diameter, D	100 mm	No. of rubber layer	15
Design bearing capacity	75 kN	Total rubber thickness, T_r	22.5 mm
Standard displacement ($d_{\max}$)	55 mm	Thick laminated sheet	1.5 mm
maximum displacement	23 mm	Thickness of connecting plate	12 mm
Layers of laminated sheet	14	Sealing plate thickness	10 mm
Design surface pressure	10 MPa	-	-

In order to test the actual performance of the rubber pad, the test of 100% horizontal shear deformation for the rubber bearings was performed. The pressure shear test for the rubber bearing was carried out under the compressive stress level of 10 MPa, while the horizontal shear deformations were 50% and 100%, respectively. The test setup is shown in Figure 4.

Figure 4. Shear test of rubber bearing under axial pressures.

Vertical stress check: The upper structure was approximately 10 tons. The calculated compressive stress for each bearing was around 3.2 MPa. This met the allowable stress limit of 10 MPa.

Horizontal deformation check: for the second category site within the eighth degree zone (rare earthquakes), calculation values for the equivalent load (kN) was:

$$F_{ek} = \alpha \times G_{eq} = 0.3471 \times 0.85 \times 100 = 31.8 \tag{7}$$

Maximum deformation of the isolation layer in mm was:

$$D_m = F_{ek}/K_d = 31.8/1112 = 28.6 \tag{8}$$

For the fourth category site within the eighth degree zone (rare earthquakes), the calculated value for the equivalent load (kN) was:

$$F_{ek} = \alpha \times G_{eq} = 0.507 \times 0.85 \times 100 = 43.08 \tag{9}$$

In the above formula F_{ek} is the standard value of total horizontal seismic force; α is the seismic influence coefficient; G_{eq} is the equivalent Total Gravity Load.

Maximum deformation of the isolation layer in mm was:

$$D_m = 43.08/1112 = 38.7 \tag{10}$$

According to the seismic code, horizontal deformation of the rubber bearing was:

$$\min(0.55 \times D = 55 \text{ mm} \mid 3 \times T_r = 67.5 \text{ mm}) \tag{11}$$

In the above formula T_r is total thickness of rubber layer; D is the diameter of rubber bearing.

Therefore, the allowable deformation of the bearing was 55 mm. The structure of the base isolation bearing was used to satisfy the deformation constraints for the second category site condition; however, for the fourth category site condition, the maximum deformation of the rubber bearing was noticeably too large, having exceeded the deformation limit value of 1.5 times the rubber layer thickness. The experimental results also prove that the rubber pad is not stable and an additional damper must be used to control the significant deformation requirement.

2.4. MR Damper

The maximum output of the MR damper in the experiment was 20 kN. The minimum output was 2.5 kN. The adjustable damping force ratio was 8. The stroke range of the damper was ±80 mm. The working excitation current was 0–4 A. Numerical simulation results indicated that the required maximum output was 1×10^3 kg for the selected test model. The maximum deformation of the isolation layer was less than 60 mm. Selection of the larger output and stroke damper was to meet the needs of other test models, but also left a certain degree of security and safety for further test verifications.

Prior to the shaking table test, the performance test for the damper was carried out. The test results are shown in Figures 5 and 6. Those are hysteresis curves for the damper, respectively, from the inside to the outside with 0.1 A as the excitation incremental step ranging from 0.2 to 1 A as the input current.

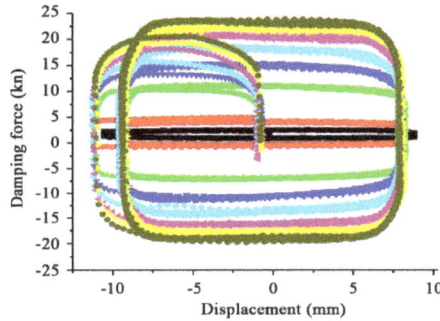

Figure 5. Damper hysteretic curve under peak value of deflection 10 mm and 1-Hz sine wave.

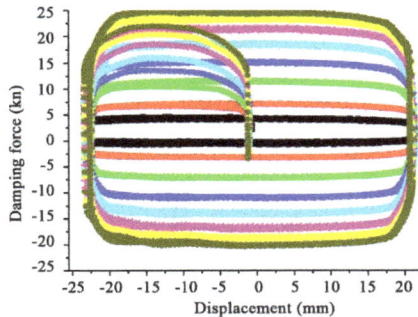

Figure 6. Damper hysteretic curve under peak value of deflection 20 mm and 0.5-Hz sine wave.

As can be seen from Figures 5 and 6, the damper hysteresis curves were sufficiently fully expanded. The maximum and minimum outputs satisfy the design requirements. In the frequency domain of the shaking table test at 1 Hz, the performance of the damper force was demonstrated to be stable.

The curve above the working conditions can be obtained. The relationship can be given as shown in Figure 7 between the current and the damping force. The regression equation for the curve is shown in Equation (12). In the formula, v is the shear velocity of MR damper and I is the applied current for the MR damper.

$$F = C_0 v + F(I) = 10629v + 1500 + 10891I + 2291I^2 - 1468I^3 \tag{12}$$

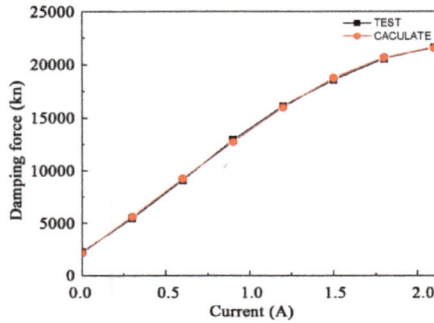

Figure 7. Relation diagram for current and damping force.

3. Test Scheme

3.1. Loading Pattern

In order to examine the control effect of the MR damper embedded base isolation system for structural vibration control under different earthquake intensities, the loading pattern was designed and described as follows. The input earthquake excitation magnitude was the fortification intensity and rare intensity corresponding to the code defined in the eighth degree zone's regulations [30], whereas the El Centro wave, Taft wave and Tianjin wave were used in this study. The selected earthquake waves in the experiment are shown in Table 3.

Table 3. Earthquake waves table in the experiment.

Group		Duration (s)	Record Interval (s)	Peak Acceleration (gal)	Earthquake Occurrence Time (s)	Seismic Wave Description
El-Centro wave	X direction	53	0.02	341.695 (NS)	2.12	The seismic wave is recorded in El Centro city, California, USA.
	Y direction			210.142 (EW)	11.44	
	Z direction			−206.34 (UD)	0.98	
Taft wave	X direction	54	0.02	175.9 (NS)	9.3	The seismic wave is recorded in Kern County, California, USA.
	Y direction			152.7 (EW)	9.1	
	Z direction			102.9 (UD)	9.76	
Tianjing wave	X direction	19	0.01	145.805 (NS)	7.65	The seismic wave is recorded in Tianjin city, PRC.
	Y direction			104.18 (EW)	7.59	
	Z direction			73.14 (UD)	9.03	

3.2. Measurement Scheme

In order to carry out the measurement and feedback of the state vector for the control system, the sensors of displacement, velocity, acceleration and force were arranged on the test model. Figure 8 shows the sensor arrangement for the experimental model. Moreover, in order to measure the actual working performance of the MR damper, a force transducer with a measuring range of 2 tons was setup in the experiment. It was directly connected with the damper embedded within the isolation layer.

3.3. Feedback Control System Scheme

The control system exhibited a full state feedback scheme. The controller host computer was installed with a PCI 8335 data acquisition board. It can complete the whole process of data collection, calculation, and instruction. The data collection board using Visual Basic (VB) compiler transforms the two control algorithm into VB executable programs for structural signal collection, online calculation and control signal generation and communication. The interface of the control operating system is shown in Figure 8.

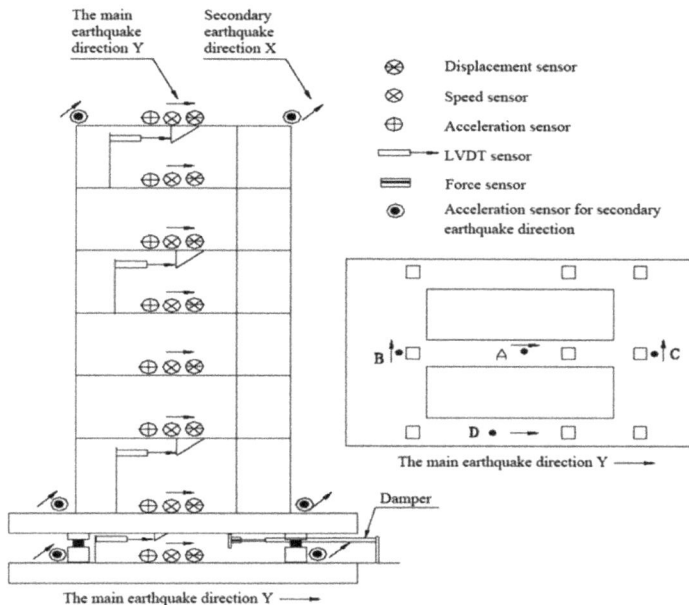

Figure 8. Sensors allocation scheme.

4. Analysis of Experimental Results

In order to proceed with the test smoothly, relevant numerical simulation for the control system was carried out prior to executing the experiments. Test results were also analyzed for the control system carried out by using HSA approach with two control strategies, namely, the continuity mode (CM) and switch mode (SM).

4.1. Structural Displacement Response

The peak value of storey drift response for each working condition is summarized in Table 4. The values in brackets are the results from numerical simulation. Figure 9 shows the time history curve for the isolation layer displacement response in the test.

Table 4. Peak value of story drift under different intensities and seismic waves (mm).

Earthquake Magnitude	Fortification Intensity					Rare Intensity			
Earthquake Wave	El Centro Wave		Taft Wave		Tianjin Wave	El Centro Wave		Taft Wave	
Control Strategy	CM	SM	CM	SM	CM	CM	SM	CM	SM
Isolation Layer	3.80 (3.6)	2.87 (2.93)	7.04 (8.52)	5.61 (9.3)	30.52 (25.1)	15.18 (15.2)	10.4 (11.7)	14.5 (18.5)	13.0 (15.3)
1st storey	2.14	1.75	2.23	3.5	4.59	3.53	3.83	3.31	4.21
2nd storey	1.38	1.22	1.03	1.74	3.91	2.76	2.53	1.57	2.18
3rd storey	1.18	1.09	1.20	1.62	3.63	2.16	2.45	1.90	2.39
4th storey	1.01	0.90	1.10	1.45	2.37	1.48	1.95	1.54	2.03
5th storey	0.86	0.75	0.87	1.41	1.49	1.88	1.95	1.47	1.87
6th storey	0.72	0.84	0.55	0.84	1.46	1.00	1.42	0.89	1.35

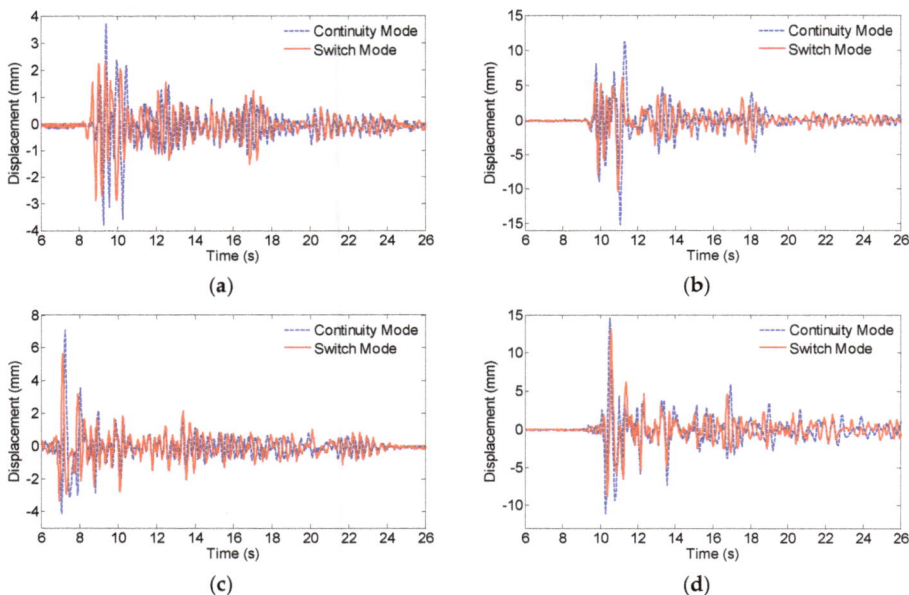

Figure 9. Displacement time history of isolation layer under two earthquake waves and control modes: (**a**) E1-Centro wave under fortification intensity; (**b**) E1-Centro wave under rare intensity; (**c**) Taft wave under fortification intensity; (**d**) Taft wave under rare intensity.

As seen from Table 4, the maximum displacement of the isolation layer (30.52 mm) is only about 55% of displacement limit value for the rubber bearing (55 mm). Hence, the control system has the capacity to sustain greater seismic loading. In addition, the numerical calculation is shown to be in good accordance with the test measurements. It lays out a good foundation for further research on the design method associated with the control system.

When the switch mode control is employed, the displacement response of the isolation layer is relatively small compared to continuity control mode. This is because the damper force outputs change only between the maximum and minimum values. When the isolation layer has a larger displacement response, the damper is directly applied with the maximum control force under the switch control mode, rather than continuously changing its value between the maximum and minimum damping force. The displacement of the isolation layer is small under the switch control mode, but the storey displacement corresponding to all the upper floors of the structure is relatively large due to the constrained base isolator deformations.

4.2. Structural Acceleration Response

The acceleration time history curves for the fifth storey of the structure under different control modes are compared and displayed in Figure 10. It is observed that the continuity control strategy outperforms the switch mode control, as the acceleration values in all scenarios have been suppressed lower than that from the switch mode control. Moreover, the control effect of continuity mode under rare intensity is found to be better than that under fortification intensity.

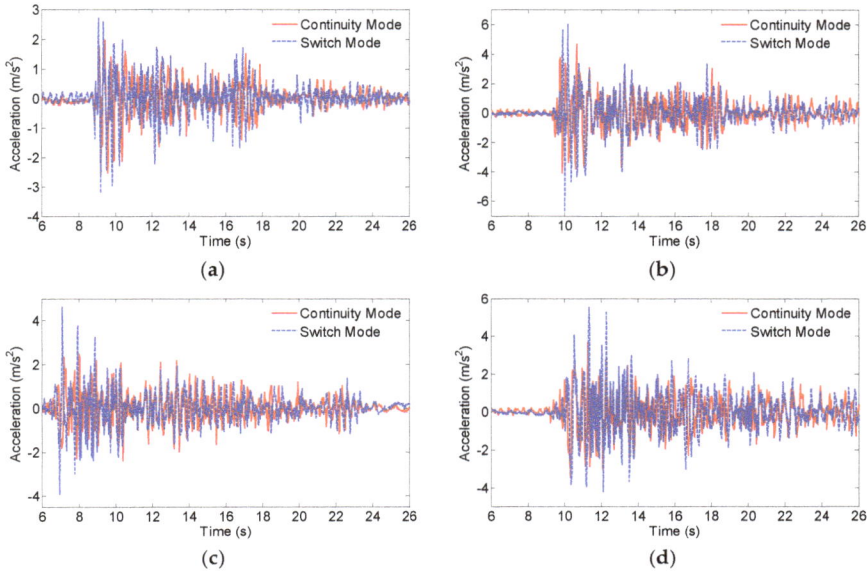

Figure 10. Acceleration time history of the fifth floor under continuity and switch modes: (**a**) El Centro wave under fortification intensity; (**b**) El Centro wave under rare intensity; (**c**) Taft wave under fortification intensity; (**d**) Taft wave under rare intensity.

Figures 11 and 12 show the peak acceleration responses of each structure storey under different earthquake profiles and amplitudes, for both the continuity control mode and switch control mode. Control effects for the El Centro wave and Taft wave are observed to be more significant than the Tianjin wave. In the rare intensity with continuity control mode for El Centro wave and Taft wave, the peak acceleration responses of the fifth storey are reduced by half of the acceleration peak of the input seismic waves. The structure vibration control effect is obvious, and structure response is reduced by 1/3 under the fortification intensity.

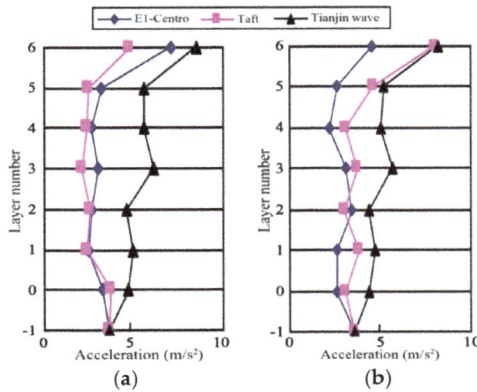

Figure 11. Acceleration peak curves under fortification intensity: (**a**) By continuity control mode; (**b**) By switch control mode.

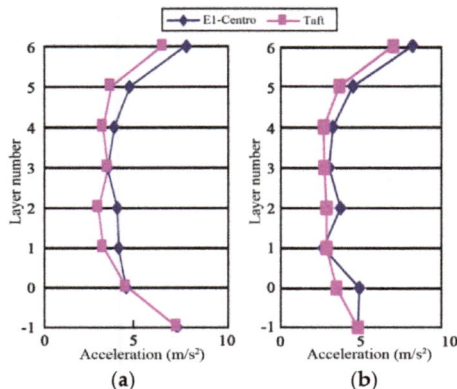

Figure 12. Acceleration peak curves under rare intensity. (**a**) By continuity control mode; (**b**) By switch control mode.

The reasons for the continuity control being better than the switch control can be attributed to the following displacement response analysis. Under switch control mode the isolation layer displacement is much smaller than that of the continuity control mode. The decrease of displacement makes the energy dissipation of the damper relatively limited to a smaller range. Thus the energy transferred to the upper structure increases, and the resulting acceleration response becomes larger. Meanwhile, the control effect under rare intensity is better than that of fortification intensity. This is because the initial viscous coefficient of the damper is larger. In rare intensity, a greater isolation layer displacement can completely overcome the initial viscous force. The dampers dissipate more energy, and hence, the acceleration response of the upper structure is significantly reduced.

It is worth noting that the peak acceleration response for the top storey of the structure model in various conditions, compared with other layers, has a large amplification. This does not conform with the control effect of the overall structure [31]. This may be due to the top storey of the structure model having been destroyed in the previous test, thereby the stiffness of the structure is reduced, and the a whiplash effect occurred in the test. The observation after the experiment also verified this hypothesis. Therefore, when the acceleration response is analyzed, the acceleration peak value of the fifth storey is adopted. The peak acceleration response of each structure storey is similar. The displacement response of the superstructure is small compared with the displacement of the isolation layer. This structure of the displacement reaction mainly occurred in the isolation layer, the upper structure close to the overall translational motion.

4.3. MR Damper Response

Figures 13 and 14 illustrate the damper hysteresis curves with continuity control mode under the El Centro wave and Taft wave with fortification and rare intensities, respectively. It is known that the area enclosed by the hysteretic curve represents the energy dissipated by the MR damper. As observed from those curves shown in Figures 13 and 14, the proposed control system is proven to effectively suppress the energy propagation upward under the different seismic scenarios. From the curve of damping force, it can be seen that the MR damper provides maximum damping force moment and seismic input peak close to the hysteretic curve of the damper to reduce energy consumption in order to reach the maximum response of the superstructure, namely, the El Centro wave and Taft wave with fortification and rare intensities. As a result, the energy transferred to the upper levels causing the secondary damages can be significantly mitigated.

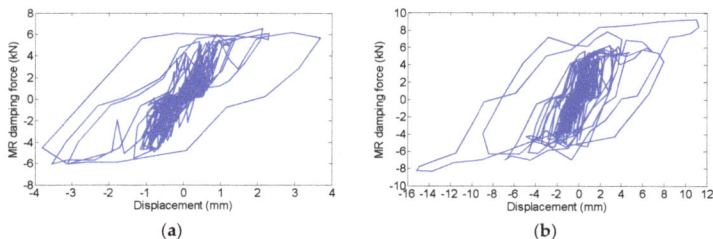

Figure 13. Damper hysteretic curves for El Centro wave with (**a**) fortification intensity, and (**b**) rare intensity of earthquake.

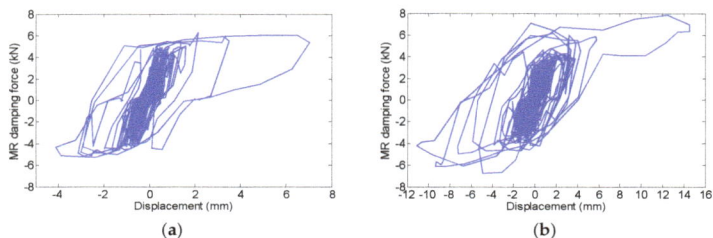

Figure 14. Damper hysteretic curves for Taft wave with (**a**) fortification intensity, and (**b**) rare intensity of earthquake.

4.4. Comparative Analysis of MR Isolation and LRB Isolation

For the same model, the isolation test with lead rubber bearing is completed in the literature [32]. Four lead rubber bearings (diameter 100 mm) were used in that shaking table experiment. General yield force for bearings is 8.652 kN. The shaking table test results are shown in Table 5 under different conditions.

Table 5. Peak value for structure acceleration and isolation layer displacement response under lead rubber bearing (LRB) isolation and seismic level 8.

	El-Centro Wave		Taft Wave
The Earthquake Wave	**Fortification Intensity**	**Rare Intensity**	**Fortification Intensity**
Table acceleration peak	−0.351	0.727	−0.368
Acceleration peak value of the first layer	−0.281	−0.596	−0.503
Acceleration peak value for fifth layer	−0.440	−0.662	−0.662
Acceleration amplification factor for structure fifth layers(LRB)	1.254	0.91	1.79
Acceleration amplification factor for structure fifth layers(MR)	0.89	0.64	0.69
Isolation layer displacement(LRB)	8.38	24.15	14.36
Isolation layer displacement(MR)	3.80	15.18	7.04

Comparison with the displacement response of the isolation layer can be seen, wherein the isolation with MR dampers is significantly less than the passive isolation with the lead rubber bearing, both displaying maximum displacement of the isolation layer and the amplification factor of the upper acceleration. In this way the isolation system with MR dampers has the ability to resist the larger earthquakes. This shows that the MR isolation system has better vibration control performance with the passive lead rubber isolation.

5. Conclusions

In this paper, the numerical simulation and shaking table test are carried out for an isolation system with MR damper and rubber bearing by using a high-order single-step algorithm. Both the continuity and switch control strategies are examined. The main findings are summarized as follows:

(1) The proposed isolation system can be applied to different site categories. The deformation corresponding to isolation layer can be effectively limited. Simultaneously, acceleration response of the superstructure can be reduced. Therefore, the smart isolation control system is shown to exhibit better control performance as compared with the traditional passive control method.

(2) Energy dissipation paths can be adapted by MR dampers in accordance with varying earthquake input excitations, and variable damping can be provided for the isolation layer. From the acceleration reductions and the hysteresis curves, the seismic energy can be found to be effectively dissipated. Therefore, the design objective concerning the mitigation of secondary damages can also be fulfilled.

(3) Restoring force can be provided by the rubber bearings to the isolation layer, thus the requirement of deformation limit of the isolation layer can be achieved under rare earthquakes.

(4) The high-order single-step algorithm has the capability of real-time calculation of structural response. Therefore, time delay issues can be effectively addressed. The entire control process, including data acquisition, real-time calculation and the results updating can be achieved by the control system. The control effect is not affected by changes associated with external input.

(5) The switch control strategy is found to be simple and reliable. The displacement response of isolation layer can be effectively controlled. Moreover, continuous regulated damping force can be achieved through a continuity control strategy. It is appropriate to consider the characteristics of MR dampers when instantaneous variable damping is needed.

Acknowledgments: The research is supported by the National Natural Science Foundation of China (Project No. 51678322, 51650110509 and 51578347), Natural Science Foundation of Heilongjiang Province (Project No. E2016053) and the Taishan Scholar Priority Discipline Talent Group program funded by the Shan Dong Province.

Author Contributions: Weiqing Fu and Chunwei Zhang developed the algorithm, conceived and designed the experimental works; Weiqing Fu, Chunwei Zhang, Li Sun, Mohsen Askari, Bijan Samali and Pezhman Sharafi investigated the control strategy and experimental implementation, Chunwei Zhang and Kwok Lun Chung analyzed the experimental data; Weiqing Fu drafted the manuscript, Chunwei Zhang, Li Sun and Kwok Lun Chung undertook the revision and editing.

References

1. Kelly, J.M.; Leitmann, G.; Soldatos, A.G. Robust control of base-isolated structures under earthquake excitation. *J. Optim. Theory Appl.* **1987**, *53*, 159–180. [CrossRef]
2. Jung, H.J.; Jang, D.D.; Lee, H.J.; Lee, I.W.; Cho, S.W. Feasibility test of adaptive passive control system using MR fluid damper with electromagnetic induction part. *J. Eng. Mech.* **2000**, *136*, 254–259. [CrossRef]
3. Tsai, H.C.; Kelly, J.M. Seismic response of the superstructure and attached equipment in a base-isolated building. *Earthq. Eng. Struct. Dyn.* **1989**, *18*, 551–564. [CrossRef]
4. Inaudi, J.A.; Kelly, J.M. Hybrid isolation systems for equipment protection. *Earthq. Eng. Struct. Dyn.* **1993**, *22*, 297–313. [CrossRef]
5. Wu, M.; Samali, B. Shake table testing of a base isolated model. *Eng. Struct.* **2002**, *24*, 1203–1215. [CrossRef]
6. Castaldo, P.; Palazzo, B.; Della, V.P. Seismic reliability of base-isolated structures with friction pendulum bearings. *Eng. Struct.* **2015**, *95*, 80–93. [CrossRef]
7. Wang, Y.; McFarland, D.M.; Vakakis, A.F.; Bergman, L.A. Seismic base isolation by nonlinear mode localization. *Arch. Appl. Mech.* **2005**, *74*, 387–414. [CrossRef]
8. Lu, L.Y.; Lin, G.L.; Kuo, T.C. Stiffness controllable isolation for near-fault seismic isolation. *Eng. Struct.* **2008**, *30*, 747–765. [CrossRef]

9. Huang, B.; Zhang, H.; Wang, H.; Song, G. Passive base isolation with superelastic nitinol SMA helical springs. *Smart Mater. Struct.* **2014**, *23*, 065009. [CrossRef]
10. Li, H.N.; Chang, Z.G.; Song, G. Studies on structural vibration control with MR dampers using mGA. *Earthq. Eng. Eng. Vib.* **2005**, *4*, 301–304. [CrossRef]
11. Zhang, C. Control force characteristics of different control strategies for the wind-excited 76-story benchmark building structure. *Adv. Struct. Eng.* **2014**, *17*, 543–560.
12. Zhang, C.; Ou, J. Control structure interaction of electromagnetic mass damper system for structural vibration control. *ASCE Eng. Mech.* **2008**, *134*, 428–437. [CrossRef]
13. Ying, Z.G.; Ni, Y.Q.; Ko, J.M. A semi-active stochastic optimal control strategy for nonlinear structural systems with MR dampers. *Smart Struct. Syst.* **2009**, *5*, 69–79. [CrossRef]
14. Yang, G.; Spencer, B.F.; Carlson, J.D.; Sain, M.K. Largescale MR fluid dampers: Modeling and dynamic performance considerations. *Eng. Struct.* **2002**, *24*, 309–323. [CrossRef]
15. Choi, S.B.; Nam, M.H.; Lee, B.K. Vibration control of a MR seat damper for commercial vehicles. *J. Intell. Mater. Syst. Struct.* **2000**, *11*, 936–944. [CrossRef]
16. Pierrick, J.; Ohayon, R.; Bihan, D. Semi-active control using magneto-rheological dampers for payload launch vibration isolation. In Proceedings of the Smart Structures and Materials 2006: Damping and Isolation SPIE 2006, San Diego, CA, USA, 26 February 2016.
17. Bhaskararao, A.V.; Jangid, R.S. Seismic analysis of structures connected with friction dampers. *Eng. Struct.* **2006**, *28*, 690–703. [CrossRef]
18. Bharti, S.D.; Dumne, S.M.; Shrimali, M.K. Seismic response analysis of adjacent buildings connected with MR dampers. *Eng. Struct.* **2010**, *32*, 2122–2133. [CrossRef]
19. Calabrese, A.; Spizzuoco, M.; Serino, G.; Della Corte, G.; Maddaloni, G. Shake table investigation of a novel, low cost, base isolation technology using recycled rubber. *Struct. Control Health Monit.* **2014**, *22*, 107–122. [CrossRef]
20. Caterino, N.; Spizzuoco, M.; Occhiuzzi, A. Promptness and dissipative capacity of MR dampers: Experimental investigations. *Struct. Control Health Monit.* **2013**, *20*, 1424–1440. [CrossRef]
21. Maddaloni, G.; Caterino, N.; Occhiuzzi, A. Shake table investigation of a structure isolated by recycled rubber. *Struct. Control Health Monit.* **2016**. [CrossRef]
22. Lin, P.Y.; Roschke, P.N.; Loh, C.H. Hybrid base isolation with magnetorheological damper and fuzzy control. *Struct. Control Health Monit.* **2007**, *14*, 384–405. [CrossRef]
23. Yoshioka, H.; Ramallo, J.C.; Spencer, B.F. "Smart" Base isolation strategies employing magnetorheological dampers. *J. Eng. Mech.* **2002**, *128*, 540–551. [CrossRef]
24. Matsagar, V.A.; Jangid, R.S. Viscoelastic damper connected to adjacent structures involving seismic isolation. *J. Civ. Eng. Manag.* **2005**, *11*, 309–322. [CrossRef]
25. Wang, H.; Zhang, Y.; Wang, W. The high order single step method for seismic response analysis of non-linear structures. *Earthq. Eng. Eng. Vib.* **1996**, *16*, 48–54.
26. Wang, H.D.; Zhang, Y.S.; Wang, W. A high order single step-β method for nonlinear structural dynamic analysis. *J. Harbin Inst. Technol.* **2003**, *10*, 113–119.
27. Zhang, C.; Ou, J. Evaluation Indices and Numerical Analysis on Characteristic of Active Control Force in Structural Active Mass Driver Control System. *Pac. Sci. Rev.* **2007**, *9*, 115–122.
28. Zhang, C.; Ou, J.; Zhang, J. Parameter Optimization and Analysis of Vehicle Suspension System Controlled by Magnetorheological Fluid Dampers. *Struct. Control Health Monit.* **2006**, *13*, 885–896. [CrossRef]
29. Zhang, C.; Ou, J. Improved semi-active control algorithm and simulation analysis for vibration reduction of structures using MR dampers. *World Inf. Earthq. Eng.* **2003**, *19*, 37–43.
30. Code of China. *Chinese Code for Seismic Design of Buildings*; GB 50011-2010; Code of China: Beijing, China, 2010.
31. Zhang, C.; Ou, J. Modeling and dynamical performance of the electromagnetic mass driver system for structural vibration control. *Eng. Struct.* **2015**, *82*, 93–103. [CrossRef]
32. Wang, T. Research on the Overturn Effect of Rubber Isolation Structure. Ph.D. Thesis, Harbin Institute of Technology, Harbin, China, 2004.

applied sciences

MDPI

Article

Integrated Design of Hybrid Interstory-Interbuilding Multi-Actuation Schemes for Vibration Control of Adjacent Buildings under Seismic Excitations †

Francisco Palacios-Quiñonero [1,*], Josep Rubió-Massegú [1], Josep Maria Rossell [1] and Hamid Reza Karimi [2]

[1] Department of Mathematics, Universitat Politècnica de Catalunya, Av. Bases de Manresa 61–73, 08242 Manresa, Barcelona, Spain; josep.rubio@upc.edu (J.R.-M.); josep.maria.rossell@upc.edu (J.M.R.)
[2] Politecnico di Milano, Department of Mechanical Engineering, via La Masa 1, 20156 Milan, Italy; hamidreza.karimi@polimi.it
* Correspondence: francisco.palacios@upc.edu; Tel.: +34-938-777-302
† This paper is an extended version of our paper published in MOVIC2016 & RASD2016. Advanced design of integrated vibration control systems for adjacent buildings under seismic excitations. *J. Phys. Conf. Ser.* **2016**, 744, 1–12.

Academic Editors: Gangbing Song, Steve C.S. Cai and Hong-Nan Li
Received: 30 January 2017; Accepted: 22 March 2017; Published: 25 March 2017

Abstract: The design of vibration control systems for the seismic protection of closely adjacent buildings is a complex and challenging problem. In this paper, we consider distributed multi-actuation schemes that combine interbuilding linking elements and interstory actuation devices. Using an advanced static output-feedback H_∞ approach, active and passive vibration control systems are designed for a multi-story two-building structure equipped with a selected set of linked and unlinked actuation schemes. To validate the effectiveness of the obtained controllers, the corresponding frequency responses are investigated and a proper set of numerical simulations is conducted using the full scale North–South El Centro 1940 seismic record as ground acceleration disturbance. The observed results indicate that using combined interstory-interbuilding multi-actuation schemes is an effective means of mitigating the vibrational response of the individual buildings and, simultaneously, reducing the risk of interbuilding pounding. These results also point out that passive control systems with high-performance characteristics can be designed using damping elements.

Keywords: structural vibration control; multi-building systems; output-feedback control; seismic protection; passive control; pounding

1. Introduction

In the design of vibration control systems for the seismic protection of closely adjacent buildings, a twofold objective has to be considered: (i) to mitigate the vibrational response of the individual structures and (ii) to provide a proper protection against interbuilding impacts (pounding) [1]. A common strategy to meet these objectives consists in connecting the adjacent buildings by linking elements, which can help both to dissipate the structural vibration energy and to keep the interbuilding gap within safe limits. Over the last years, a significant research effort has been made in this field. Recent works on the seismic response of adjacent structures include nonlinear dynamical models for pounding events [2–5], stochastic assessment of pounding risk [6], structural analysis of large buildings connected by sky-bridge links [7], seismic response of adjacent nonstructural components [8], and reduced-order models for the study of the analytical characteristics of linked structures [9–12]. Also, a wide variety of control strategies has been recently proposed, including passive linking systems with viscous and viscoelastic elements [13–18], nonlinear linking dampers [19,20], semiactive magnetorheological linking dampers [21–23], shared Tuned-Mass-Dampers [24–26], and active linking devices [27,28]. Improved performance and robustness can be attained by hybrid control schemes that make a combined use of interbuilding linking elements and vibration control systems implemented in the individual buildings. Positive results are reported by recent works in this line, which combine the connected control method with base-isolation systems [29–33], multi-isolation systems [34], and interstory multi-actuation systems [35–37].

Encouraged by the aforementioned statements, in this paper two different kinds of force actuation devices are considered: *interstory actuators*, which are implemented between consecutive stories of the same building and exert structural forces restricted to this building, and *interbuilding actuators*, which are implemented between stories located at the same level of adjacent buildings and produce structural forces affecting both buildings (see Figure 1b). The main objective is to design effective vibration control systems with hybrid interstory-interbuilding multi-actuation schemes. More specifically, we are interested in designing: (i) active control systems with reduced and realistic feedback information, and (ii) passive control systems with high-performance characteristics. To meet the former objective, we assume that the relative velocities associated to the actuation devices are measurable and compute static velocity-feedback H_∞ controllers following an advanced linear matrix inequality (LMI) approach [38, 39]. While, in the latter, the actuation devices are assumed to be passive viscous dampers and the corresponding damping capacities are computed by designing a fully decentralized velocity-feedback H_∞ controller [40]. The main problem is described by means of a particular two-building system equipped with different linked and unlinked actuation schemes. To assess the effectiveness of the proposed active and passive control strategies, the corresponding frequency responses are investigated. Also, a proper set of numerical simulations is conducted using the full scale North–South El Centro 1940 seismic record as ground acceleration disturbance.

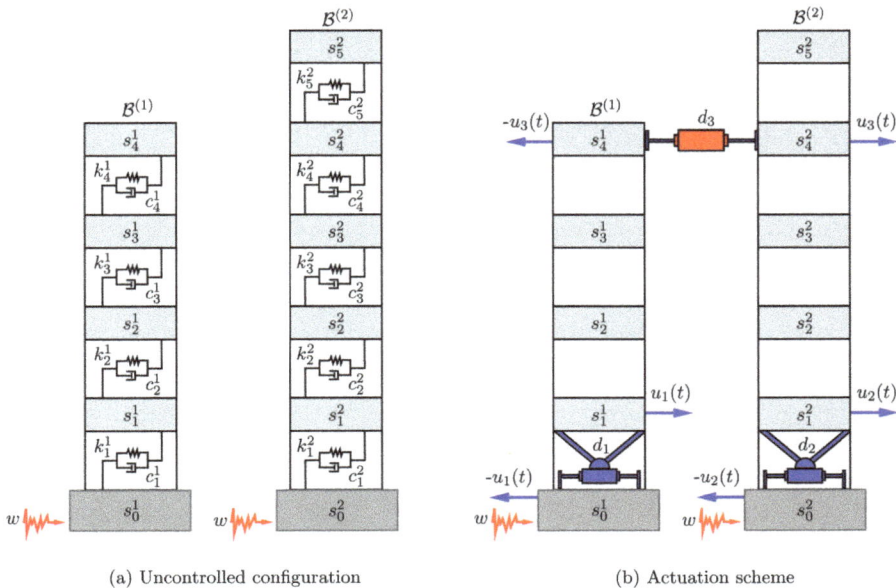

(a) Uncontrolled configuration   (b) Actuation scheme

Figure 1. Two-building model: (**a**) Uncontrolled configuration. (**b**) Linked actuation scheme formed by two interstory actuators (d_1 and d_2) and one interbuilding actuator (d_3).

The rest of the paper is organized as follows: In Section 2, state-space models for the uncontrolled two-building system and the different linked and unlinked controlled configurations are provided. In Section 3, the centralized and decentralized static velocity-feedback H_∞ controllers are designed and the characteristics of the corresponding frequency responses are discussed. In Section 4, numerical simulations of the time responses are conducted and illustrative peak-value plots of the interstory drifts, interbuilding approaches and control efforts are provided and compared. In Section 5, some conclusions and future research directions are briefly presented. Additionally, the particular parameter values of the two-building system used in the controller designs and numerical simulations are included in Appendix A, and the main elements of the LMI-based static output-feedback H_∞ controller design methodology are summarized in Appendix B.

2. Two-Building Mathematical Models

2.1. Uncontrolled Configuration

Let us consider a two-building system formed by a four-story building adjacent to a five-story building as schematically depicted in Figure 1a. In this uncontrolled configuration, the lateral motion of the buildings can be described by the second-order differential equation

$$\mathbf{M}\,\ddot{\mathbf{q}}(t) + \mathbf{C}_{\mathrm{d}}\,\dot{\mathbf{q}}(t) + \mathbf{K}_{\mathrm{s}}\,\mathbf{q}(t) = \mathbf{T}_{\mathrm{w}}\,w(t), \tag{1}$$

where $\mathbf{q}(t)$ is the vector of story displacements with respect to the ground, $\mathbf{M}$ is the mass matrix, $\mathbf{C}_{\mathrm{d}}$ is the damping matrix, $\mathbf{K}_{\mathrm{s}}$ is the stiffness matrix, $w(t)$ is the ground acceleration disturbance and $\mathbf{T}_{\mathrm{w}}$ is the disturbance input matrix. The vector of story displacements can be written in the form

$$\mathbf{q}(t) = \left[q_1^1(t),\, q_2^1(t),\, q_3^1(t),\, q_4^1(t),\, q_1^2(t),\, q_2^2(t),\, q_3^2(t),\, q_4^2(t),\, q_5^2(t) \right]^T, \tag{2}$$

where $q_i^j(t)$ represents the displacement of the ith story in the building $\mathcal{B}^{(j)}$ (denoted as s_i^j in Figure 1) with respect to the building's ground level s_0^j. The mass matrix is a diagonal matrix

$$\mathbf{M} = \mathrm{diag}(m_1^1,\, m_2^1,\, m_3^1,\, m_4^1,\, m_1^2,\, m_2^2,\, m_3^2,\, m_4^2,\, m_5^2), \tag{3}$$

where m_i^j denotes the mass of the ith story in the building $\mathcal{B}^{(j)}$. The damping and stiffness matrices have a similar block diagonal structure

$$\mathbf{K}_{\mathrm{s}} = \begin{bmatrix} \mathbf{K}_{\mathrm{s}}^{(1)} & [\mathbf{0}]_{4\times5} \\ [\mathbf{0}]_{5\times4} & \mathbf{K}_{\mathrm{s}}^{(2)} \end{bmatrix}, \quad \mathbf{C}_{\mathrm{d}} = \begin{bmatrix} \mathbf{C}_{\mathrm{d}}^{(1)} & [\mathbf{0}]_{4\times5} \\ [\mathbf{0}]_{5\times4} & \mathbf{C}_{\mathrm{d}}^{(2)} \end{bmatrix}, \tag{4}$$

where $\mathbf{C}_{\mathrm{d}}^{(j)}$ and $\mathbf{K}_{\mathrm{s}}^{(j)}$ represent the local damping and stiffness matrices, respectively, corresponding to the building $\mathcal{B}^{(j)}$, and $[\mathbf{0}]_{r\times s}$ is a zero matrix of dimensions $r \times s$. Typically, the local stiffness matrices have the following tridiagonal structure:

$$\mathbf{K}_{\mathrm{s}}^{(1)} = \begin{bmatrix} k_1^1 + k_2^1 & -k_2^1 & 0 & 0 \\ -k_2^1 & k_2^1 + k_3^1 & -k_3^1 & 0 \\ 0 & -k_3^1 & k_3^1 + k_4^1 & -k_4^1 \\ 0 & 0 & -k_4^1 & k_4^1 \end{bmatrix}, \quad \mathbf{K}_{\mathrm{s}}^{(2)} = \begin{bmatrix} k_1^2 + k_2^2 & -k_2^2 & 0 & 0 & 0 \\ -k_2^2 & k_2^2 + k_3^2 & -k_3^2 & 0 & 0 \\ 0 & -k_3^2 & k_3^2 + k_4^2 & -k_4^2 & 0 \\ 0 & 0 & -k_4^2 & k_4^2 + k_5^2 & -k_5^2 \\ 0 & 0 & 0 & -k_5^2 & k_5^2 \end{bmatrix}, \tag{5}$$

where k_i^j denotes the stiffness coefficient of the ith story in the building $\mathcal{B}^{(j)}$ (see Figure 1a). When the values of the damping coefficients c_i^j are known, the local damping matrices $\mathbf{C}_{\mathrm{d}}^{(j)}$ can be obtained by replacing the stiffness coefficients k_i^j in Equation (5) by the corresponding damping coefficients. Frequently, however, the values of the damping coefficients cannot be properly determined and other computational methods are used to obtain the matrices $\mathbf{C}_d^{(j)}$ [41]. Finally, the disturbance input matrix has the following form:

$$\mathbf{T}_w = -\mathbf{M}\,[\mathbf{1}]_{9\times1} \tag{6}$$

where $[\mathbf{1}]_{n\times1}$ is a vector of dimension n with all its entries equal to one.

In order to describe the vibrational response of the two-building system, we consider two different types of output variables: *interstory drifts* and *interbuilding approaches*. The interstory drift $r_i^j(t)$ is the relative displacement between the consecutive stories s_i^j and s_{i-1}^j of the building $\mathcal{B}^{(j)}$, and can be defined as

$$\begin{cases} r_1^j(t) = q_1^j(t), \\ r_i^j(t) = q_i^j(t) - q_{i-1}^j(t), \quad 1 < i \leq n_j, \end{cases} \tag{7}$$

where n_j represents the number of stories of the building $\mathcal{B}^{(j)}$. The overall vector of interstory drifts

$$\mathbf{r}(t) = \left[r_1^1(t),\, r_2^1(t),\, r_3^1(t),\, r_4^1(t),\, r_1^2(t),\, r_2^2(t),\, r_3^2(t),\, r_4^2(t),\, r_5^2(t) \right]^T \tag{8}$$

can be computed as

$$\mathbf{r}(t) = \widetilde{\mathbf{C}}_r\, \mathbf{q}(t), \tag{9}$$

with

$$\widetilde{\mathbf{C}}_r = \begin{bmatrix} \mathbf{C}_r^{(1)} & [\mathbf{0}]_{4\times5} \\ [\mathbf{0}]_{5\times4} & \mathbf{C}_r^{(2)} \end{bmatrix}, \quad \mathbf{C}_r^{(1)} = \begin{bmatrix} 1 & 0 & 0 & 0 \\ -1 & 1 & 0 & 0 \\ 0 & -1 & 1 & 0 \\ 0 & 0 & -1 & 1 \end{bmatrix}, \quad \mathbf{C}_r^{(2)} = \begin{bmatrix} 1 & 0 & 0 & 0 & 0 \\ -1 & 1 & 0 & 0 & 0 \\ 0 & -1 & 1 & 0 & 0 \\ 0 & 0 & -1 & 1 & 0 \\ 0 & 0 & 0 & -1 & 1 \end{bmatrix}. \tag{10}$$

The interbuilding approach $a_i(t)$ describes the approaching between the stories s_i^1 and s_i^2 placed at the same level in the adjacent buildings, and can be defined as

$$a_i(t) = -\left(q_i^2(t) - q_i^1(t) \right), \; 1 \leq i \leq \min(n_1, n_2). \tag{11}$$

For the considered two-building system, the vector of interbuilding approaches

$$\mathbf{a}(t) = [a_1(t),\, a_2(t),\, a_3(t),\, a_4(t)]^T \tag{12}$$

can be computed as

$$\mathbf{a}(t) = \widetilde{\mathbf{C}}_a\, \mathbf{q}(t), \tag{13}$$

with

$$\widetilde{\mathbf{C}}_a = \begin{bmatrix} \mathbf{I}_4 & -\mathbf{I}_4 & [\mathbf{0}]_{4\times1} \end{bmatrix}. \tag{14}$$

Next, by introducing the state vector

$$\mathbf{x}(t) = \begin{bmatrix} \mathbf{q}(t) \\ \dot{\mathbf{q}}(t) \end{bmatrix}, \tag{15}$$

we obtain a first-order state-space model

$$\dot{\mathbf{x}}(t) = \mathbf{A}\,\mathbf{x}(t) + \mathbf{E}\,w(t), \tag{16}$$

with system matrices

$$\mathbf{A} = \begin{bmatrix} [\mathbf{0}]_{9 \times 9} & \mathbf{I}_9 \\ -\mathbf{M}^{-1}\mathbf{K} & -\mathbf{M}^{-1}\mathbf{C} \end{bmatrix}, \quad \mathbf{E} = \begin{bmatrix} [\mathbf{0}]_{9 \times 1} \\ -[\mathbf{1}]_{9 \times 1} \end{bmatrix}, \tag{17}$$

where $\mathbf{I}_n$ denotes an identity matrix of dimension n. The vectors of interstory drifts and interbuilding approaches can be computed in the form

$$\mathbf{r}(t) = \mathbf{C}_r\,\mathbf{x}(t), \quad \mathbf{a}(t) = \mathbf{C}_a\,\mathbf{x}(t), \tag{18}$$

using the output matrices

$$\mathbf{C}_r = \begin{bmatrix} \tilde{\mathbf{C}}_r & [\mathbf{0}]_{9 \times 9} \end{bmatrix}, \quad \mathbf{C}_a = \begin{bmatrix} \tilde{\mathbf{C}}_a & [\mathbf{0}]_{4 \times 9} \end{bmatrix}. \tag{19}$$

2.2. Controlled Configurations

In order to mitigate the vibrational response of the adjacent buildings, we consider two different kinds of force actuation devices: (i) *interstory actuators*, which are implemented between consecutive stories of the same building, and (ii) *interbuilding actuators*, which are implemented between stories located at the same level in the adjacent buildings. In both cases, the actuation device produces a pair of opposite structural forces on the corresponding stories. An actuation scheme with two interstory actuators (d_1 and d_2) located at the buildings' lowest level, and one interbuilding actuator (d_3) implemented at the fourth-story level is schematically depicted in Figure 1b, where $u_j(t)$ denotes the actuation force produced by the actuation device d_j. For this control configuration, the lateral motion of the buildings can be described by the second-order differential equation

$$\mathbf{M}\,\ddot{\mathbf{q}}(t) + \mathbf{C}_d\,\dot{\mathbf{q}}(t) + \mathbf{K}_s\,\mathbf{q}(t) = \mathbf{T}_u\,\mathbf{u}(t) + \mathbf{T}_w\,w(t), \tag{20}$$

where

$$\mathbf{u}(t) = \begin{bmatrix} u_1(t), \ldots, u_{n_d}(t) \end{bmatrix}^T \tag{21}$$

is the vector of control forces, n_d is the total number of actuation devices and $\mathbf{T}_u$ is the control location matrix, which models the overall effect of the actuation system. By considering the state vector $\mathbf{x}(t)$, we obtain the state-space model

$$\dot{\mathbf{x}}(t) = \mathbf{A}\,\mathbf{x}(t) + \mathbf{B}\,\mathbf{u}(t) + \mathbf{E}\,w(t), \tag{22}$$

with the control input matrix

$$\mathbf{B} = \begin{bmatrix} [\mathbf{0}]_{9 \times 1} \\ \mathbf{M}^{-1}\mathbf{T}_u \end{bmatrix}. \tag{23}$$

For the control configuration presented in Figure 1b, the control location matrix is

$$\mathbf{T_u} = \begin{bmatrix} 1 & 0 & 0 \\ 0 & 0 & 0 \\ 0 & 0 & 0 \\ 0 & 0 & -1 \\ 0 & 1 & 0 \\ 0 & 0 & 0 \\ 0 & 0 & 0 \\ 0 & 0 & 1 \\ 0 & 0 & 0 \end{bmatrix}. \tag{24}$$

In this work, we consider a set of six different control configurations, defined by a particular actuation scheme. The configurations that contain interbuilding actuation devices are called *linked*, and those that only contain interstory actuators are called *unlinked*. The *control configuration I* (CC1), presented in Figure 2a, is an unlinked configuration that includes two interstory actuators (d_1 and d_2) located at the buildings' lowest level. The *control configuration II* (CC2), presented in Figure 2b, is a linked configuration that includes an interstory actuator (d_1) located at the lowest level of the building $\mathcal{B}^{(2)}$, and an interbuilding actuator (d_2), located at the fourth-story level. Control configurations with three and four actuation devices are presented in Figures 3 and 4, respectively. The control location matrices corresponding to the configurations CC1, CC2 and CC3 are, respectively,

$$\mathbf{T_u^I} = \begin{bmatrix} 1 & 0 \\ 0 & 0 \\ 0 & 0 \\ 0 & 0 \\ 0 & 1 \\ 0 & 0 \\ 0 & 0 \\ 0 & 0 \\ 0 & 0 \end{bmatrix}, \quad \mathbf{T_u^{II}} = \begin{bmatrix} 0 & 0 \\ 0 & 0 \\ 0 & 0 \\ 0 & -1 \\ 1 & 0 \\ 0 & 0 \\ 0 & 0 \\ 0 & 1 \\ 0 & 0 \end{bmatrix}, \quad \mathbf{T_u^{III}} = \begin{bmatrix} 1 & 0 & 0 \\ 0 & 0 & 0 \\ 0 & 0 & 0 \\ 0 & 0 & 0 \\ 0 & 1 & -1 \\ 0 & 0 & 1 \\ 0 & 0 & 0 \\ 0 & 0 & 0 \\ 0 & 0 & 0 \end{bmatrix}, \tag{25}$$

the matrix $\mathbf{T_u^{IV}}$ corresponding to the configuration CC4 has been previously presented in Equation (24), and the control location matrices corresponding to the configurations CC5 and CC6 are, respectively,

$$\mathbf{T_u^V} = \begin{bmatrix} 1 & -1 & 0 & 0 \\ 0 & 1 & 0 & 0 \\ 0 & 0 & 0 & 0 \\ 0 & 0 & 0 & 0 \\ 0 & 0 & 1 & -1 \\ 0 & 0 & 0 & 1 \\ 0 & 0 & 0 & 0 \\ 0 & 0 & 0 & 0 \\ 0 & 0 & 0 & 0 \end{bmatrix}, \quad \mathbf{T_u^{VI}} = \begin{bmatrix} 1 & 0 & 0 & 0 \\ 0 & 0 & 0 & 0 \\ 0 & 0 & 0 & 0 \\ 0 & 0 & 0 & -1 \\ 0 & 1 & -1 & 0 \\ 0 & 0 & 1 & 0 \\ 0 & 0 & 0 & 0 \\ 0 & 0 & 0 & 1 \\ 0 & 0 & 0 & 0 \end{bmatrix}. \tag{26}$$

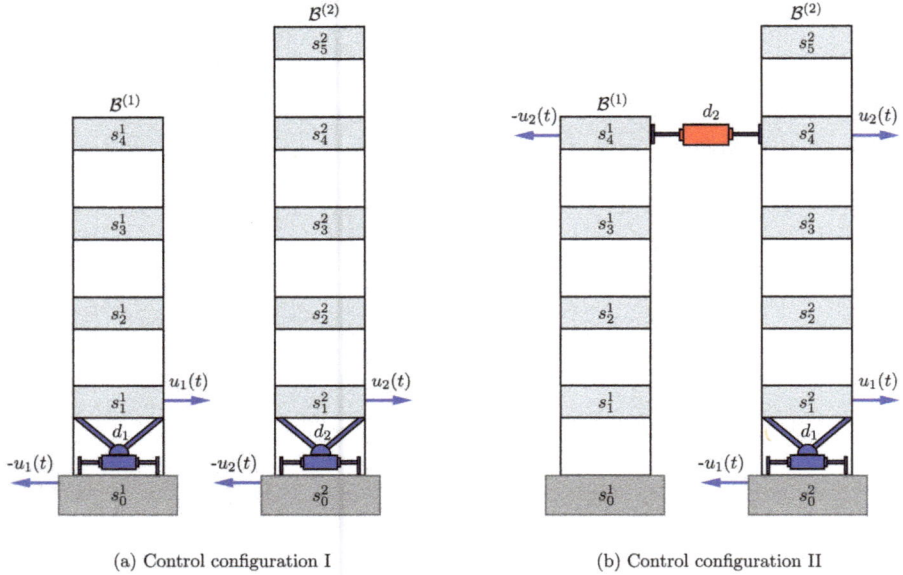

(a) Control configuration I

(b) Control configuration II

Figure 2. Actuation schemes with two actuation devices. (**a**) Unlinked configuration CC1, with two interstory actuators (d_1 and d_2). (**b**) Linked configuration CC2, with an interstory actuator (d_1) and an interbuilding actuator (d_2).

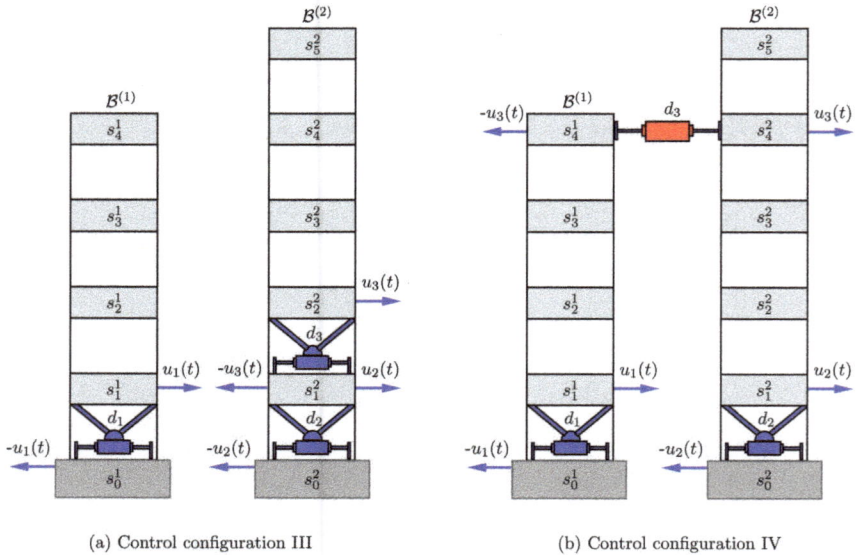

(a) Control configuration III

(b) Control configuration IV

Figure 3. Actuation schemes with three actuation devices. (**a**) Unlinked configuration CC3, with three interstory actuators (d_1, d_2 and d_3). (**b**) Linked configuration CC4, with two interstory actuators (d_1 and d_2) and an interbuilding actuator (d_3).

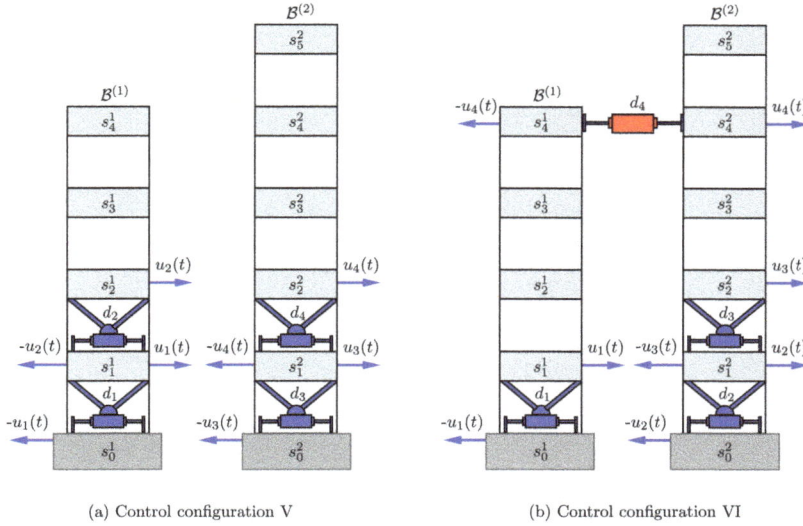

(a) Control configuration V

(b) Control configuration VI

Figure 4. Actuation schemes with four actuation devices. (**a**) Unlinked configuration CC5, with four interstory actuators (d_1, d_2, d_3 and d_4). (**b**) Linked configuration CC6, with three interstory actuators (d_1, d_2 and d_3) and an interbuilding actuator (d_4).

3. Controllers Design

3.1. Static Output-Feedback H_∞ Centralized Controllers

Static output-feedback controllers can be designed to perform a fast and effective computation of the control actions from the available feedback information. Typically, these kind of controllers can be written in the form

$$\mathbf{u}(t) = \mathbf{K}\mathbf{y}(t), \tag{27}$$

where $\mathbf{u}(t)$ is the vector of control actions, $\mathbf{K}$ is a constant control gain matrix and $\mathbf{y}(t)$ is a vector of measured outputs. According to the results summarized in Appendix B, a suboptimal static output-feedback H_∞ controller can be computed for the linear system in Equation (22) by solving the LMI optimization $\mathcal{P}$ given in Equation (A8). The design procedure involves a measured-output vector of the form

$$\mathbf{y}(t) = \mathbf{C}_y\mathbf{x}(t), \tag{28}$$

that models the available feedback information, and a vector of controlled outputs

$$\mathbf{z}(t) = \mathbf{C}_z\,\mathbf{x}(t) + \mathbf{D}_z\,\mathbf{u}(t), \tag{29}$$

that allows computing the overall cost of the system response and the control action. In this work, we assume that the relative velocities associated to the actuation devices are measurable and consider the vector of measured outputs:

$$\mathbf{y}(t) = [y_1(t), \dots, y_{n_d}(t)]^T, \tag{30}$$

where n_d is the total number of actuation devices and $y_j(t)$ represents the relative velocity corresponding to the actuation device d_j. For the control configuration ℓ, the vector of measured outputs can be written as a linear combination of the state variables defined in Equation (15) using the measured-output matrix

$$\mathbf{C}_y^\ell = \left[\ [0]_{n_d \times 9} \quad \left(\mathbf{T}_u^\ell \right)^T \ \right], \tag{31}$$

where $\mathbf{T}_u^\ell$ is the corresponding control location matrix. Thus, for the configuration CC6 displayed in Figure 4b, the number of actuation devices is $n_d = 4$ and the measured-output vector has the form $\mathbf{y}(t) = [y_1(t), y_2(t), y_3(t), y_4(t)]^T$, where $y_1(t)$ is the interstory velocity at the first-story level in building $\mathcal{B}^{(1)}$, $y_2(t)$ and $y_3(t)$ denote the interstory velocity at the first and second story levels in building $\mathcal{B}^{(2)}$, respectively, and $y_4(t)$ is the interbuilding velocity at the four-story level. In this case, by considering the matrix $\mathbf{T}_u^{VI}$ given in Equation (26), we obtain the measured-output matrix

$$\mathbf{C}_y^{VI} = \begin{bmatrix} 0 & 0 & 0 & 0 & 0 & 0 & 0 & 0 & 0 & 1 & 0 & 0 & 0 & 0 & 0 & 0 & 0 & 0 \\ 0 & 0 & 0 & 0 & 0 & 0 & 0 & 0 & 0 & 0 & 0 & 0 & 0 & 1 & 0 & 0 & 0 & 0 \\ 0 & 0 & 0 & 0 & 0 & 0 & 0 & 0 & 0 & 0 & 0 & 0 & 0 & -1 & 1 & 0 & 0 & 0 \\ 0 & 0 & 0 & 0 & 0 & 0 & 0 & 0 & 0 & 0 & 0 & 0 & -1 & 0 & 0 & 0 & 1 & 0 \end{bmatrix}. \tag{32}$$

To define the controlled-output vector $\mathbf{z}(t)$, it should be noted that large interstory drifts and interbuilding approaches must both be avoided in order to prevent buildings' structural damage and interbuilding collisions. Additionally, moderate control efforts are also convenient. To this end, we consider the controlled-output matrices

$$\mathbf{C}_z = \begin{bmatrix} \alpha_r \mathbf{C}_r \\ \alpha_a \mathbf{C}_a \\ [0]_{n_d \times 18} \end{bmatrix}, \quad \mathbf{D}_z = \begin{bmatrix} [0]_{13 \times n_d} \\ \alpha_u \mathbf{I}_{n_d} \end{bmatrix}, \tag{33}$$

where α_r, α_a and α_u are scaling coefficients that compensate the different magnitude of interstory drifts, interbuilding approaches and control forces, respectively, $\mathbf{C}_r$ and $\mathbf{C}_a$ are the matrices given in Equation (19), and n_d is the total number of actuation devices of the considered control configuration. With this choice, the controlled-output vector satisfies

$$\|\mathbf{z}(t)\|_2^2 = \alpha_r^2 \|\mathbf{r}(t)\|_2^2 + \alpha_a^2 \|\mathbf{a}(t)\|_2^2 + \alpha_u^2 \|\mathbf{u}(t)\|_2^2. \tag{34}$$

To obtain a velocity-feedback H_∞ controller for the control configuration ℓ, we consider the system matrices $\mathbf{A}$, $\mathbf{B}$ and $\mathbf{E}$ corresponding to the control location matrix $\mathbf{T}_u^\ell$, the mass and stiffness values given in Table A1 (see Appendix A) and the damping matrices in Equations (A1) and (A2); the measured-output matrix $\mathbf{C}_y^\ell$ in Equation (31); and the controlled-output matrices $\mathbf{C}_z$ and $\mathbf{D}_z$ in Equation (33) defined by the corresponding control dimension n_d and the scaling coefficients

$$\alpha_r = 5, \quad \alpha_a = 1, \quad \alpha_u = 10^{-7.4}. \tag{35}$$

Next, by applying the computational procedure described in Appendix B, we obtain a velocity-feedback control gain matrix $\mathbf{K}^\ell$ and an upper bound $\widetilde{\gamma}_\ell$ of the corresponding H_∞-norm γ_ℓ, which can be computed by considering the closed-loop transfer function $\mathbf{T}_{K\ell}(2\pi f j)$ in Equation (A17) and solving the optimization problem given in Equation (A16). For the control configurations CC1 and CC2, we obtain the following velocity-feedback control gain matrices

$$\mathbf{K}^I = 10^7 \times \begin{bmatrix} -0.7525 & 0.0142 \\ -0.8003 & -1.1890 \end{bmatrix}, \quad \mathbf{K}^{II} = 10^7 \times \begin{bmatrix} -1.0176 & 0.0748 \\ 0.1392 & -0.4387 \end{bmatrix}, \tag{36}$$

and the γ-value upper bounds

$$\widetilde{\gamma}_I = 0.5973, \quad \widetilde{\gamma}_{II} = 0.8851. \tag{37}$$

64

The actual γ-values are

$$\gamma_{\mathrm{I}} = 0.4967, \quad \gamma_{\mathrm{II}} = 0.7833. \tag{38}$$

To illustrate the frequency behavior of the velocity-feedback controllers defined by the gain matrices $\mathbf{K}^{\mathrm{I}}$ and $\mathbf{K}^{\mathrm{II}}$, the maximum singular values of the closed-loop transfer functions $\mathbf{T}_{\mathrm{KI}}(2\pi fj)$, $\mathbf{T}_{\mathrm{KII}}(2\pi fj)$ and the open-loop transfer function

$$\mathbf{T}(2\pi fj) = \mathbf{C}_z(2\pi fj\mathbf{I} - \mathbf{A})^{-1}\mathbf{E} \tag{39}$$

are presented in Figure 5. In this figure, the thin black solid line displays the maximum singular values of the open-loop transfer function and shows the frequency response characteristics of the uncontrolled two-building structure. The peaks in this plot are associated to the natural resonant frequencies of the individual buildings, which are presented in Table 1. The main peak-value is associated to the frequency 1.0082 Hz and has a magnitude

$$\gamma_0 = 2.0479, \tag{40}$$

which corresponds to the H_∞-norm of the uncontrolled configuration. The thick blue solid line presents the frequency response of the configuration CC1 with the velocity-feedback control gain matrix $\mathbf{K}^{\mathrm{I}}$, and the red dash-dotted line shows the frequency response of the configuration CC2 with the velocity-feedback control gain matrix $\mathbf{K}^{\mathrm{II}}$. As indicated in Equation (A16), the largest peak-values in these plots correspond to the controllers H_∞-norms given in Equation (38). Following the same design procedure, we obtain the velocity-feedback control gain matrices

$$\mathbf{K}^{\mathrm{III}} = 10^7 \times \begin{bmatrix} -1.4671 & 0.3369 & 0.5595 \\ 0.0936 & -0.6758 & -0.4428 \\ 0.1090 & -0.5970 & -0.5382 \end{bmatrix}, \quad \mathbf{K}^{\mathrm{IV}} = 10^6 \times \begin{bmatrix} -6.8093 & -3.7631 & -0.3129 \\ -1.2553 & -8.6084 & 0.3653 \\ 0.5425 & 0.9675 & -2.9665 \end{bmatrix}, \tag{41}$$

for the control configurations CC3 and CC4, respectively, the control gain matrix

$$\mathbf{K}^{\mathrm{V}} = 10^7 \times \begin{bmatrix} -1.0323 & 0.4066 & 0.2143 & -0.4538 \\ -0.3656 & -0.5823 & -0.3065 & -0.0591 \\ 0.9548 & -1.7601 & -1.5193 & 0.2758 \\ 0.3918 & 0.3666 & -0.5359 & -0.6302 \end{bmatrix} \tag{42}$$

for the configuration CC5, and finally, the control gain matrix

$$\mathbf{K}^{\mathrm{VI}} = 10^6 \times \begin{bmatrix} -4.7133 & -4.9748 & -3.9433 & 1.9188 \\ -4.1747 & -5.9723 & -2.0281 & -1.4118 \\ -1.6926 & -4.8034 & -4.5185 & 0.1502 \\ -4.2366 & 3.2550 & 6.0731 & -6.8744 \end{bmatrix} \tag{43}$$

for the control configuration CC6. The corresponding γ-values and upper bounds $\widetilde{\gamma}_\ell$ are collected in Table 2. The frequency responses are presented in Figures 6 and 7, using a blue solid line for the unlinked configurations (CC3 and CC5), and a red dash-dotted line for the linked configurations (CC4 and CC6). According to the γ-values in Table 2 and the value γ_0 in Equation (40), three facts can be noted: (i) all the controllers obtained with the proposed design methodology provide a good level of reduction in the H_∞-norm value, (ii) an increasing effectiveness is attained with a larger number of actuation devices, and (iii) smaller γ-values are obtained by the unlinked control configurations. All these facts can also be clearly appreciated by observing the peak-values in the frequency plots presented in Figures 5–7. These frequency plots also show that, in all the cases, a significant reduction of the vibrational response is additionally attained in the secondary resonant peaks.

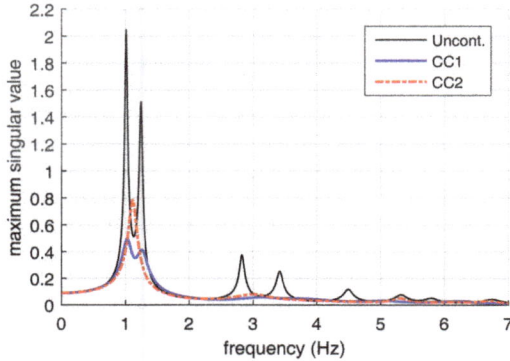

Figure 5. Frequency response corresponding to the centralized velocity-feedback H_∞ controllers with active implementation defined by the control gain matrices $\mathbf{K}^{\mathrm{I}}$ and $\mathbf{K}^{\mathrm{II}}$. Maximum singular values of the closed-loop transfer function $\mathbf{T}_{\mathbf{K}^{\mathrm{I}}}(2\pi f j)$ (thick blue solid line), the closed-loop transfer function $\mathbf{T}_{\mathbf{K}^{\mathrm{II}}}(2\pi f j)$ (red dash-dotted line) and the open-loop transfer function $\mathbf{T}(2\pi f j)$ (thin black solid line).

Table 1. Resonant natural frequencies of the unlinked buildings.

	Frequency (Hz)				
Building $\mathcal{B}^{(1)}$	1.2404	3.4161	5.3160	6.7227	
Building $\mathcal{B}^{(2)}$	1.0082	2.8246	4.4929	5.7974	6.7735

Table 2. Active configurations: Actual γ-values and γ-value upper bounds corresponding to the centralized velocity-feedback controllers defined by the gain matrices $\mathbf{K}^\ell$.

	CC1	CC2	CC3	CC4	CC5	CC6
Upper bound $\widetilde{\gamma}$	0.5973	0.8851	0.3856	0.6528	0.3396	0.4333
γ-value	0.4967	0.7833	0.3395	0.5037	0.2855	0.3277

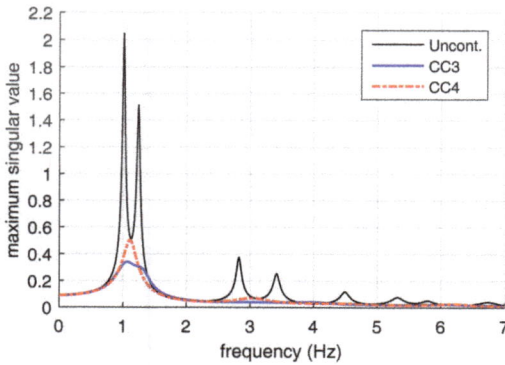

Figure 6. Frequency response corresponding to the centralized velocity-feedback H_∞ controllers with active implementation defined by the control gain matrices $\mathbf{K}^{\mathrm{III}}$ and $\mathbf{K}^{\mathrm{IV}}$. Maximum singular values of the closed-loop transfer function $\mathbf{T}_{\mathbf{K}^{\mathrm{III}}}(2\pi f j)$ (thick blue solid line), the closed-loop transfer function $\mathbf{T}_{\mathbf{K}^{\mathrm{IV}}}(2\pi f j)$ (red dash-dotted line) and the open-loop transfer function $\mathbf{T}(2\pi f j)$ (thin black solid line).

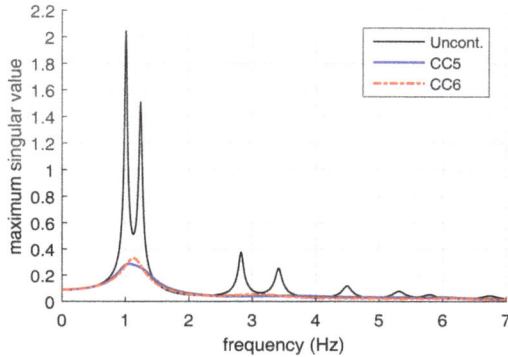

Figure 7. Frequency response corresponding to the centralized velocity-feedback H_∞ controllers with active implementation defined by the control gain matrices $\mathbf{K}^{\mathrm{V}}$ and $\mathbf{K}^{\mathrm{VI}}$. Maximum singular values of the closed-loop transfer function $\mathbf{T}_{\mathbf{K}^{\mathrm{V}}}(2\pi f j)$ (thick blue solid line), the closed-loop transfer function $\mathbf{T}_{\mathbf{K}^{\mathrm{VI}}}(2\pi f j)$ (red dash-dotted line) and the open-loop transfer function $\mathbf{T}(2\pi f j)$ (thin black solid line).

3.2. Static Velocity-Feedback H_∞ Decentralized Controllers

From a practical point of view, the controllers defined by the velocity-feedback gain matrices $\mathbf{K}^\ell$ computed in the previous section have the important advantage of using a reduced system of sensors which are naturally associated to the actuation devices. However, they also present two serious drawbacks. Firstly, the complete vector of measured outputs is used to compute the control actions and, consequently, a wide communication system would be necessary in the controller implementation. Secondly, producing the corresponding actuation forces would require active devices with a large power consumption and potential reliability issues. These two disadvantages, typically present in active vibration control of large structures, can be properly overcome by considering fully decentralized velocity-feedback controllers with control gain matrices of the form

$$\widehat{\mathbf{K}}^\ell = \mathrm{diag}(\hat{k}_1^\ell, \ldots, \hat{k}_{n_d}^\ell), \tag{44}$$

which can be obtained by solving the LMI optimization problem $\mathcal{P}$ in Equation (A8) with the same matrices used in the previous controller designs and constraining the LMI variable matrices $\mathbf{X}_R$ and $\mathbf{Y}_R$ to a diagonal form. As indicated in [17,40], if the gain matrix elements $\hat{k}_i^\ell$ are all negative, then this kind of controllers admit a passive implementation using linear viscous dampers. Thus, for instance, by applying this design methodology to the control configuration CC1, we obtain the following diagonal gain matrix:

$$\widehat{\mathbf{K}}^{\mathrm{I}} = 10^7 \times \begin{bmatrix} -0.8543 & 0 \\ 0 & -1.2968 \end{bmatrix}. \tag{45}$$

Hence, the decentralized velocity-feedback controller defined by the diagonal gain matrix $\widehat{\mathbf{K}}^{\mathrm{I}}$ can be implemented using two interstory linear damping devices d_1 and d_2 with respective damping constants $0.8543 \times 10^7 \, \mathrm{Ns/m}$ and $1.2968 \times 10^7 \, \mathrm{Ns/m}$. The frequency response characteristics of this passive control system are displayed in Figure 8 using a red dash-dotted line. The frequency response corresponding to the uncontrolled configuration (thin black solid line) and the centralized controller defined by the gain matrix $\mathbf{K}^{\mathrm{I}}$ (thick blue solid line) are also included as a reference. The plots in the figure clearly show the good behavior of the obtained passive controller, which practically matches the performance of the active centralized controller over most of the frequency range, and produces a small increment (of about 10%) in the peak-value corresponding to the main resonant frequency.

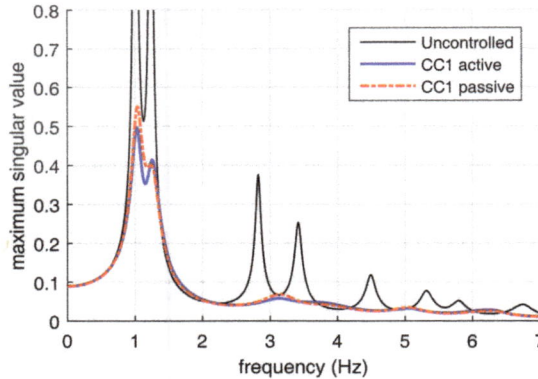

Figure 8. Frequency response corresponding to the decentralized velocity-feedback H_∞ controller with passive implementation defined by the diagonal control gain matrix $\widehat{\mathbf{K}}^!$. Maximum singular values of the closed-loop transfer function $\mathbf{T}_{\widehat{\mathbf{K}}^!}(2\pi fj)$ (red dash-dotted line), the closed-loop transfer function $\mathbf{T}_{\mathbf{K}^!}(2\pi fj)$ (thick blue solid line) and the open-loop transfer function $\mathbf{T}(2\pi fj)$ (thin black solid line).

Using the same design strategy to compute decentralized velocity-feedback controllers for the other control configurations considered in this paper, we have obtained diagonal gain matrices with the values presented in Table 3. Comparing the corresponding γ-values collected in Table 4 and the value γ_0 in Equation (40), it can be appreciated that the same three facts observed in the centralized controller designs also apply to the decentralized controllers: (i) all the controllers obtained with the proposed design methodology provide a good level of reduction in the H_∞-norm value, (ii) an increasing effectiveness is attained with a larger number of actuation devices, and (iii) smaller γ-values are obtained by the unlinked control configurations. However, two new elements appear in the decentralized design: (iv) the risk of failure of the controller design procedure is higher in the decentralized designs, and (v) the γ-value attained by some decentralized controllers is lower than the one obtained by the associated centralized controller.

Table 3. Coefficients of the passive controllers $\widehat{\mathbf{K}}^\ell = \mathrm{diag}\big(\hat{k}_1^\ell, \ldots, \hat{k}_{n_d}^\ell\big)$ obtained for the different control configurations ($\times 10^7$ Ns/m). The values corresponding to the configuration CC5 are missing due to the feasibility issues encountered in the associated LMI optimization problem.

	CC1	CC2	CC3	CC4	CC5	CC6
$\hat{k}_1^\ell$	−0.8543	−0.9442	−1.5106	−0.6138	–	−2.2631
$\hat{k}_2^\ell$	−1.2968	−0.2483	−0.7879	−0.9207	–	−2.2309
$\hat{k}_3^\ell$			−1.1282	−0.3257	–	−0.9255
$\hat{k}_4^\ell$					–	−0.3883

Table 4. Passive configurations: Actual γ-values and γ-value upper bounds corresponding to the decentralized velocity-feedback controllers defined by the diagonal gain matrices $\widehat{\mathbf{K}}^\ell$. The values of the control configuration CC5 are missing due to the feasibility issues encountered in the associated linear matrix inequality (LMI) optimization problem.

	CC1	CC2	CC3	CC4	CC5	CC6
Upper bound $\widetilde{\gamma}$	0.6537	0.8865	0.5745	0.7657	–	0.6079
γ-value	0.5500	0.7646	0.3241	0.5987	–	0.3058

The increased risk of failure in the decentralized design procedure can be explained by the additional structure constraints introduced in the LMI optimization problem, which can produce feasibility issues. This kind of numerical problems are poorly understood and sometimes depend on the particular numerical solver and the options used in the LMI optimization procedure. In our case, the feasibility issues have only been encountered in the design of a diagonal gain matrix for the control configuration CC5. Regarding the γ-value produced by some decentralized controllers, by comparing the γ-values in Tables 2 and 4, it can be seen that a lower γ-value is attained by the decentralized controllers for the control configurations CC2, CC3 and CC6. A clear view of this situation can be obtained in Figure 9, which presents the frequency response corresponding to the passive control system defined by the diagonal gain matrix $\widehat{\mathbf{K}}^{\mathrm{VI}}$ (red dash-dotted line) and the corresponding centralized controller defined by the gain matrix $\mathbf{K}^{\mathrm{VI}}$ given in Equation (43) (thick blue solid line). To explain this unexpected fact, it should be noted that the LMI optimization procedure is based on the upper bound $\widetilde{\gamma}$ and produces a suboptimal H_∞ controller. Obviously, the upper bound $\widetilde{\gamma}(\mathbf{K}^\ell)$ corresponding to the unstructured gain matrix $\mathbf{K}^\ell$ must be inferior to the upper bound $\widetilde{\gamma}(\widehat{\mathbf{K}}^\ell)$ obtained for the diagonal gain matrix $\widehat{\mathbf{K}}^\ell$, which has been computed by solving a constrained version of the same LMI optimization problem. Looking at the data in Tables 2 and 4, we can see that the inequality $\widetilde{\gamma}(\mathbf{K}^\ell) \leq \widetilde{\gamma}(\widehat{\mathbf{K}}^\ell)$ certainly holds for all the control configurations. Additionally, the actual γ-values $\gamma(\mathbf{K}^\ell)$ and $\gamma(\widehat{\mathbf{K}}^\ell)$ must satisfy $\gamma(\mathbf{K}^\ell) \leq \widetilde{\gamma}(\mathbf{K}^\ell)$ and $\gamma(\widehat{\mathbf{K}}^\ell) \leq \widetilde{\gamma}(\widehat{\mathbf{K}}^\ell)$. However, these inequalities do not exclude the observed fact that $\gamma(\widehat{\mathbf{K}}^\ell) < \gamma(\mathbf{K}^\ell)$. That is, they do not exclude the unexpected possibility of obtaining a passive controller with a better performance than the corresponding active controller.

Remark 1. *In this paper, all the computations have been carried out using Matlab® R2015b on a regular laptop with an Intel® Core™ i7-2640M processor at 2.80 GHz. The LMI optimization problems corresponding to the different controller designs have been solved with the function* mincx() *included in the Robust Control Toolbox™. A relative accuracy of 10^{-7} has been set in the solver options.*

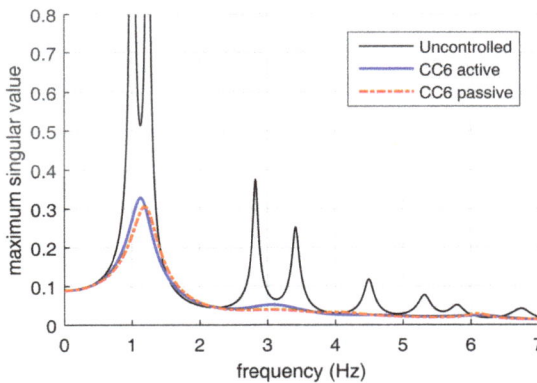

Figure 9. Frequency response corresponding to the decentralized velocity-feedback H_∞ controller with passive implementation defined by the diagonal control gain matrix $\widehat{\mathbf{K}}^{\mathrm{VI}}$. Maximum singular values of the closed-loop transfer function $\mathbf{T}_{\widehat{\mathbf{K}}^{\mathrm{VI}}}(2\pi f j)$ (red dash-dotted line), the closed-loop transfer function $\mathbf{T}_{\mathbf{K}^{\mathrm{VI}}}(2\pi f j)$ (thick blue solid line) and the open-loop transfer function $\mathbf{T}(2\pi f j)$ (thin black solid line).

4. Numerical Simulations

In this section, a proper set of numerical simulations is conducted to investigate the vibrational time-response of the two-building system for the considered control configurations. The simulations include the active controllers designed in Section 3.1, the passive controllers computed in Section 3.2 and the response of the uncontrolled buildings, which is taken as a natural reference in the performance assessment. In all the cases, the full-scale *North–South El Centro 1940* ground acceleration seismic record (see Figure 10) has been used as external disturbance. To describe the vibrational response of the individual buildings and the buildings interactions, the vectors of interstory drifts $\mathbf{r}(t)$ and interbuilding approaches $\mathbf{a}(t)$ have been computed. Additionally, the vector of control efforts $\mathbf{u}(t)$ has also been computed for the controlled configurations. Overall, the simulations include eleven different control configurations plus the uncontrolled case. To provide an intuitive and effective summary of this complex set of numerical results, the control configurations have been grouped by the number of actuation devices. Thus, the peak-values of the absolute interstory drifts and interbuilding approaches corresponding to the control configurations with two actuation devices (CC1 and CC2) are presented in Figure 11, where the interstory drift peak-values corresponding to the four-story building $\mathcal{B}^{(1)}$ are presented in the left-hand-side graphic, the interbuilding approach peak-values are shown in the central graphic, and the interstory drift peak-values corresponding to the five-story building $\mathcal{B}^{(2)}$ are displayed in the right-hand-side graphic. In the graphics of this section, the following colors and line styles have been used: blue lines present the values of the unlinked control configurations, red lines represent the linked control configurations, and black lines correspond to the uncontrolled configuration; additionally, active controllers are represented by solid lines, and passive controllers by non-solid lines. Detailed legends and captions have also been included in the figures to facilitate an unambiguous interpretation of the graphics. Specifically, the following colors, line styles and symbols have been used in the plots of Figure 11: black solid line with squares for the uncontrolled configuration, blue solid line with circles for the unlinked control configuration CC1 with the active controller defined by the control gain matrix $\mathbf{K}^{\mathrm{I}}$ given in Equation (36), blue dashed line with asterisks for the unlinked control configuration CC1 with the passive controller defined by the control gain matrix $\hat{\mathbf{K}}^{\mathrm{I}}$ given in Equation (45), red solid line with triangles for the linked control configuration CC2 with the active controller defined by the control gain matrix $\mathbf{K}^{\mathrm{II}}$ given in Equation (36), and red dotted line with hexagrams for the linked control configuration CC2 with the passive controller defined by the control gain matrix $\hat{\mathbf{K}}^{\mathrm{II}}$ given in Table 3. The peak-values of the corresponding absolute control efforts are presented in Figure 12a. Looking at graphics in Figures 11 and 12a, the following facts can be clearly appreciated: (i) all the controllers provide a significant level of reduction in the interstory drift and interbuilding approach peak-values when compared with the uncontrolled response; (ii) the unlinked control configuration CC1 is more effective in mitigating the interstory drift response; (iii) the linked control configuration CC2 attains better results in reducing the interbuilding approaches; (iv) smaller control-effort peak-values are produced by the linked configuration CC2; and (v) the levels of performance attained by the passive controllers are quite similar to those achieved by the corresponding active controllers, especially in the case of the linked control configuration CC2.

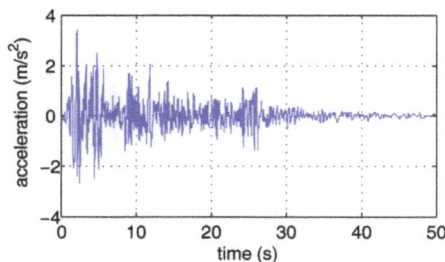

Figure 10. Full-scale North–South El Centro 1940 ground acceleration seismic record.

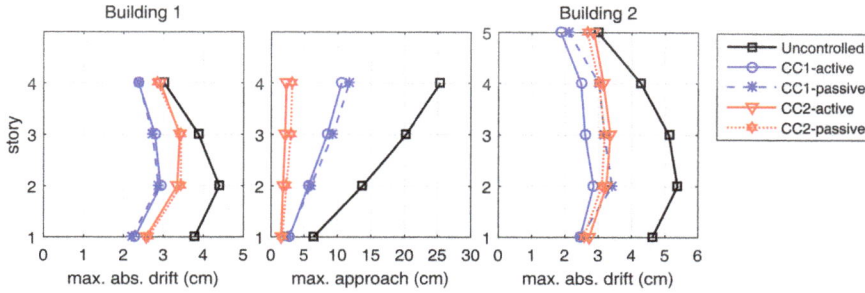

Figure 11. *Interstory drift and interbuilding approach peak-values for the control configurations CC1 and CC2.* Maximum absolute interstory drifts and maximum interbuilding approaches corresponding to the uncontrolled configuration (black solid line with squares), the active controller defined by the control gain matrix $\mathbf{K}^{\mathrm{I}}$ (blue solid line with circles), the passive controller defined by the control gain matrix $\widehat{\mathbf{K}}^{\mathrm{I}}$ (blue dashed line with asterisks), the active controller defined by the control gain matrix $\mathbf{K}^{\mathrm{II}}$ (red solid line with triangles) and the passive controller defined by the control gain matrix $\widehat{\mathbf{K}}^{\mathrm{II}}$ (red dotted line with hexagrams).

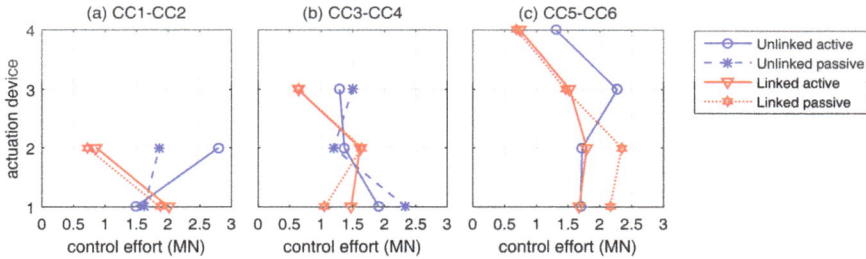

Figure 12. *Maximum absolute control efforts.* (**a**) Configurations with two actuation devices: CC1 (unlinked) and CC2 (linked). (**b**) Configurations with three actuation devices: CC3 (unlinked) and CC4 (linked). (**c**) Configurations with four actuation devices: CC5 (unlinked) and CC6 (linked).

For the control configurations with three actuation devices CC3 and CC4, the plots of interstory drift and interbuilding approach peak-values presented in Figure 13 show that the best overall behavior corresponds to the linked control configuration CC4 with the active controller defined by the control gain matrix $\mathbf{K}^{\mathrm{IV}}$ given in Equation (41) (red solid line with triangles). Also remarkable is the overall performance of the linked control configuration CC4 with the passive controller defined by the control gain matrix $\widehat{\mathbf{K}}^{\mathrm{IV}}$ given in Table 3 (red dotted line with hexagrams). Moreover, looking at the plots shown in Figure 12b, it can be appreciated that smaller control-effort peak-values are required by the controllers corresponding to the linked control configuration CC4. For the control configurations with four actuation devices CC5 and CC6, the plots of interstory drift and interbuilding approach peak-values presented in Figure 14 and the control-effort peak-values displayed in Figure 12c indicate the superior performance of the linked control configuration CC6 with the active controller defined by the control gain matrix $\mathbf{K}^{\mathrm{VI}}$ given in Equation (43). The good properties of this linked control configuration with the passive controller $\widehat{\mathbf{K}}^{\mathrm{VI}}$ can also be clearly appreciated. It should be recalled that feasibility issues appeared in the design of a fully decentralized velocity-feedback controller for the unlinked control configuration CC5 and, consequently, no passive controller is available for this case.

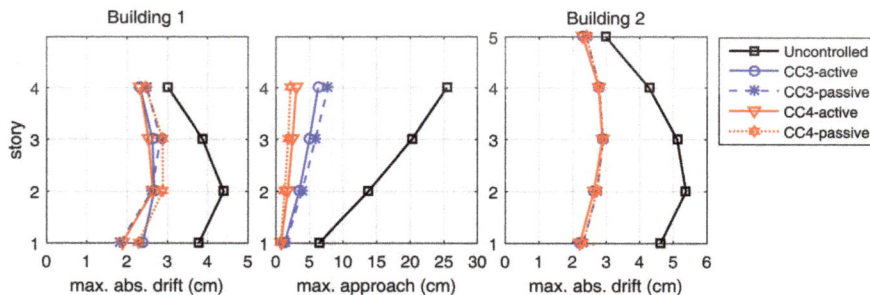

Figure 13. *Interstory drift and interbuilding approach peak-values for the control configurations CC3 and CC4.* Maximum absolute interstory drifts and maximum interbuilding approaches corresponding to the uncontrolled configuration (black solid line with squares), the active controller defined by the control gain matrix $\mathbf{K}^{\mathrm{III}}$ (blue solid line with circles), the passive controller defined by the control gain matrix $\hat{\mathbf{K}}^{\mathrm{III}}$ (blue dashed line with asterisks), the active controller defined by the control gain matrix $\mathbf{K}^{\mathrm{IV}}$ (red solid line with triangles) and the passive controller defined by the control gain matrix $\hat{\mathbf{K}}^{\mathrm{IV}}$ (red dotted line with hexagrams).

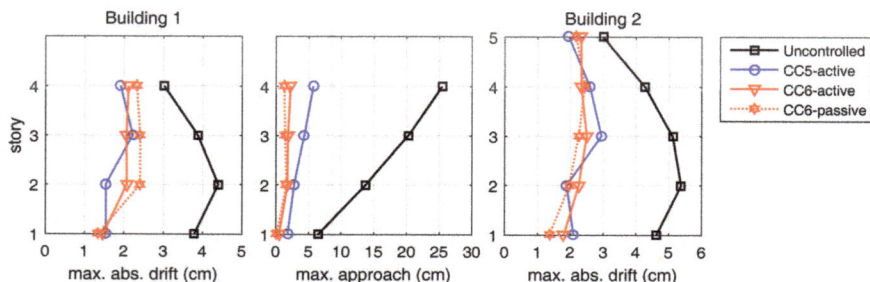

Figure 14. *Interstory drift and interbuilding approach peak-values for the control configurations CC5 and CC6.* Maximum absolute interstory drifts and maximum interbuilding approaches corresponding to the uncontrolled configuration (black solid line with squares), the active controller defined by the control gain matrix $\mathbf{K}^{\mathrm{V}}$ (blue solid line with circles), the active controller defined by the control gain matrix $\mathbf{K}^{\mathrm{VI}}$ (red solid line with triangles) and the passive controller defined by the control gain matrix $\hat{\mathbf{K}}^{\mathrm{VI}}$ (red dotted line with hexagrams).

To complement the information supplied by the interstory drift and the interbuilding approach time responses, the absolute acceleration peak-values of the buildings' top-level stories corresponding to the proposed active and passive velocity-feedback controllers are presented in Tables 5 and 6, respectively. Additionally, to provide a wider vision of the acceleration response characteristics, the story absolute acceleration peak-values corresponding to the uncontrolled configuration and the active and passive controllers proposed for the control configurations CC3 and CC4 are displayed in Figure 15. Looking at the values of building $\mathcal{B}^{(1)}$ in Table 5, it can be appreciated that better results are attained by the linked control configurations. In this case, the best performance is achieved by the linked configuration CC6, which produces a 50.6% of reduction with respect to the uncontrolled response. In contrast, the values of building $\mathcal{B}^{(2)}$ indicate a better behavior of the unlinked control configurations. For this building, the best performance corresponds to the unlinked control configuration CC3, which produces a 45.5% of reduction with respect to the uncontrolled response. Looking at the data in Table 6, a similar pattern can be appreciated in the values produced by the passive controllers. In this case, the best results are attained by the linked configuration CC4 in building $\mathcal{B}^{(1)}$, which produces a 39.7%

of reduction with respect to the uncontrolled response, and by the unlinked configuration CC3 in building $\mathcal{B}^{(2)}$, with a relative reduction of 43.1%. These acceleration response characteristics are further illustrated by the plots in Figure 15, where the following facts can be clearly appreciated: (i) all the proposed controllers produce positive results in reducing the story absolute acceleration peak-values when compared with the uncontrolled response; (ii) the unlinked control configuration CC3 is more effective in mitigating the acceleration response of the taller building $\mathcal{B}^{(2)}$; (iii) the linked control configuration CC4 attains better results in reducing the acceleration response of the shorter building $\mathcal{B}^{(1)}$; and (iv) the levels of performance of the passive controllers are similar to those attained by the corresponding active controllers, especially in the case of the linked control configuration CC4.

Table 5. Active controllers. Absolute acceleration peak-values (m/s²) of the buildings' top-level stories corresponding to the uncontrolled response and the active velocity-feedback controllers defined by the gain matrices $\mathbf{K}^\ell$.

	CC1	CC2	CC3	CC4	CC5	CC6	Uncontrolled
Building $\mathcal{B}^{(1)}$ (4th story)	7.5070	6.4787	7.3414	5.7286	6.0715	4.7087	9.5253
Building $\mathcal{B}^{(2)}$ (5th story)	5.9233	9.0380	5.2443	7.2716	6.1648	7.3445	9.6258

Table 6. Passive controllers. Absolute acceleration peak-values (m/s²) of the buildings' top-level stories corresponding to the uncontrolled response and the passive velocity-feedback controllers defined by the gain matrices $\widehat{\mathbf{K}}^\ell$. The values of the control configuration CC5 are missing due to the feasibility issues encountered in the associated LMI optimization problem.

	CC1	CC2	CC3	CC4	CC5	CC6	Uncontrolled
Building $\mathcal{B}^{(1)}$ (4th story)	7.5438	6.7569	7.8870	5.7426	–	5.9200	9.5253
Building $\mathcal{B}^{(2)}$ (5th story)	6.7197	8.4009	5.4798	7.6020	–	6.8801	9.6258

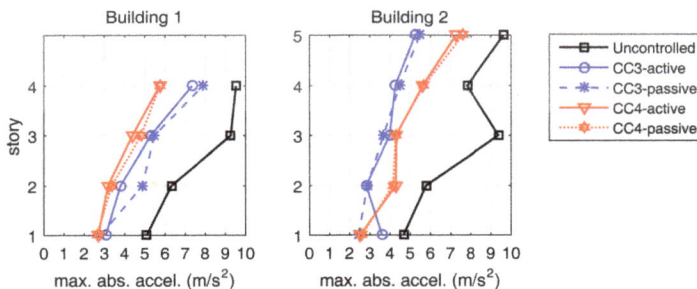

Figure 15. *Story absolute acceleration peak-values for the control configurations CC3 and CC4.* Maximum story absolute acceleration corresponding to the uncontrolled configuration (black solid line with squares), the active controller defined by the control gain matrix $\mathbf{K}^{\mathrm{III}}$ (blue solid line with circles), the passive controller defined by the control gain matrix $\widehat{\mathbf{K}}^{\mathrm{III}}$ (blue dashed line with asterisks), the active controller defined by the control gain matrix $\mathbf{K}^{\mathrm{IV}}$ (red solid line with triangles) and the passive controller defined by the control gain matrix $\widehat{\mathbf{K}}^{\mathrm{IV}}$ (red dotted line with hexagrams).

Remark 2. *In order to avoid the modeling and simulation difficulties associated to interbuilding collisions, the numerical simulations have been carried out assuming that the interbuilding spacing is large enough to prevent pounding events. In this case, the maximum interbuilding approaches can be understood as lower bounds of safe interbuilding distances. Thus, for example, the central plots in Figure 14 point out that, for the considered seismic event, an interbuilding distance of 2.5 cm can be considered safe for the active and passive controllers of the linked control configuration CC6 while, in contrast, an interbuilding separation of 5 cm would produce* *interbuilding collisions for the active controller of the unlinked control configuration CC5. For the uncontrolled configuration, a interbuilding gap of more than 25 cm would have been necessary to avoid pounding events.*

Remark 3. *It should be noted that the linked control configuration CC2 is an incomplete actuation scheme, in the sense that it does not include any interstory actuation device in building $\mathcal{B}^{(1)}$. This element can possibly help to explain its reduced effectiveness in mitigating the interstory drift response. In contrast, all the other control configurations include a more balanced actuation scheme with interstory actuation devices in both buildings.*

Remark 4. *Although no general conclusions can be drawn from the considered particular configurations, the obtained numerical results seem to indicate that the presence of interbuilding actuation devices can produce opposite effects in the acceleration response of linked buildings with a different height. This fact certainly opens a number of interesting questions related to the acceleration response characteristics of linked buildings, such as the relevance of the height difference, the behavior of linked buildings with the same height, and the effect of multiple linking devices.*

5. Conclusions and Future Directions

In this paper, the design of advanced structural vibration control systems for the seismic protection of adjacent multi-story buildings has been investigated. The proposed approach considers multi-actuation schemes that combine interstory and interbuilding force actuation devices implemented at different locations of the structure. Using an advanced static output-feedback H_∞ controller design methodology, active and passive vibration control systems have been obtained for a multi-story two-building structure equipped with a selected set of linked and unlinked actuation schemes. After studying the corresponding frequency and time responses, the following positive points can be highlighted: (i) The proposed design methodology allows dealing with a wide variety of actuation schemes. (ii) The obtained control systems provide a significant reduction of the frequency response in the main and secondary resonant modes. Moreover, they also produce a significant reduction of the maximum interbuilding approaches and the absolute interstory drift peak-values in both buildings. (iii) The obtained control systems produce positive results in reducing the buildings acceleration responses. (iv) Control configurations with interbuilding linking devices provide a higher protection against pounding events and produce lower control-effort peak-values. (v) Lower interstory drift peak-values are attained in buildings equipped with interstory actuation devices. (vi) A remarkable performance level is achieved by the passive control systems. (vii) In general, more effective and better balanced results are obtained with a larger number of actuation devices. Additionally, the following negative aspects can be pointed out: (viii) The design procedure can fail due to feasibility issues in the associated LMI optimization problem, specially in the constrained passive designs. (ix) The controller H_∞-norm is a suitable index to obtain effective controllers, but it cannot be used to identify optimal configurations of the actuation system. (x) Due to the computational cost and complexity, the proposed design methodology is only effective for structures with a moderate number of stories.

In summary, the observed results indicate the convenience of using multi-actuation systems that combine interbuilding linking devices and interstory actuators implemented in both buildings. The good behavior exhibited by the obtained passive control systems is a fact of singular relevance, which certainly deserves a deeper investigation. Further research effort should also be invested in studying the acceleration response of linked buildings, and in performing more realistic numerical simulations, which should include the effect of pounding events and other nonlinear aspects. Finally, it is worth mentioning the important open problem of finding a suitable methodology to determine the optimal configuration of distributed multi-actuation systems in large-scale structures.

Acknowledgments: This work was partially supported by the Spanish Ministry of Economy and Competitiveness under Grant DPI2015-64170-R/FEDER.

Author Contributions: All the authors contributed to the modeling and the scientific elaboration. Moreover, J. M. Rossell and J. Rubió-Massegú designed the centralized controllers; H. R. Karimi designed the decentralized controllers and computed the frequency responses; F. Palacios-Quiñonero and J. Rubió-Massegú performed the numerical simulations; F. Palacios-Quiñonero, in collaboration with the other authors, wrote the paper.

Conflicts of Interest: The authors declare no conflict of interest.

Appendix A. Buildings Parameters

Table A1. Mass and stiffness coefficient values corresponding to the two-building system.

Story	Building $\mathcal{B}^{(1)}$				Building $\mathcal{B}^{(2)}$				
	1	2	3	4	1	2	3	4	5
mass ($\times 10^5$ Kg)	2.152	2.092	2.070	2.661	2.152	2.092	2.070	2.048	2.661
stiffness ($\times 10^8$ N/m)	1.470	1.130	0.990	0.840	1.470	1.130	0.990	0.890	0.840

In this appendix, the particular parameter values of the two-building system used in the controller designs and numerical simulations are presented. The mass and stiffness coefficients are collected in Table A1. These values are similar to those presented in [42]. Approximate damping matrices $\mathbf{C}_d^{(1)}$ and $\mathbf{C}_d^{(2)}$ have been computed following a Rayleigh damping approach [41], by setting a 2% of relative damping on the corresponding smallest and largest modes. The obtained particular values (in Ns/m) are the following:

$$\mathbf{C}_d^{(1)} = 10^5 \times \begin{bmatrix} 2.6450 & -0.9034 & 0 & 0 \\ -0.9034 & 2.2455 & -0.7915 & 0 \\ 0 & -0.7915 & 2.0078 & -0.6715 \\ 0 & 0 & -0.6715 & 1.3719 \end{bmatrix}, \tag{A1}$$

$$\mathbf{C}_d^{(2)} = 10^5 \times \begin{bmatrix} 2.6017 & -0.9244 & 0 & 0 & 0 \\ -0.9244 & 2.1958 & -0.8099 & 0 & 0 \\ 0 & -0.8099 & 1.9946 & -0.7281 & 0 \\ 0 & 0 & -0.7281 & 1.8670 & -0.6872 \\ 0 & 0 & 0 & -0.6872 & 1.2741 \end{bmatrix}. \tag{A2}$$

Appendix B. Static Output-Feedback H_∞ Controller Design

This appendix provides a brief summary of the static output-feedback H_∞ controller design methodology presented in [38,39]. Let us consider the linear model

$$\begin{cases} \dot{x}(t) = \mathbf{A}\,x(t) + \mathbf{B}\,u(t) + \mathbf{E}\,w(t) \\ z(t) = \mathbf{C}_z\,x(t) + \mathbf{D}_z\,u(t) \\ y(t) = \mathbf{C}_y x(t) \end{cases} \tag{A3}$$

where $x(t)$ is the state, $u(t)$ is the control action, $w(t)$ is the external disturbance, $z(t)$ is the controlled output, and $y(t)$ is the measured output. The design objective is to obtain an optimal static output-feedback H_∞ controller

$$u(t) = \mathbf{K}y(t) \tag{A4}$$

that produces an asymptotically stable closed-loop matrix

$$\mathbf{A}_K = \mathbf{A} + \mathbf{B}\mathbf{K}\mathbf{C}_y \tag{A5}$$

and, simultaneously, minimizes the associated H_∞-norm

$$\gamma_K = \sup_{\|\mathbf{w}\|_2 \neq 0} \frac{\|\mathbf{z}\|_2}{\|\mathbf{w}\|_2}, \tag{A6}$$

where $\|\mathbf{f}\|_2 = \left[\int_0^\infty \mathbf{f}^T(t)\,\mathbf{f}(t)\,dt\right]^{1/2}$ denotes the usual continuous 2-norm. Computing this kind of optimal controllers is a challenging problem that still remains open. However, according to the results in [38,39], a suboptimal static output-feedback H_∞ controller

$$\mathbf{u}(t) = \widetilde{\mathbf{K}}\mathbf{y}(t) \tag{A7}$$

can be computed by solving the following LMI optimization problem:

$$\mathcal{P} : \begin{cases} \text{maximize } \eta \\ \text{subject to } \mathbf{X_Q} > 0, \ \mathbf{X_R} > 0, \ \eta > 0 \text{ and the LMI in (A9),} \end{cases} \tag{A8}$$

$$\begin{bmatrix} \mathbf{AQX_QQ}^T + \mathbf{QX_QQ}^T\mathbf{A}^T + \mathbf{ARX_RR}^T + \mathbf{RX_RR}^T\mathbf{A}^T + \mathbf{BY_RR}^T + \mathbf{RY_R^T B}^T + \eta\mathbf{EE}^T & * \\ \mathbf{C_zQX_QQ}^T + \mathbf{C_zRX_RR}^T + \mathbf{D_zY_RR}^T & -\mathbf{I} \end{bmatrix} < 0, \tag{A9}$$

where $*$ denotes the transpose of the symmetric entry, $\mathbf{X_Q}$, $\mathbf{X_R}$ and $\mathbf{Y_R}$ are the optimization variables, $\mathbf{Q}$ is a matrix whose columns contain a basis of $\text{Ker}(\mathbf{C_y})$, and the matrix $\mathbf{R}$ has the following form:

$$\mathbf{R} = \mathbf{C_y^\dagger} + \mathbf{Q}\widetilde{\mathbf{L}}, \quad \widetilde{\mathbf{L}} = \mathbf{Q}^\dagger \widetilde{\mathbf{X}}\mathbf{C_y}^T (\mathbf{C_y}\widetilde{\mathbf{X}}\mathbf{C_y}^T)^{-1}, \tag{A10}$$

where

$$\mathbf{C_y^\dagger} = \mathbf{C_y}^T (\mathbf{C_y}\,\mathbf{C_y}^T)^{-1}, \quad \mathbf{Q}^\dagger = (\mathbf{Q}^T\mathbf{Q})^{-1}\mathbf{Q}^T \tag{A11}$$

are the *Moore-Penrose* pseudoinverses of $\mathbf{C_y}$ and $\mathbf{Q}$, respectively, and $\widetilde{\mathbf{X}}$ is the optimal X-matrix of the auxiliary LMI optimization problem

$$\mathcal{P}_a : \begin{cases} \text{maximize } \eta_a \\ \text{subject to } \mathbf{X} > 0, \ \eta_a > 0 \text{ and the LMI in (A13),} \end{cases} \tag{A12}$$

$$\begin{bmatrix} \mathbf{AX} + \mathbf{XA}^T + \mathbf{BY} + \mathbf{Y}^T\mathbf{B}^T + \eta_a\mathbf{EE}^T & * \\ \mathbf{C_zX} + \mathbf{D_zY} & -\mathbf{I} \end{bmatrix} < 0. \tag{A13}$$

If an optimal value $\widetilde{\eta}$ is attained in $\mathcal{P}$ for the triplet $(\widetilde{\mathbf{X}}_Q, \widetilde{\mathbf{X}}_R, \widetilde{\mathbf{Y}}_R)$, then the output gain matrix $\widetilde{\mathbf{K}}$ can be written in the form

$$\widetilde{\mathbf{K}} = \widetilde{\mathbf{Y}}_R (\widetilde{\mathbf{X}}_R)^{-1}. \tag{A14}$$

Moreover, the value

$$\widetilde{\gamma}_{\widetilde{K}} = (\widetilde{\eta})^{-1/2} \tag{A15}$$

provides an upper bound of the associated H_∞-norm $\gamma_{\widetilde{K}}$, which can be computed by solving the optimization problem

$$\gamma_{\widetilde{K}} = \sup_f \sigma_{\max}\left[\mathbf{T}_{\widetilde{K}}(2\pi f j)\right], \tag{A16}$$

where $j = \sqrt{-1}$, f is the frequency in hertz, $\sigma_{\max}[\cdot]$ denotes the maximum singular value and

$$\mathbf{T}_{\tilde{K}}(s) = \mathbf{C}_{\tilde{K}}(s\mathbf{I} - \mathbf{A}_{\tilde{K}})^{-1}\mathbf{E}, \tag{A17}$$

with

$$\mathbf{A}_{\tilde{K}} = \mathbf{A} + \mathbf{B}\tilde{\mathbf{K}}\mathbf{C}_y, \quad \mathbf{C}_{\tilde{K}} = \mathbf{C}_z + \mathbf{D}_z\tilde{\mathbf{K}}\mathbf{C}_y, \tag{A18}$$

is the closed-loop transfer function from the disturbance input to the controlled output.

References

1. Abdel Raheem, S.E. Mitigation measures for earthquake induced pounding effects on seismic performance of adjacent buildings. *Bull. Earthq. Eng.* **2014**, *12*, 1705–1724.
2. Kandemir-Mazanoglu, E.C.; Mazanoglu, K. An optimization study for viscous dampers between adjacent buildings. *Mech. Syst. Signal Process.* **2017**, *89*, 88–96.
3. Kumar, P.; Karuna, S. Effect of seismic pounding between adjacent buildings and mitigation measures. *Int. J. Res. Eng. Technol.* **2015**, *4*, 208–216.
4. Pawar, P.D.; Murnal, P.B. Effect of seismic pounding on adjacent buildings considering soil-structure interaction. *Int. J. Adv. Found. Res. Sci. Eng.* **2015**, *2*, 286–294.
5. Sorace, S.; Terenzi, G. Damped interconnection-based mitigation of seismic pounding between adjacent R/C buildings. *LACSIT Int. J. Eng. Technol.* **2013**, *5*, 406–412.
6. Tubaldi, E.; Freddi, F.; Barbato, M. Probabilistic seismic demand model for pounding risk assessment. *Earthq. Eng. Struct. Dyn.* **2016**, *45*, 1743–1758.
7. McCall, A.J.; Balling, R.J. Structural analysis and optimization of tall buildings connected with skybridges and atria. *Struct. Multidiscip. Optim.* **2016**, *55*, 583–600.
8. Pardalopoulos, S.I.; Pantazopoulou, S.J. Seismic response of nonstructural components attached on multistorey buildings. *Earthq. Eng. Struct. Dyn.* **2015**, *44*, 139–158.
9. Behnamfar, F.; Dorafshan, S.; Taheri, A.; Hashemi, B.H. A method for rapid estimation of dynamic coupling and spectral responses of connected adjacent structures. *Struct. Des. Tall Spec. Build.* **2016**, *25*, 605–625.
10. Tubaldi, E. Dynamic behavior of adjacent buildings connected by linear viscous/viscoelastic dampers. *Struct. Control Health Monit.* **2015**, *22*, 1086–1102.
11. Patel, C.C.; Jangid, R.S. Dynamic response of identical adjacent structures connected by viscous damper. *Struct. Control Health Monit.* **2014**, *21*, 205–224.
12. Richardson, A.; Walsh, K.K.; Abdullah, M.M. Closed-form equations for coupling linear structures using stiffness and damping elements. *Struct. Control Health Monit.* **2013**, *20*, 259–281.
13. Jankowski, R.; Mahmoud, S. Linking of adjacent three-storey buildings for mitigation of structural pounding during earthquakes. *Bull. Earthq. Eng.* **2016**, *14*, 3075–3097.
14. Bigdeli, K.; Hare, W.; Nutini, J.; Tesfamariam, S. Optimizing damper connectors for adjacent buildings. *Optim. Eng.* **2016**, *17*, 47–75.
15. Greco, R.; Marano, G.C. Multi-objective optimization of a dissipative connection for seismic protection of wall-frame structures. *Soil Dyn. Earthq. Eng.* **2016**, *87*, 151–163.
16. Yang, Z.D.; Lam, E.S.S. Dynamic responses of two buildings connected by viscoelastic dampers under bidirectional earthquake excitations. *Earthq. Eng. Eng. Vib.* **2014**, *13*, 137–150.
17. Palacios-Quiñonero, F.; Rubió-Massegú, J.; Rossell, J.M.; Karimi, H.R. Vibration control for adjacent structures using local state information. *Mechatronics* **2014**, *24*, 336–344.
18. Huang, X.; Zhu, H. Optimal arrangement of viscoelastic dampers for seismic control of adjacent shear-type structures. *J. Zhejiang Univ. Sci. A* **2013**, *14*, 47–60.
19. Tehrani, M.G.; Gattulli, V. Vibration control using nonlinear damped coupling. *J. Phys. Conf. Ser.* **2016**, *744*, 1–9.
20. Kasagi, M.; Fujita, K.; Tsuji, M.; Takewaki, I. Effect of non-linearity of connecting dampers on vibration control of connected building structures. *Front. Built Environ.* **2016**, *1*, 1–9.
21. Uz, M.E.; Sharafi, P. Investigation of the optimal semi-active control strategies of adjacent buildings connected with magnetorheological dampers. *Int. J. Optim. Civ. Eng.* **2016**, *6*, 523–547.

22. Abdeddaim, M.; Ounis, A.; Djedoui, N.; Shrimali, M.K. Pounding hazard mitigation between adjacent planar buildings using coupling strategy. *J. Civ. Struct. Health Monit.* **2016**, *6*, 603–617.
23. Uz, M.E.; Hadi, M.N.S. Optimal design of semi active control for adjacent buildings connected by MR damper based on integrated fuzzy logic and multi-objective genetic algorithm. *Eng. Struct.* **2014**, *69*, 135–148.
24. Kim, H.S. Seismic response control of adjacent buildings coupled by semi-active shared TMD. *Int. J. Steel Struct.* **2016**, *16*, 647–656.
25. Sun, H.; Liu, M.; Zhu, H. Connecting parameters optimization on unsymmetrical twin-tower structure linked by sky-bridge. *J. Cent. South Univ.* **2014**, *21*, 2460–2468.
26. Kim, H.S.; Kim, Y.J. Control performance evaluation of shared tuned mass damper. *Adv. Sci. Technol. Lett.* **2014**, *69*, 1–4.
27. Gao, H.; Zhan, W.; Karimi, H.R.; Yang, X.; Yin, S. Allocation of actuators and sensors for coupled-adjacent-building vibration attenuation. *IEEE Trans. Ind. Electron.* **2013**, *60*, 5792–5801.
28. Park, K.S.; Ok, S.Y. Optimal design of actively controlled adjacent structures for balancing the mutually conflicting objectives in design preference aspects. *Eng. Struct.* **2012**, *45*, 213–222.
29. Dumne, S.M.; Shrimali, M.K.; Bharti, S.D. Earthquake performance of hybrid controls for coupled buildings with MR dampers and sliding base isolation. *Asian J. Civ. Eng.* **2017**, *18*, 63–97.
30. Fathi, F.; Bahar, O. Hybrid coupled building control for similar adjacent buildings. *KSCE J. Civ. Eng.* **2017**, *21*, 265–273.
31. Kasagi, M.; Fujita, K.; Tsuji, M.; Takewaki, I. Automatic generation of smart earthquake-resistant building system: Hybrid system of base-isolation and building-connection. *Helyon* **2016**, *2*, 1–21.
32. Shrimali, M.K.; Bharti, S.D.; Dumne, S.M. Seismic response analysis of coupled building involving MR damper and elastomeric base isolation. *Ain Shams Eng. J.* **2015**, *6*, 457–470.
33. Murase, M.; Tsuji, M.; Takewaki, I. Smart passive control of buildings with higher redundancy and robustness using base-isolation and inter-connection. *Earthq. Struct.* **2013**, *4*, 649–670.
34. Taniguchi, M.; Fujita, K.; Tsuji, M.; Takewaki, I. Hybrid control system for greater resilience using multiple isolation and building connection. *Front. Built Environ.* **2016**, *2*, 1–10.
35. Tubaldi, E.; Barbato, M.; Dall'Asta, A. Efficient approach for the reliability-based design of linear damping devices for seismic protection of buildings. *ASCE-ASME J. Risk Uncertain. Eng. Syst. Part A Civ. Eng.* **2016**, *2*, 1–10.
36. Park, K.S.; Ok, S.Y. Optimal design of hybrid control system for new and old neighboring buildings. *J. Sound Vib.* **2015**, *336*, 16–31.
37. Park, K.S.; Ok, S.Y. Hybrid control approach for seismic coupling of two similar adjacent structures. *J. Sound Vib.* **2015**, *349*, 1–17.
38. Rubió-Massegú, J.; Rossell, J.M.; Karimi, H.R.; Palacios-Quiñonero, F. Static output-feedback control under information structure constraints. *Automatica* **2013**, *49*, 313–316.
39. Palacios-Quiñonero, F.; Rubió-Massegú, J.; Rossell, J.M.; Karimi, H.R. Feasibility issues in static output-feedback controller design with application to structural vibration control. *J. Frankl. Inst.* **2014**, *351*, 139–155.
40. Palacios-Quiñonero, F.; Rubió-Massegú, J.; Rossell, J.M.; Karimi, H.R. Optimal passive-damping design using a decentralized velocity-feedback H_∞ approach. *Model. Identif. Control* **2012**, *33*, 87–97.
41. Chopra, A. *Dynamics of Structures. Theory and Applications to Earthquake Engineering*, 3rd ed.; Prentice Hall: Upper Saddle River, NJ, USA, 2007.
42. Kurata, N.; Kobori, T.; Takahashi, M.; Niwa, N.; Midorikawa, H. Actual seismic response controlled building with semi-active damper system. *Earthq. Eng. Struct. Dyn.* **1999**, *28*, 1427–1447.

applied
sciences

MDPI

Article

Active Vibration Suppression of a Motor-Driven Piezoelectric Smart Structure Using Adaptive Fuzzy Sliding Mode Control and Repetitive Control

Chi-Ying Lin * and Hong-Wu Jheng

Department of Mechanical Engineering, National Taiwan University of Science and Technology, No. 43, Keelung Rd., Sec. 4, Taipei 106, Taiwan; kingiffe@gmail.com
* Correspondence: chiying@mail.ntust.edu.tw; Tel.: +886-02-2737-6484

Academic Editors: Gangbing Song, Steve C.S. Cai and Hong-Nan Li
Received: 28 December 2016; Accepted: 28 February 2017; Published: 4 March 2017

Abstract: In this paper, we report on the use of piezoelectric sensors and actuators for the active suppression of vibrations associated with the motor-driven rotation of thin flexible plate held vertically. Motor-driven flexible structures are multi-input multi-output systems. The design of active vibration-suppression controllers for these systems is far more challenging than for flexible structures with a fixed end, due to the effects of coupling and nonlinear vibration behavior generated in structures with poor damping. To simplify the design of the controller and achieve satisfactory vibration suppression, we treated the coupling of vibrations caused by the rotary motion of the thin flexible plate as external disturbances and system uncertainties. We employed an adaptive fuzzy sliding mode control algorithm in the design of a single-input–single-output controller for the suppression of vibrations using piezoelectric sensors and actuators. We also used a repetitive control system to reduce periodic vibrations associated with the repetitive motions induced by the motor. Experimental results demonstrate that the hybrid intelligent control approach proposed in this study can suppress complex vibrations caused by modal excitation, coupling effects, and periodic external disturbances.

Keywords: flexible structure; piezoelectric materials; active vibration control; adaptive fuzzy sliding mode control; repetitive control

1. Introduction

Flexible structure based positioning is generally achieved using motors to expand the range of strokes and applications. Examples include the grabbing motion of flexible robot manipulators and the lifting operations of cranes. Due to their poor damping, the coupling of the entire structures, and various nonlinear effects, the rotary motions of flexible structures produce complex vibrations that can severely affect the precision of positioning. A popular and effective research approach to address this issue in recent years has been the use of piezoelectric materials as actuators and sensors in conjunction with active suppression control algorithms to create smart active suppression and control designs for structures [1–3].

Motor-driven flexible structures generally have two control objectives: (1) motor tracking control; and (2) active suppression control of vibrations in flexible structures. Most of these control systems are multi-input–multi-output (MIMO) systems in which achieving the high-performance motion control of these two objectives makes for a certain level of difficulty in controller design. However, discussion regarding these systems usually focuses on vibration-suppression control for the flexible structures. Thus, the majority of existing literature examines the effectiveness of various control methods in active vibration suppression [4–8]. Tso et al. [4] employed proportional-derivative (PD) control to actively suppress the vibrations in a single-axis flexible robot manipulator. Shan et al. [5] used positive position feedback (PPF) control for the active suppression of vibrations in a flexible robot manipulator with one degree of freedom. Ahmad et al. [6] combined fuzzy control with PD control and input shaping control for the tracking and control of deflection in flexible robot manipulators. Park et al. [7] combined fuzzy controller design with H-infinity control to enhance system robustness in reducing vibrations in a single-axis flexible robot manipulator during positioning. Lin and Chao [8] combined a flexible beam with a beam-cart system to investigate vibration responses during cart motion. They also employed piezoelectric transducers and adaptive neuro-fuzzy control for the active suppression of vibrations in the flexible beam.

Aside from the common single-axis flexible robot manipulators, flexible structures with two or more degrees of freedom enable greater work space and more diverse actions. However, coupling effects and nonlinearities greatly complicate vibration behavior [9–13]. Moudgal et al. [10] took into account the coupling effects between the axes in the design of a fuzzy controller aimed at reducing vibrations in a dual-link flexible robot manipulator. Lin et al. [11] designed a piezoelectric vibration absorber for a beam–cart–seesaw system and suppressed vibration responses in the system by pairing passive mechanical components with active piezoelectric proportional-integral-derivative (PID) feedback control. Lin and Zheng [12] designed a piezoelectric truss structure driven by two motors to enable rotation in two directions. They combined neural-fuzzy control with a genetic algorithm to achieve active vibration suppression control; however, the system was limited to static positioning and did not take into account the vibration responses created by dynamic tracking [13]. Lin et al. established a mathematical model to facilitate modal analysis of the large flexible structure in [12]. They also applied a hybrid PD/repetitive control framework [14] for the active suppression of vibrations produced in a thin plate undergoing periodic movement. Their simulation results demonstrated the effectiveness of a hybrid repetitive control (RC) algorithm in suppressing vibration responses in a motor-driven structure. Nonetheless, they treated the entire system and the coupling behavior as linear terms; i.e., they did not take into account the influence of complex coupling terms of the nonlinear terms created by external disturbances in actual flexible structures. Determining whether the same vibration-suppression effects could be obtained in actual rotary systems will require further investigation.

Horizontal thin-plate rotors have been proposed by previous researchers [12,13]. Our primary objective in this study was to determine the effectiveness of an active vibration-suppression control system based on the motor-driven rotation of thin flexible plate held vertically. This system was developed in conjunction with a dynamic model that takes into account nonlinear coupling effects as well as nonlinear terms. We attached piezoelectric devices to the thin flexible plate, which underwent excitation through the periodic motion of the motor. Feedback signals from the piezoelectric sensors were used to examine active vibration-suppression control in the flexible structure. The simulation results in [13] indicate that taking system coupling into consideration during the controller design stage and applying a multi-input-multi-output controller will result in better suppression effects on the vibrations created by periodic disturbances. However, this approach will also greatly increase the complexity of the overall controller design. We sought to overcome these difficulties by assuming that the coupling of vibrations caused by rotary motion could be regarded as external disturbances and systemic uncertainties associated with the thin flexible plate. This led us to design single-input–single-output controllers for the motor-driven rotation system as well as for the piezoelectric vibration suppression system. Adaptive fuzzy sliding mode control (AFSMC) was used in the latter to deal with complex vibrations generated by the rotation of the thin flexible plate. This control method can effectively handle nonlinear systems with uncertain parameters and achieve the expected control performance [15–17]. To minimize the influence of the periodic disturbances caused by the periodic motions of the motor, we adopted a hybrid RC framework [13,14] to combine RC with AFSMC and enhance piezoelectric vibration-suppression control. Experimental results demonstrate the effectiveness of the proposed hybrid intelligent control method in suppressing the complex vibrations generated in flexible structures subjected to periodic motor-driven rotations.

2. System Modeling

We developed a mathematical model for the large flexible structure used in this study to facilitate subsequent analysis. The structure comprises a large and flexible truss-like thin plate with a servomotor as the driving force. A transmission mechanism enables the rotation of the thin plate around the roll and yaw axes. Piezoelectric actuators and piezoelectric sensors were placed at the root of the vertical thin plate, as shown in Figure 1.

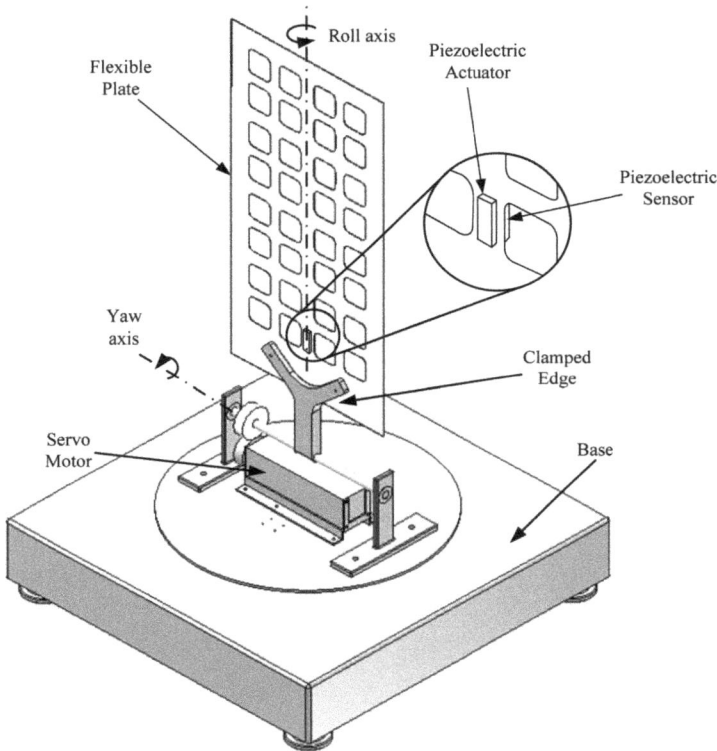

Figure 1. Motor-driven system for the rotation of thin flexible plate held in a vertical position.

In formulating the dynamic equations, we took into account the transverse vibrations in the thin flexible plate. The dynamic equations used for the overall system can be regarded as a combination of the dynamic equations pertaining to a cantilevered thin flexible plate and a servo motor. The thin plate system is treated as linear and the rotation system is assumed to be rigid. The assumptions and derivation procedures used in modeling the overall structure are similar to those in [13]; however, we also included nonlinearities and coupling terms in the current model. When the motor introduces rotary motions, coupling generates disturbances in the thin flexible plate. Vibrations in the thin plate also influence the rotary motions. Based on the effects of coupling, the dynamic equations of the overall system can be written as follows:

$$\mathbf{M}\ddot{\boldsymbol{\eta}} + \mathbf{C}^{D}\dot{\boldsymbol{\eta}} + \mathbf{K}\boldsymbol{\eta} = \mathbf{B}^{P}u_{p} + \boldsymbol{\Phi}_{\theta} + \mathbf{F} \tag{1}$$

$$\mathbf{L_a}\dot{\mathbf{i}}_a + \mathbf{R_a}\mathbf{i}_a = \mathbf{e_a} - \mathbf{K_b}\dot{\theta}_m \tag{2}$$

$$\mathbf{J_m}\ddot{\theta}_m + \mathbf{B_m}\dot{\theta}_m + \mathbf{\Phi}_\eta = \mathbf{K_i}\mathbf{i}_a \tag{3}$$

$$
\mathbf{L_a} = \begin{bmatrix} L_{ar} & 0 \\ 0 & L_{ay} \end{bmatrix}; \quad
\mathbf{R_a} = \begin{bmatrix} R_{ar} & 0 \\ 0 & R_{ay} \end{bmatrix}; \quad
\mathbf{e_a} = \begin{bmatrix} e_{ar} & 0 \\ 0 & e_{ay} \end{bmatrix}; \quad
\mathbf{K_b} = \begin{bmatrix} K_{br} & 0 \\ 0 & K_{by} \end{bmatrix}
$$
$$
\mathbf{J_m} = \begin{bmatrix} J_{mr} & 0 \\ 0 & J_{my} \end{bmatrix}; \quad
\mathbf{B_m} = \begin{bmatrix} B_{mr} & 0 \\ 0 & B_{my} \end{bmatrix}; \quad
\mathbf{K_i} = \begin{bmatrix} K_{ir} & 0 \\ 0 & K_{iy} \end{bmatrix}; \quad
\theta_m = \begin{bmatrix} \theta_{mr} & 0 \\ 0 & \theta_{my} \end{bmatrix}
\tag{4}
$$

where $\mathbf{M}$ and $\mathbf{K}$ denote the mass and stiffness coefficient matrices, respectively; both are ij × ij diagonal matrices with no coupling between the modes; η is a vector containing modal participation factors with dimensions ij × 1; $\mathbf{C^D}$ is the damping coefficient matrix; $\mathbf{B^P}$ signifies the piezoelectric coefficient matrix; u_p indicates the control input from the piezoelectric actuator; and $\mathbf{F}$ represents the external disturbance. The parameters associated with motor dynamics are as follows. $\mathbf{L_a}$ denotes the inductance constant matrix; $\mathbf{R_a}$ indicates the resistance matrix; $\mathbf{i}_a$ is the current vector; $\mathbf{e_a}$ signifies the motor input voltage vector; $\mathbf{K_b}$ is the back-emf constant matrix; $\mathbf{J_m}$ denotes the inertia matrix; $\mathbf{B_m}$ is the viscous friction coefficient matrix; $\mathbf{K_i}$ is the torque constant matrix; and θ_m denotes the motor angle vector. The subscripts r and y in the above matrices, respectively, refer to servomotor parameters of the roll and yaw axes. Finally, $\mathbf{\Phi}_\theta$ and $\mathbf{\Phi}_\eta$ are the coupling function matrices defined as

$$\mathbf{\Phi}_\theta = \alpha_c \mathbf{N^T P}_\theta(\theta, \dot{\theta}, \ddot{\theta}) \tag{5}$$

$$\mathbf{\Phi}_\eta = \beta_c \mathbf{N P}_\eta(\eta, \dot{\eta}, \ddot{\eta}) \tag{6}$$

where $\mathbf{P}_\theta$ is a 2 × 1 nonlinear thin plate displacement function matrix, which originates from motor inertia and friction; $\mathbf{P}_\eta$ is a ij × 1 nonlinear motor rotation function matrix, in which the influences of the nonlinear terms originate from the flexible vibrations in the lightweight thin plate; α_c and β_c are weighting values in the coupling function; generally, $\alpha_c \gg \beta_c$; and $\mathbf{N}$ is the rigid body coupling matrix defined as

$$\mathbf{N} = \begin{bmatrix} N_r \\ N_y \end{bmatrix} \tag{7}$$

Furthermore, the relationship between the rotation angles of the cantilever thin plate (θ_r and θ_y) and those of the servomotor (θ_{mr} and θ_{my}) can be expressed as

$$\theta = \begin{bmatrix} \theta_r \\ \theta_y \end{bmatrix} = \begin{bmatrix} g_r\theta_{mr} \\ g_y\theta_{my} \end{bmatrix} \tag{8}$$

3. Controller Design for Active Vibration Suppression

In this section, we introduce the control algorithm used in the vibration-suppression control experiments. The AFSMC processes complex vibrations in the thin flexible plate generated in the application of rotary motion. We used a hybrid RC framework [14] incorporating RC to reduce the influence of periodic disturbances created by the periodic motions introduced by the motor.

3.1. Adaptive Fuzzy Sliding Mode Control

The nonlinearity of the mathematical model in the previous section makes it difficult to obtain the relevant state variables. Generally, the format of the fuzzy rule bank in a fuzzy controller is fixed, and the parameters are not updated when the system changes. Consequently, it is difficult to predict control performance in the event of changes to the system parameters. This led researchers to develop AFSMC [15], which combines the advantages of fuzzy control, sliding mode control, and adaptive control. In a control system with unknown parameters, analyzing the input and output data can further understanding of the control system in order to establish language rules. Furthermore, adaptive control means that the rule bank can be adjusted online, and the robustness can be enhanced via the sliding mode. Figure 2 displays the AFSMC framework, in which gs and gu denote the input and output scaling factors, respectively.

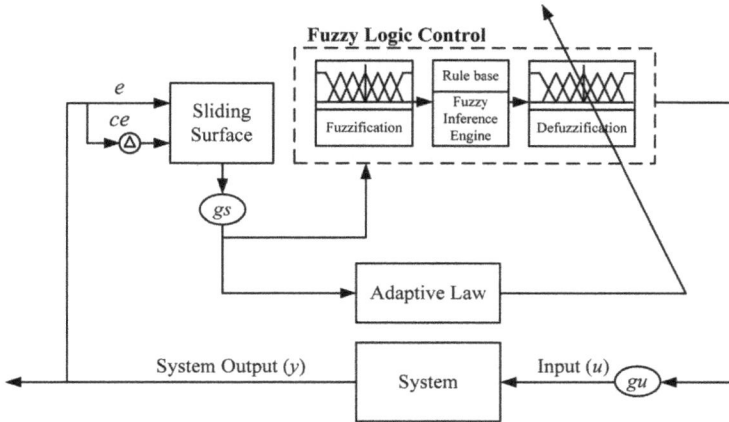

Figure 2. Block diagram of adaptive fuzzy sliding mode control system.

During the design process of the controller, the sliding surface must be defined. In this study, we used a second-order sliding surface, which can be mathematically presented as

$$S(e) = ce + \lambda e \tag{9}$$

where λ is the weight of the sliding mode. Using the sliding surface, we can convert the two-dimensional input variables (e, ce) of the original fuzzy controller into a one-dimensional variable (s). With appropriate scaling based on gs, the input variable can then be input into the fuzzy controller as follows:

$$s(e) = S(e) \times gs \tag{10}$$

As shown in Figure 2, the concept of the controller is to describe physical quantities using linguistic variables and identify the corresponding membership grade. With the adaptive law, the rule table can be adjusted and defuzzified online, which enables the rule bank to begin learning from zero rules. This in turn revises the defuzzified control rules. The control variable can then be obtained using the center of gravity defuzzification method, and then the control input variable is adjusted based on output scaling factor gu.

With regard to fuzzy control, we employed a triangular membership function with seven one-dimensional fuzzy variables for the input variable s and defuzzified variable u, as shown in Figure 3, where NB is the negative big, NM is the negative medium, NS is the negative small, ZO is the zero, PS is the positive small, PM is the positive medium, and PB is the positive big.

In sliding model control, the sliding surface reaching condition is

$$s\dot{s} < 0 \tag{11}$$

When $s > 0$, the control variable must be increased to reduce $s\dot{s}$; if $s < 0$, then the control variable must be decreased to reduce $s\dot{s}$. Input variable s and control variable u can thus be designed to satisfy the reaching condition, where rule i can be written as

$$R_i: IF \ s = S_i \ THEN \ U_i = C_i \tag{12}$$

where S_i is the rule i of the component in question prior to fuzzification; U_i denotes the rule i of the component following defuzzification; and C_i indicates the central position of rule table i of the component following defuzzification. Using the defuzzified U_i corresponding to the ith membership grade obtained from component S before fuzzification, the control variable can be written as

$$u = \frac{\sum_{i=1}^{m} \mu_i U_i}{\sum_{i=1}^{m} \mu_i} = \frac{\sum_{i=1}^{m} \mu_i C_i}{\sum_{i=1}^{m} \mu_i} \tag{13}$$

where m is the number of rules, with i equaling 1, 2, ... , and m; the initial values of C_i are all 0, and they are therefore referred to as zero rules. During the control process, we can use the adaptive law to update C_i, the mathematical formula of which is

$$\dot{C}_i(t) = -\gamma \frac{\partial s(t)\dot{s}(t)}{\partial C_i(t)} \tag{14}$$

where γ denotes the adaptive rate. Using the chain rule, we can rewrite Equation (14) into

$$\begin{aligned}\dot{C}_i(t) &= -\gamma \frac{\partial s(t)\dot{s}(t)}{\partial u(t)}\frac{\partial u(t)}{\partial C_i(t)} \\ &= -\gamma c(e)s(t)\frac{\partial u(t)}{\partial C_i(t)} \\ &= \gamma_a s(t)\frac{\partial u(t)}{\sum_{i=1}^{m}\mu_i(t)}\end{aligned} \tag{15}$$

where $c(e)$ is the direction of the control variable, and γ_a is the learning rate parameter. When errors and error variations exist in the system, C_i increases continually, which causes the system to diverge. To resolve this issue, we can use the adaptive law of e-modification [18] to revise Equation (15) to

$$\dot{C}_i(t) = \gamma_a s(t)\frac{\partial u(t)}{\sum_{i=1}^{m}\mu_i(t)} - \Lambda|s(t)|C_i(t) \tag{16}$$

where Λ can be adjusted based on system stability, and the learning rate exerts direct impact on the central position of defuzzification. This gives the AFSMC the functions of online learning and updating the defuzzification rule bank. For details on the design process and theory of AFSMC, please refer to [15].

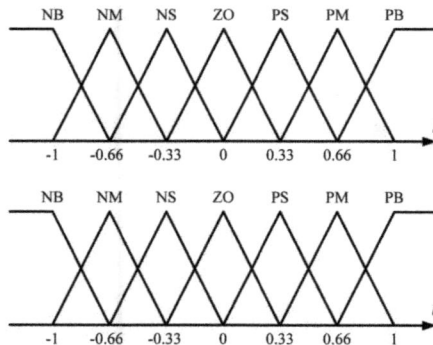

Figure 3. Membership function of AFSMC.

3.2. Hybrid Repetitive Control

Figure 4 presents the hybrid RC framework adopted in this study, where C_1 and C_2 respectively represent the AFSMC controller and the repetitive controller. The sum of the control variable of the repetitive controller, u_{RC}, and the control variable of the AFSMC controller, u_{AFSMC}, is the total control input variable of the system, u_{sum}. In the hybrid control framework used in this study, we designed the controllers separately with the aim of having them work synergistically. The controller designs are similar to those in [14]; however, we replace PD control with AFSMC control to handle the complex vibration behaviors found in actual flexible structures more effectively.

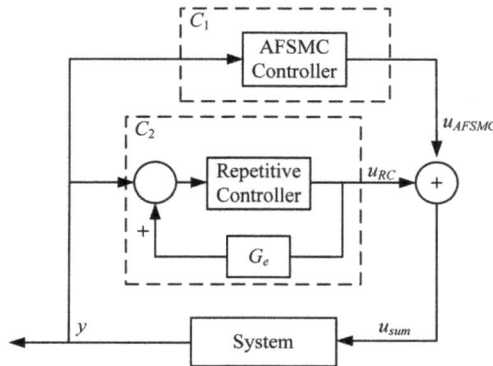

Figure 4. Block diagram of the proposed hybrid AFSMC/RC system.

Periodic disturbances from the external environment or the motions of mechanical structures can damage the system via resonance. Repetitive controllers have proven highly effective in handling periodic signal trajectories and disturbances [19]. Thus, we employed a repetitive controller to eliminate disturbances caused by the periodic motions of the motor. The design principle of a repetitive controller involves adding the internal model of exogenous signals into the feedback control system. According to the internal model principle, asymptotic error tracking can be achieved as long as the closed loop system remains stable [19]. Figure 5 presents a block diagram of the RC adopted in this study.

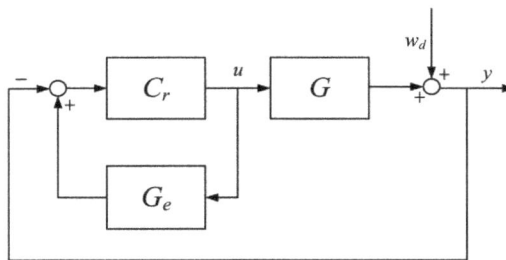

Figure 5. Block diagram of the applied repetitive control system.

In Figure 5, C_r denotes the repetitive controller; G represents the actual system, and G_e symbolizes the mathematical model of the system. As can be seen, the relationship between disturbance w_d and output y in the system of the RC framework is

$$y = \frac{1 - C_r G_e}{1 - C_r (G_e - G)} w_d \tag{17}$$

When the mathematical model G_e approximates the actual system G, Equation (17) can be rewritten as

$$y = (1 - C_r G_e) w_d \tag{18}$$

In this way, the original feedback control issue can be converted into a feedforward control issue. To mitigate the impact of disturbance w_d, a cost function can be established:

$$J_{rc} = 1 - C_r G_e \tag{19}$$

The design objective of this control problem is to derive C_r, the solution to stable plant inversion [20], while satisfying the constraint conditions that accompany periodic signals serving as the exogenous input. For systems with repetitive signals, the internal model D_r can be expressed as $D_r = 1 - z^{-N_d}$ where N_d is the length of the time delay, which is determined by the period of the input signal. If the cost function J_{rc} is written as

$$1 - C_r G_e = R_r D_r \quad or \quad C_r G_e + R_r D_r = 1 \tag{20}$$

then Equation (20) becomes the renowned "Bezout Identity" [21], of which (R_r, C_r) is a solution, and G_e are C_r are coprime. As the piezoelectric vibration suppression system in this study adopts a single-input–single-output control design, the mathematical model of this system, G_e, can be written as

$$G_e = G_d G_i \tag{21}$$

where G_d is the minimal phase of G_e, and G_i denotes the non-minimal phase of G_e.

Substituting Equation (21) into Equation (20) gives

$$\begin{aligned} R_r D_r + C C_i &= 1 \\ C &= C_r G_d \end{aligned} \tag{22}$$

Based on the concept of stable plant inversion in [20], a solution (R_r, C_r) represented in discrete-time can be obtained to fulfill Equation (22):

$$\begin{aligned} R_r &= \frac{1}{1 - (1 - k_r G_i^* G_i) q z^{-N_d}} \\ C &= k_r G_i^* q z^{-N_d} R_r \\ C_r &= C G_d^{-1} \end{aligned} \tag{23}$$

where k_r is the learning gain that can be used to adjust the convergence speed of the repetitive controller, and q is a zero-phase low-pass filter that can enhance the robustness of the repetitive controller. Introducing a narrower bandwidth for filter q results in a repetitive controller of greater robustness but diminishes performance when processing periodic signals, and vice versa. In accordance with Equation (23), the repetitive controller can be expressed as follows:

$$C_r = \frac{k_r G_i^* G_d^{-1} q z^{-N_d}}{1 - (1 - k_r G_i^* G_i) q z^{-N_d}} \tag{24}$$

4. System Setup

The proposed hybrid RC algorithm was implemented in Matlab at a sampling frequency of 1000 Hz for the active suppression of vibrations in a flexible structure. Figures 6 and 7, respectively, present a schematic diagram and photo of the hardware used in the vibration control system. Measurements were obtained using a data acquisition card (NI PCI-6259). After sending control commands to a power amplifier, they were respectively used to drive a servomotor (MHMD-042P1S from Panasonic, Osaka, Japan) and piezoelectric ceramic transducer (SB4020008 from SINOCERAMICS, Shanghai, China). Motor encoder values were sent back to another data acquisition card (MCC PCI QUADO4). Table 1 lists the properties of the thin flexible plate and Table 2 lists the properties of the piezoelectric actuator and piezoelectric thin film sensor used in this study. A piezoelectric amplifier (VP7206 from PiezoMaster, Marlboro, MA, USA) was used to amplify the piezoelectricity from the original piezoelectric actuator by 20 times, thereby generating sufficient bending force to suppress vibrations in the thin flexible plate.

We conducted modal analysis using finite element software to optimize the placement of sensors for subsequent experiments on the suppression of vibration. Due to the large structural mass of the system, we considered only the first bending resonance. In simulations, we applied a 1 Hz sinusoidal moment to the root of the structure and excited the structure for 10 s. We recorded strain values at various locations (A–E in Figure 8) to determine the best location for the sensors used in feedback control. Figures 8 and 9 present the modal analysis associated with various time responses and Fast Fourier transform (FFT) results. The symmetric structure of the device allows us to present strain values on only one side of the thin plate. Our results clearly indicate that vibration was dominated by persistent disturbances with the first vibration mode appearing at approximately 1.5 Hz. Time plots obtained from every location present similar composite periodic responses, whereas the highest strain occurred at the root of the structure (location E), as confirmed by the peak values in the FFT results. We therefore placed the piezoelectric sensor (Model NO. LDT0-028K/L from MEAS, Hampton, VA, USA) at the fixed end of the thin plate, in order to obtain measurements of high sensitivity for use in vibration control. The frequency bandwidth of bandpass filter in the coordinating piezoelectric sensor amplifier (Piezo Film Lab Amplifier from MEAS) was set between 0.1 Hz and 10 Hz. The feedback signals of the piezoelectric vibration suppression system were amplified 10 times.

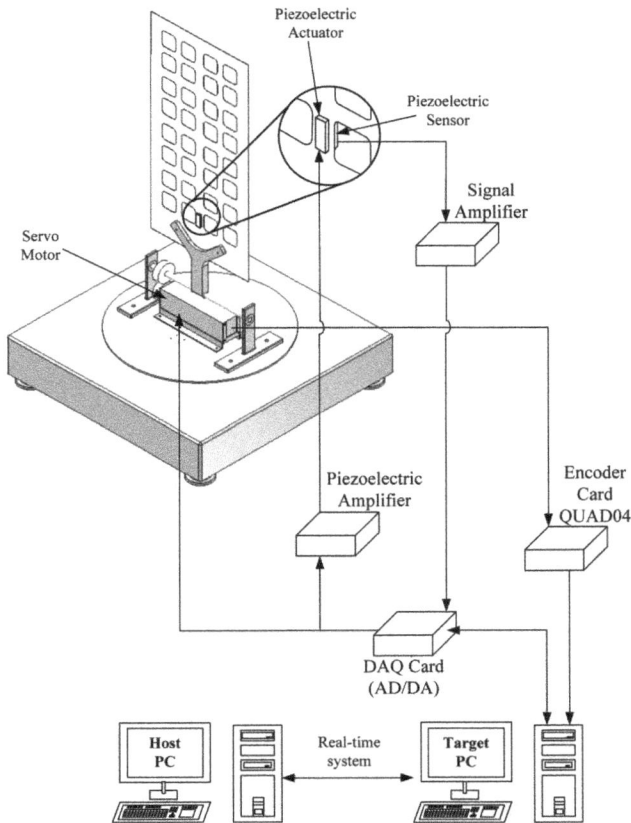

Figure 6. Schematic diagram showing vibration control system for large flexible structures.

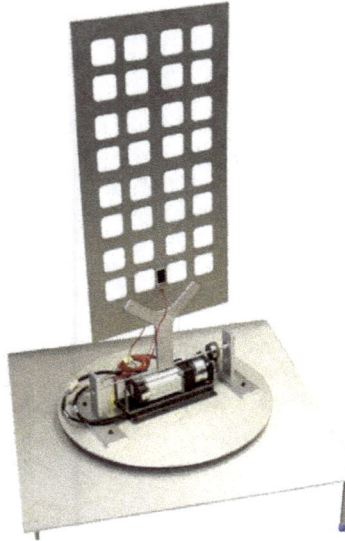

Figure 7. Hardware photograph representing a large flexible structure in experiments.

Table 1. Properties of the flexible plate.

Symbol	Description	Flexible Plate	Units
L_{plate}	Plate Length	700	mm
h_{plate}	Plate Thickness	1.58	mm
W_{plate}	Plate Width	350	mm
ρ_{plate}	Plate Density	2700	Kg/m^3
E_{plate}	Young's Modulus	7.0×10^{10}	N/m^2
ν_{plate}	Poisson Ratio	0.359	N/A

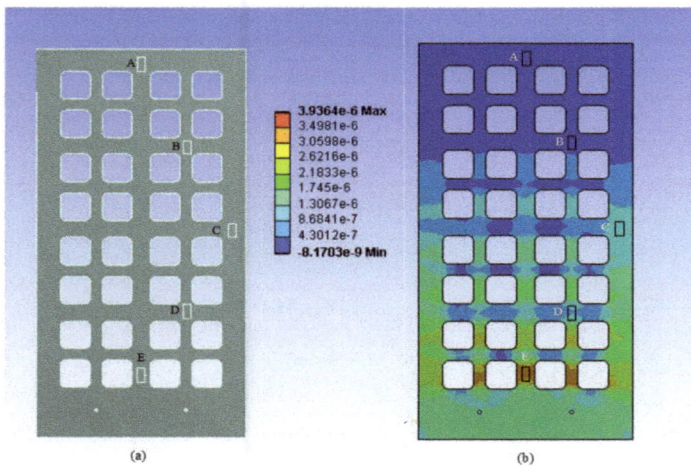

Figure 8. Modal analysis of flexible structure under persistent excitation: (**a**) illustration of the structure used for modal analysis; and (**b**) strain distribution throughout entire structure. Capital letters refer to the five locations used in evaluating the placement of piezoelectric sensors.

Table 2. Properties of the applied piezoelectric sensor and actuator.

Symbol	Description	PZT Actuator	PVDF Sensor	Units
L_{px}	Length	40	25	mm
h_p	Thickness	0.8	0.2	mm
L_{py}	Width	20	13	mm
ρ_p	Density	7.4×10^3	1.78×10^3	Kg/m^3
g_{31}	Stress Constant	-8.2×10^{-3}	0.216	Vm/N
d_{31}	Stress Constant	-3.2×10^{-11}	2.3×10^{-11}	C/N
E_a	Young's Modulus	7.1×10^{10}	0.2×10^{10}	N/m^2

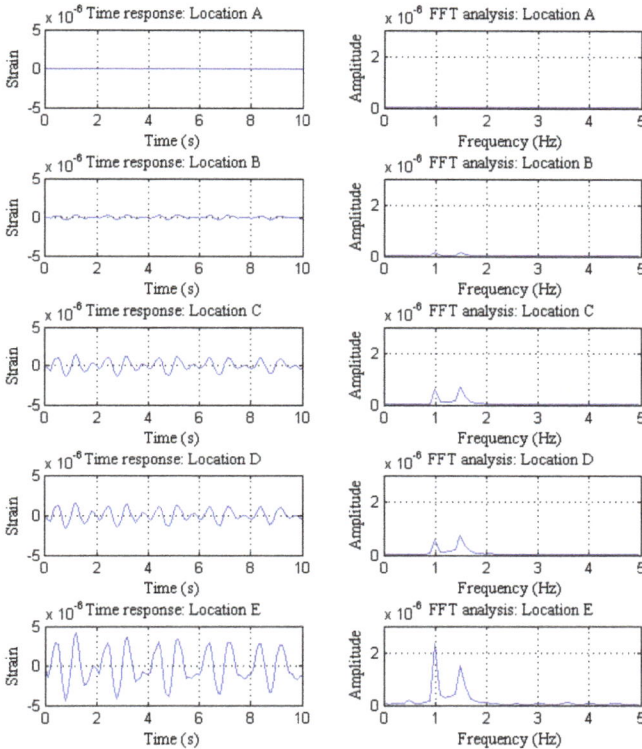

Figure 9. Modal analysis of flexible structure under persistent excitation. Subplots in the left column show the time responses at each location; subplots in the right column show the corresponding FFT results.

5. Results and Discussion

In the following two sections, we discuss the results of the experiments. In the first section, we discuss the output responses of the motor tracking system without vibration suppression. A proportional-integral (PI) controller was used to track periodic motions introduced by the motor around the yaw axis, and the piezoelectric sensor was used to measure vibrations produced by periodic disturbances in the thin plate. We used this data to identify the primary frequency determining the vibration response, the results of which were verified based on analysis in the frequency domain. We then applied active piezoelectric vibration-suppression control under the same excitation conditions in order to elucidate the control performance of RC, AFSMC, and the proposed hybrid intelligent control.

5.1. Motor Periodic Excitation

All of the data obtained from piezoelectric sensors passed through an offline fifth-order Butterworth bandpass filter. The frequency range was set at 0.01–50 Hz, and the purpose was to filter out the disturbances from the DC and 60 Hz power source for clearer result analysis. We used a PI controller to perform the closed loop tracking control of the motor around the yaw axis, causing the motor to perform periodic motions at 20 s with amplitude 0.05 rad and frequency 1 Hz so that the thin plate swung back and forth. From the perspective of the thin flexible plate, this motion can be considered a periodic disturbance. Figure 10 displays the tracking error and control input of the motor around the yaw axis, which produced periodic steady-state error of approximately 4%. Figure 11 lists the measurement results obtained from the piezoelectric sensors without vibration suppression. As can be seen, the periodic motions of the motor induce clear periodic vibration responses at 1 Hz in the thin plate output. FFT results (Figure 11b,c) revealed that the first mode also produced vibrations at 1.4 Hz, which is in agreement with the results obtained from finite element simulations. The peak values in the FFT plots revealed that the vibration responses in the flexible structure are influenced primarily by periodic disturbances. To suppress the vibrations in the thin plate during dynamic tracking motions, we placed a piezoelectric actuator and sensor on the thin flexible plate and reduced the amount of vibration in the structure using active vibration-suppression control.

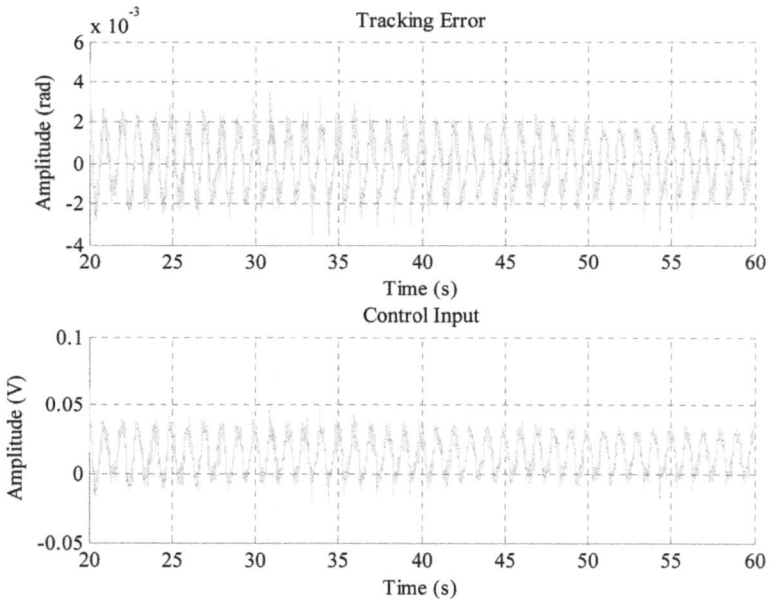

Figure 10. Periodic tracking results of the Yaw axis motor: tracking error (**top**); and control input (**bottom**).

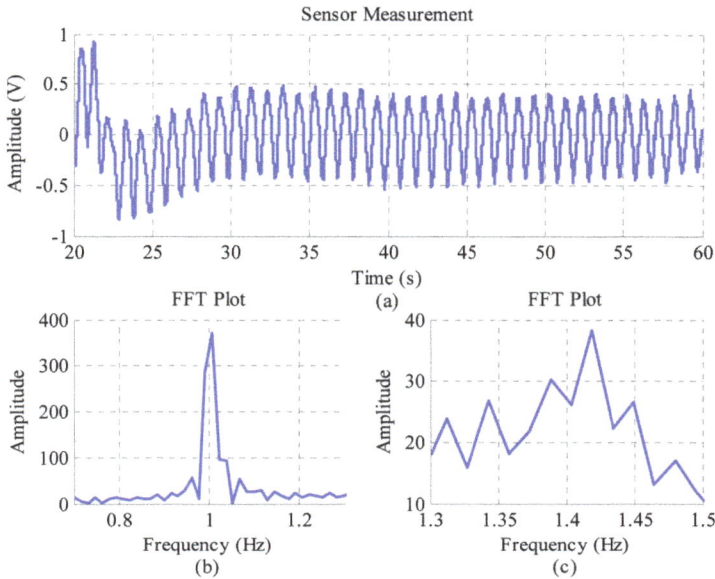

Figure 11. Flexible plate output response and FFT analysis without vibration suppression: (a) time response of uncontrolled system output, (b) FFT results at 1 Hz, and (c) FFT results at resonance frequency.

5.2. Piezoelectric Vibration Control

This study examines the active vibration suppression effects of the used three control algorithms: AFSMC, RC, and hybrid AFSMC/RC. The design parameters of the AFSMC controller were set as follows: $gu = 0.2$, $gs = 0.4$, and $\gamma_a = 0.2$. Using the time-domain system identification method, we obtained the mathematical model of the thin flexible plate system for the repetitive controller design. The relevant parameter setting was $k_r = 0.15$, and a low-pass q filter with a 10 Hz cutoff frequency was adopted to ensure the robustness of the repetitive controller. Figure 12 shows the vibration response results of the thin plate with active vibration-suppression control performed by various controllers. The AFSMC controller proved highly effective in suppressing vibrations in the thin plate; however, periodic responses remained while in steady state. RC proved more effective in reducing the periodic output in steady state; however, the initial rate of convergence was slow (40 s to reach steady state), and the amplitudes were greater. This can be attributed to an overly conservative learning gain k_r and low-pass filter q, the purpose of which was to ensure that the unmodeled dynamics in the system did not cause instability in the RC system. The hybrid AFSMC/RC control algorithm provided the benefits of AFSMC as well as RC, resulting in swift convergence (at approximately 25 s) as well as improved output responses during steady state. Figure 13 presents a comparison of the performance of the three controllers with regard to the active suppression of vibrations. The frequency band was divided into three ranges to facilitate analysis: (1) low-frequency peaks were dominated by the dynamic response of the mechanical system; (2) peak values at 1 Hz were caused by periodic excitations of the motor; and (3) peak values at 1.4 Hz were associated with the natural resonance frequency of the thin plate. AFSMC is clearly able to reduce the effects of disturbances at low frequencies and largely eliminate disturbances at 1 Hz resulting from the periodic motions of the motor. RC proved effective in suppressing vibration responses at 1 Hz; however, it was shown to amplify responses at lower frequencies as well as at natural resonance frequency of the thin plate. The hybrid AFSMC/RC control was shown to reduce output responses in all three frequency ranges.

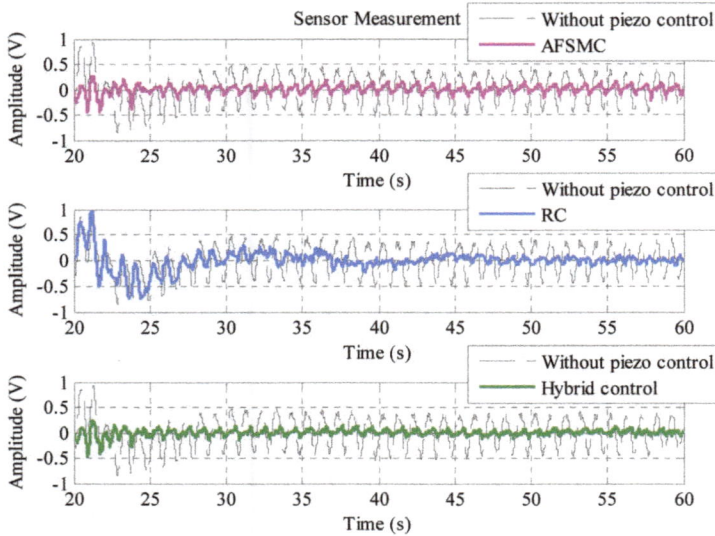

Figure 12. Piezoelectric vibration control in flexible structure: Controlled output responses obtained using: (**top**) AFSMC; (**middle**) RC; and (**bottom**) hybrid AFSMC/RC method.

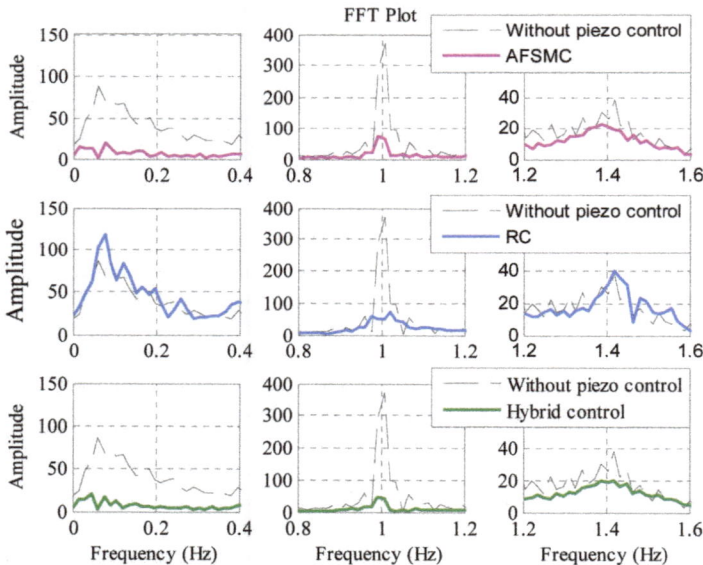

Figure 13. Piezoelectric vibration control for the flexible structure system: FFT analysis (**left**) low frequency range; (**middle**) 1 Hz; and (**right**) resonance frequency of 1.4 Hz.

Figure 14 shows the control input for each type of control. As can be seen, the control input of AFSMC shows continuous fluctuations. This is because the rule bank changes with each sampling point, and the continuous switching back and forth produces this chattering. The RC maintains periodic control input after the output converges to cope with the continuous excitation. As for the control input of the hybrid control, the graph shows overlapping effects of the swift rule bank switching in AFSMC and the learning process in RC, which also presents greater control efforts. Figures 15 and 16, respectively, present the variations in the rule banks of AFSMC and hybrid control. The resulting variations in the two rule banks are similar; i.e., they both converge to within a stable range because they use the same AFSMC design parameters. When using the hybrid control, rules C3, C4, and C5 resulted in smaller fluctuations. This is because the addition of RC reduces the periodic errors, thereby enabling the rule bank to achieve the control objective without many dramatic changes.

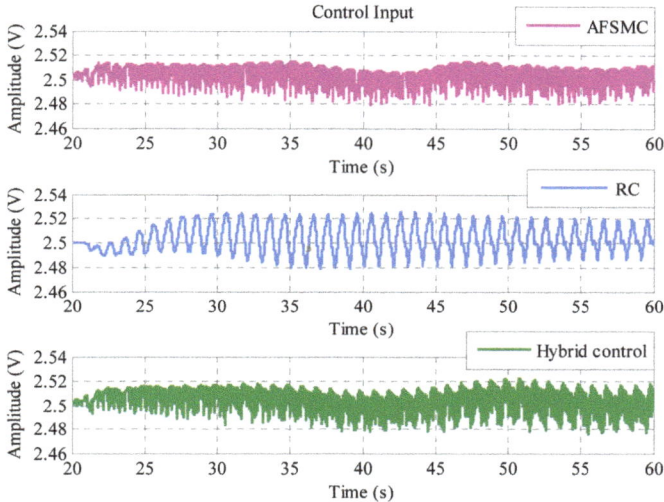

Figure 14. Piezoelectric vibration control for the flexible structure system: control input.

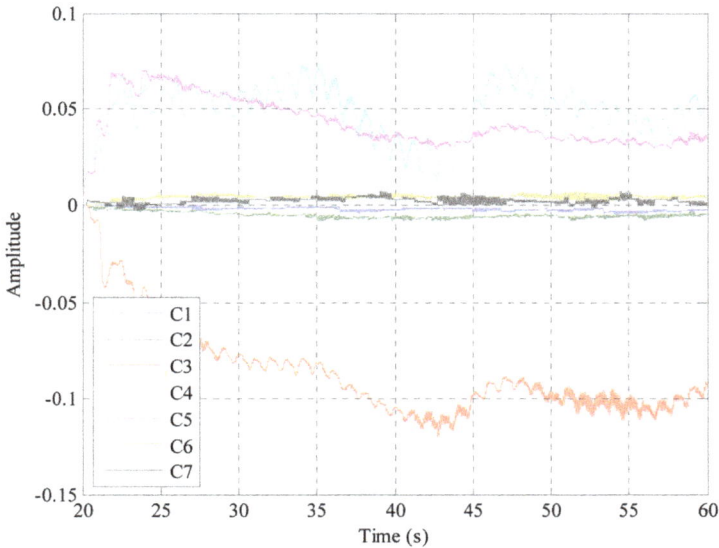

Figure 15. Fuzzy rule bank used in AFMSC vibration control system.

Appl. Sci. **2017**, *7*, 240

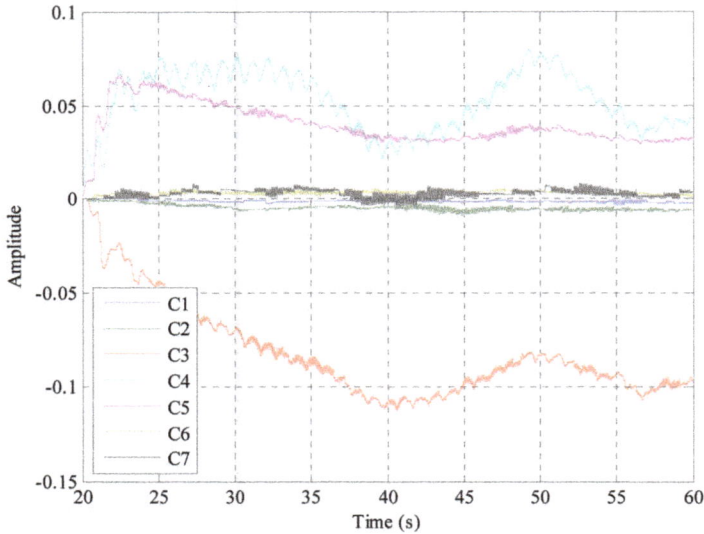

Figure 16. Fuzzy rule bank used in hybrid AFSMC/RC vibration control system.

6. Conclusions

This paper presents a novel approach to the active suppression of complex vibration responses in large flexible structures undergoing rotary motions. We applied piezoelectric materials to a thin flexible plate and then excited the plate by performing periodic motions in the motor system. The piezoelectric vibration control system was then examined using AFSMC, RC, and hybrid AFSMC/RC control algorithms. Experimental results show that AFSMC or RC can be used alone to attenuate periodic disturbances; however, the AFSMC controller was shown to reduce vibrations over a wider range of frequencies. Combining the AFSMC and RC controllers resulted in superior vibration suppression effects.

Acknowledgments: The authors would like to thank Ministry of Science and Technology, Taiwan, for providing financial support of this research under grant number MOST-101-2221-E-011-007. They are also thankful to Yu-Hsi Huang for valuable discussions and comments on this study.

Author Contributions: Chi-Ying Lin conceived and designed the active vibration control system, analyzed the results, and wrote the paper. Hong-Wu Jheng performed the vibration suppression control experiments and the data collection.

Conflicts of Interest: The authors declare no conflict of interest.

References

1. Iorga, L.; Baruh, H.; Ursu, I. A review of H_∞ robust control of piezoelectric smart structures. *Appl. Mech. Rev.* **2008**, *61*, 1–15. [CrossRef]
2. Lin, J. A vibration absorber of smart structures using adaptive networks in hierarchical fuzzy control. *J. Sound Vib.* **2005**, *287*, 683–705. [CrossRef]
3. Kang, Y.K.; Park, H.C.; Hwang, W.; Han, K.S. Optimum placement of piezoelectric sensor/actuator for vibration control of laminate beams. *AIAA J.* **1996**, *34*, 1921–1926. [CrossRef]
4. Tso, S.K.; Yang, T.W.; Xu, W.L.; Sun, Z.Q. Vibration control for a flexible-link robot arm with deflection feedback. *Int. J. Nonlinear Mech.* **2003**, *38*, 51–62. [CrossRef]
5. Shan, J.; Liu, H.T.; Sun, D. Slewing and vibration control of a single-link flexible manipulator by positive position feedback (PPF). *Mechatronics* **2005**, *15*, 487–503. [CrossRef]

6. Ahmad, M.A.; Ismail, R.R.; Ramli, M.S.; Zawawi, M.A.; Hambali, N.; Ghani, N.M.A. Vibration control of flexible joint manipulator using input shaping with PD-type fuzzy logic control. In Proceedings of the IEEE International Symposium on Industrial Electronics, Seoul, Korea, 5–8 July 2009; pp. 1184–1189.

7. Park, H.W.; Yang, H.S.; Park, Y.P.; Kim, S.H. Position and vibration control of a flexible robot manipulator using hybrid controller. *Robot. Autom. Syst.* **1999**, *28*, 31–41. [CrossRef]

8. Lin, J.; Chao, W.S. Vibration suppression control of a beam-cart system with piezoelectric transducers by decomposed parallel adaptive neuro-fuzzy control. *J. Vib. Control* **2009**, *15*, 1885–1906. [CrossRef]

9. Shin, H.C.; Choi, S.B. Position control of a two-link flexible manipulator featuring piezoelectric actuators and sensors. *Mechatronics* **2001**, *11*, 707–729. [CrossRef]

10. Moudgal, V.G.; Kwong, W.A.; Passino, K.M.; Yurkovich, S. Fuzzy learning control for a flexible-link robot. *IEEE Trans. Fuzzy Syst.* **1995**, *3*, 199–210. [CrossRef]

11. Lin, J.; Huang, C.J.; Chang, J.; Wang, S.W. Vibration suppression controller for a novel beam-cart-seesaw system. In Proceedings of the American Control Conference, Baltimore, MD, USA, 30 June–2 July 2010; pp. 1526–1531.

12. Lin, J.; Zheng, Y.B. Vibration suppression control of smart piezoelectric rotating truss structure by parallel neuro-fuzzy control with genetic algorithm tuning. *J. Sound Vib.* **2012**, *331*, 3677–3694. [CrossRef]

13. Lin, C.Y.; Chiu, W.H.; Lin, J. Rejecting multiple-period disturbances: Active vibration control of a two degree-of-freedom piezoelectric flexible structure system. *J. Vib. Control* **2015**, *21*, 3368–3382. [CrossRef]

14. Lin, C.Y.; Chang, C.M. Hybrid proportional derivative/repetitive control for active vibration control of smart piezoelectric structures. *J. Vib. Control* **2013**, *19*, 992–1003. [CrossRef]

15. Huang, S.J.; Huang, K.S. An adaptive fuzzy sliding-mode controller for servomechanism disturbance rejection. *IEEE. Trans. Ind. Electron.* **2001**, *48*, 845–852. [CrossRef]

16. Huang, S.J.; Lin, W.C. Adaptive fuzzy controller with sliding surface for vehicle suspension control. *IEEE. Trans. Fuzzy Syst.* **2003**, *11*, 550–559. [CrossRef]

17. Huang, S.J.; Shieh, H.W. Motion control of a nonlinear pneumatic actuating table by using self-adaptation fuzzy controller. In Proceedings of the IEEE International Conference on Industrial Technology (ICIT), Churchill, Victoria, Australia, 10–13 February 2009; pp. 1–6.

18. Narendra, K.; Annaswamy, A. A new adaptive law for robust adaptation without persistent excitation. *IEEE Trans. Autom. Control* **1987**, *32*, 134–145. [CrossRef]

19. Tomizuka, M.; Tsao, T.C.; Chew, K.K. Analysis and synthesis of discrete-time repetitive controllers. *J. Dyn. Syst. Meas. Contr.* **1989**, *111*, 353–358. [CrossRef]

20. Tomizuka, M. Zero phase error tracking algorithm for digital control. *J. Dyn. Syst. Meas. Contr.* **1987**, *109*, 65–68. [CrossRef]

21. Skogestad, S.; Postlethwaite, I. *Multivariable Feedback Control: Analysis and Design*; John Wiley & Sons: Hoboken, NJ, USA, 1996.

applied
sciences

MDPI

Article

Dynamic Response of a Simplified Turbine Blade Model with Under-Platform Dry Friction Dampers Considering Normal Load Variation

Bingbing He [1], Huajiang Ouyang [2,3,*], Xingmin Ren [1] and Shangwen He [4]

[1] School of Mechanics, Civil Engineering and Architecture, Northwestern Polytechnical University, Xi'an 710072, China; hebb714@gmail.com (B.H.); renxmin@nwpu.edu.cn (X.R.)
[2] State Key Laboratory of Structural Analysis for Industrial Equipment, Dalian University of Technology, Dalian 116023, China
[3] School of Engineering, University of Liverpool, Liverpool L69 3GH, UK
[4] School of Mechanics & Engineering Science, Zhengzhou University, Zhengzhou 450001, China; hsw2013@zzu.edu.cn
*   Correspondence: h.ouyang@liverpool.ac.uk or huajiang.ouyang@gmail.com; Tel.: +44-151-7944-815; Fax: +44-151-7944-848

Academic Editors: Gangbing Song, Steve C.S. Cai and Hong-Nan Li
Received: 1 January 2017; Accepted: 20 February 2017; Published: 1 March 2017

Abstract: Dry friction dampers are widely used to reduce vibration. The forced vibration response of a simplified turbine blade with a new kind of under-platform dry friction dampers is studied in this paper. The model consists of a clamped blade as two rigidly connected beams and two dampers in the form of masses which are allowed to slide along the blade platform in the horizontal direction and vibrate with the blade platform in the vertical direction. The horizontal and vertical vibrations of the two dampers, and the horizontal and transverse platform vibrations are coupled by friction at the contact interfaces which is assumed to follow the classical discontinuous Coulomb's law of friction. The vertical motion of the dampers leads to time-varying contact forces and can cause horizontal stick-slip motion between the contact surfaces. Due to the relative horizontal motion between the dampers and the blade platform, the vertical contact forces and the resultant friction forces act as moving loads. The Finite Element (FE) method and Modal Superposition (MS) method are applied to solve the dynamic response, together with an algorithm that can capture nonsmooth transitions from stick to slip and slip to stick. Quasi-periodic vibration is found even under harmonic excitation.

Keywords: dry friction damper; turbine blade; vibration reduction; beam; moving load; discontinuous Coulomb's law of friction

1. Introduction

Blades are a major component in aero-engines. High-cycle fatigue (HCF) failure due to high dynamic stresses caused by blade vibration is one of the main causes of aero-engine incidents [1]. A dry friction damper dissipates energy in the form of heat as a result of the relative rubbing motion at the contact surfaces, which has many advantages, for instance, having a simple structure, and being insensitive to temperature variation and easy to manufacture and install. Thus, dry friction dampers are widely used as a means of vibration suppression for turbine blades. A common damper configuration is the so-called under-platform dry friction damper in the form of a small piece of metal device located underneath the blade platform and actuated by centrifugal force against the platform due to engine shaft rotation. Figure 1 shows a Dummy bladed disk for vibration analysis. A damper could be installed between two adjacent blades under their platforms.

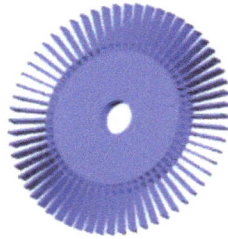

Figure 1. Dummy bladed disk for vibration analysis.

Extensive research on under-platform dry friction dampers has been carried out. A macroslip model to investigate the resonant stresses of a blade with a dry friction damper was presented by Griffin [2]. Menq et al. [3] developed a microslip model for analyzing the dynamic response of frictionally damped structures in which the friction interface was subjected to high normal loads, the microslip model derived by Menq et al. [3] was improved by Csaba [4]. A new two-dimensional model for point friction contacts was introduced by Sanliturk and Ewins [5]. Xia [6] proposed a model for investigating the stick–slip motion caused by dry friction of a two-dimensional oscillator under arbitrary excitations. A friction contact model was proposed to characterize the contact kinematics that imposed both friction nonlinearity and intermittent separation nonlinearity on structures having three-dimensional frictional constraint by Yang et al. [7]. Cigeroglu et al. [8] adopted a one-dimensional dynamic microslip friction model, including the damper inertia. This microslip friction model was further developed for a two-dimensional distributed parameter model with normal load variation induced by normal motion [9], in which they explored the use of this new model with harmonic balance method in frictionally constrained structures with a varying normal load. Further on, Cigeroglu et al. [10] implemented this model on wedge-shaped under-platform dampers in a bladed disc assembly. They allowed the wedge dampers to undergo three-dimensional translation and rotation along with elastic deformation while the damper was constrained only by friction contacts. Allara [11] proposed a model to characterize friction contact of non-spherical contact geometries obeying the Coulomb's law of friction with a constant friction coefficient and constant normal load. From this model, the effect of the main contact parameters (contact geometry, material properties, and loads) on the contact behavior could be effectively estimated.

In [3,5] the dampers were always in full contact with the blade platform. In [9,12] the dampers could partially detach from the blade platforms during vibration. In [13,14] the dampers were modeled with the finite elements. A carefully designed and constructed rotating test rig was used to make precise measurements of the forced vibration response of a bladed disk with fitted under-platform "cottage-roof" friction dampers, the corresponding numerical predictions were carried out too, and then a comparison between the measured and predicted response curves was made and the degree of correlation was discussed [15]. To what extent microslip due to the combined nonlinearities along the normal and the tangent of non-conforming contact surfaces influenced the damper behavior was investigated [16]. Ostachowicz [17] established a harmonic balance method (HBM) for forced vibration analysis of dynamic systems damped by dry friction forces. Guillen and Pierre [18] introduced a Hybrid Frequency-Time (HFT) method for analyzing the steady-state response of the large-scale dry-friction damped structural systems. The correlation of the static/dynamic coupling of the under-platform dampers was investigated by Firrone, Zucca [19]. Mathematical relationships of dry friction force versus relative velocity in friction contact of two bodies were studied by Pûst et al. [20]. Schwingshackl et al. [21] focused their research on contact interface parameters in a nonlinear dynamic analysis of assembled structures.

Gola et al. [22] studied the design and the calibration of a test rig specially developed to measure the in-plane forces transferred between the blade platforms through the under-platform damper and their relative displacement. A nonlinear analysis based on an updated explicit damper model having different levels of details was performed [23], and the results were evaluated against a newly-developed under-platform damper test rig. A standard Scanning Laser Doppler Vibrometer (SLDV) technology to measure the mistuned bladed disk vibration was put forward by Di Maio [24]. Then Zucca et al. [25] used a rotating laser Doppler Vibrometer to measure the performance of under-platform dampers for turbine blades. Nikhamkin et al. [26] developed an experimental technique for damping efficiency estimation of gas-turbine blades. The experimental investigation into the dynamic response of the blades of a gas turbine for power generation carrying asymmetric under-platform dampers was presented [27]. Quot et al. [28] proposed a novel approach which could directly measure the forces transmitted between the two platforms through the dampers. Rastogi et al. [29] utilized the Bond graph model of dry friction dampers in structural analysis of turbine blades. An FE modal analysis was described and a simplified method to evaluate the under-platform damper effects was also presented by Bessone and Traversone [30]. Based on a set of newly introduced non-dimensional parameters that ensured a similar dynamic behavior of the test rig to a real turbine blade-damper system, a new experimental damper rig was developed [31].

Friction dampers are widely used in machines and structures, including under-platform dampers studied as a particular example in this paper, shrouds in steam engines [32], those used in buildings [33], washing machines [34] and train suspensions [35]. They share the same features of stick-slip vibration and are worth studying. Stick-slip vibration is a kind of nonsmooth nonlinearity and requires special mathematical treatment. Popp et al. [36] made a comparison and classification of different contact models with friction that had been commonly used. In [37], the sticking and non-sticking orbits of a two-degree-of-freedom oscillator subjected to dry friction and a harmonic load were obtained in a closed form. Li et al. [38] developed a model of an elastic disc in sliding frictional contact with a rotating oscillator. Separation of the moving slider form the disc and its subsequent reattachment to the disc were considered, and various dynamic behaviors were discovered.

One aim of this study is to propose a more accurate model to study the motion of a friction damper. Another aim is to study the general dynamic behaviour of friction dampers in various applications (using aero engine under-platform dampers as a particular example). A third aim is to establish an effective algorithm for dealing with stick-slip vibration. In this paper, the normal contact force and the resultant friction force at the contact interface is modeled as moving loads, which is what really happens in many friction dampers but has not been modeled as such in the past. A numerical approach for solving the steady-state vibration of a simplified blade-damper system with nonsmooth friction contact is put forward and three distinct types of vibration, including stick-slip vibration between the two dampers and the blade platform surface, are studied. Each damper is partially constrained by a short cantilever beam and thus is different from those dampers reported in all of the above-mentioned papers about friction dampers.

2. Blade-Damper Model and Theoretical Development

The structural model of the blades with under-platform dry friction dampers is shown in Figure 2: the X-Y-Z coordinate system (the global cylindrical coordinate system) is defined in accordance with the convention of tangential (X), radial (Y), and axial (Z) directions; the x_1-y_1 coordinate system which is a plane Cartesian coordinate frame (called the left damper coordinate system) is defined with the local coordinate origin being at the left initial contact point, the x_2-y_2 coordinate system which is a plane Cartesian coordinate frame (called the right damper coordinate system) is defined similarly. In this paper, the blade-platform structure is simplified and regarded as two rigidly connected Euler–Bernoulli beams, as illustrated in Figure 3. Unlike a conventional under-platform damper which is a floating mass actuated by centrifugal force, the under-platform damper studied here is mounted on the free end of a small vertical cantilever beam. The mechanical models of the blades and under-platform dampers are shown in Figures 3 and 4. In Figures 3 and 4, F is the external excitation force acting on a blade, N_1 and N_2 are the normal contact forces acting on the blade platform from the dampers, f_1 and f_2 are the friction forces at the contact interface, F_1 and F_2 are the centrifugal forces of the under-platform dampers, and k is a spring constant (the equivalent lateral stiffness of the tip of the small vertical cantilever beam). $F_1 = m_1 r \Omega^2$, $F_2 = m_2 r \Omega^2$, where m_1 and m_2 are the mass of the dampers, r is the distance between the platform and the centre of the disk, and Ω is the rotating speed of the disk. The weights of the dampers are very small in comparison with the centrifugal forces and thus can be ignored.

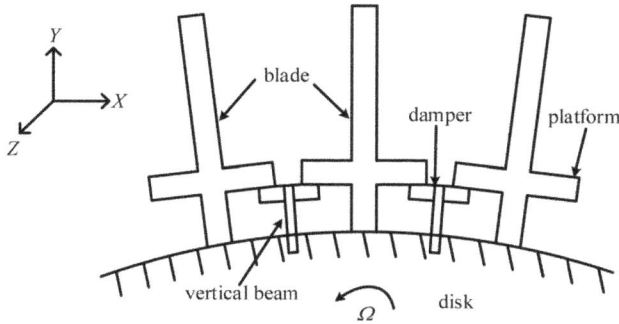

Figure 2. The structural model of the blades with under-platform dry friction dampers.

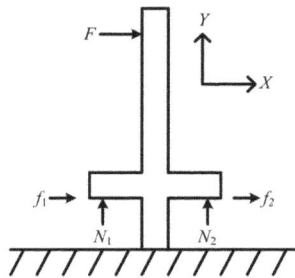

Figure 3. The mechanical model of the simplified blade-platform system.

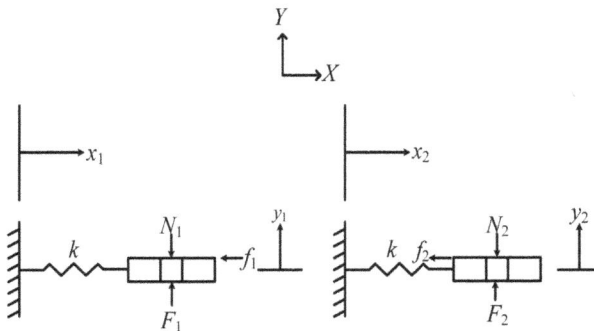

Figure 4. The mechanical model of the under-platform dampers.

In this investigation, the normal contact forces and the tangential friction forces at the contact interface are modeled as moving loads (which are a distinct feature of this paper and have not been modeled as such in previous research of blade under-platform friction dampers), and the two dampers are assumed to undergo horizontal and vertical vibrations, but no rotation. They are in point contact with the platform and do not separate with the platform during vibration excited by excitation force F. Therefore, displacements y_1 and y_2 of the dampers are equal to the local transverse deflections of the platform; but they can slide along the underside of the platform in the global X direction. The horizontal vibration and the vertical vibration of the two dampers, and the horizontal and transverse platform vibrations are coupled by friction at the contact interface.

A classical Coulomb friction model [39] is adopted in this paper: before sliding occurs, the friction force equals the shear force at the contact interface due to the applied external force and acts to resist the initiation of sliding; during sliding the friction force is proportional to the normal force at the contact interface and acts in the direction opposite to sliding. A further assumption is that the static coefficient of friction μ_s is greater than the kinetic coefficient of friction μ_k (so that stick-slip vibration of the damper is possible) and both are constant. Figure 5 describes the Coulomb friction model.

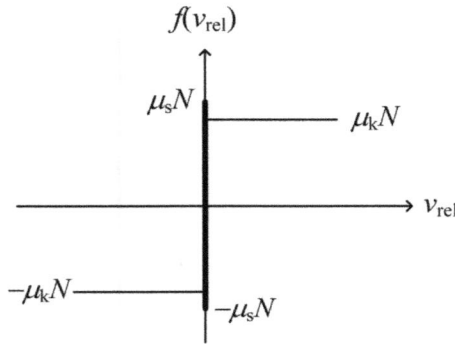

Figure 5. The classical Coulomb friction model.

Accordingly, the relationship between the friction force $f(v_{rel})$ and the relative velocity v_{rel} and the local normal force N is

$$
\begin{cases}
f(v_{rel}) = \mu_k N (v_{rel} > 0) \\
f(v_{rel}) = -\mu_k N (v_{rel} < 0) \\
-\mu_s N \leq f(v_{rel}) \leq \mu_s N (v_{rel} = 0)
\end{cases}
\tag{1}
$$

The equations of free vibration in the local transverse and longitudinal directions of the component of the blade modeled as an Euler–Bernoulli beam [40] (denoted by 1) in its local coordinate ε are

$$
\begin{cases}
\rho_1 A_1 \frac{\partial^2 w_1}{\partial t^2}(\varepsilon, t) + E_1 I_1 \frac{\partial^4 w_1}{\partial \varepsilon^4}(\varepsilon, t) = 0 \\
\rho_1 A_1 \frac{\partial^2 u_1}{\partial t^2}(\varepsilon, t) - E_1 A_1 \frac{\partial^2 u_1}{\partial \varepsilon^2}(\varepsilon, t) = 0
\end{cases}
\tag{2}
$$

In the same approach, the equations of free vibration of the platform component modeled as an Euler–Bernoulli beam (denoted by 2) are

$$
\begin{cases}
\rho_2 A_2 \frac{\partial^2 w_2}{\partial t^2}(\varepsilon, t) + E_2 I_2 \frac{\partial^4 w_2}{\partial \varepsilon^4}(\varepsilon, t) = 0 \\
\rho_2 A_2 \frac{\partial^2 u_2}{\partial t^2}(\varepsilon, t) - E_2 A_2 \frac{\partial^2 u_2}{\partial \varepsilon^2}(\varepsilon, t) = 0
\end{cases}
\tag{3}
$$

where w_1 and w_2 are the transverse displacements, and u_1 and u_2 are the longitudinal displacements, all in their local coordinates; E_1 and E_2 are Young's moduli of these two components; ρ_1 and ρ_2 are the mass densities; A_1 and A_2 are the cross-sectional areas; and I_1 and I_2 are the second moment of the areas. The w-component and u-component of the k-th beam's n-th analytical mode can be denoted as $\psi_{wkn}(\varepsilon)$ and $\psi_{ukn}(\varepsilon)$, ($k = 1, 2$) in its local coordinate ε, which consist of base functions of sine, cosine, hyperbolic sine and hyperbolic cosine functions when expressed exactly [41]. Equations (2) and (3) are solved to get the natural frequencies and modes of the cross structure, which will be used in the mode superposition method for subsequent analysis. However, when the configuration of the beam structure is complex, analytical modes are difficult to obtain accurately. In general, the FE method provides a simple and effective approach to obtain the frequencies and modes of a structure of arbitrary configuration and hence is used.

The blade-platform system is discretized with a number of two-node Euler–Bernoulli Beam elements. Each node has three degrees-of-freedom: one translation along the longitudinal axis of the beam element, denoted by u; one translation lateral to the beam axis, denoted by w; and a rotation around the axis normal to this plane, denoted by θ [42]. Denote the k-th beam's n-th numerical mode as ψ_{kn}^{FE}

$$\psi_{kn}^{FE} = \{u_{k1}, w_{k1}, \theta_{k1}, u_{k2}, w_{k2}, \theta_{k2}, \ldots u_{ki}, w_{ki}, \theta_{ki}, u_{k(i+1)}, w_{k(i+1)}, \theta_{k(i+1)}, \ldots\}_n^T$$

$$\psi_{ukn}^{FE} = \left\{u_{k1}, u_{k2}, \ldots u_{ki}, u_{k(i+1)}, \ldots\right\}_n^T, \ \psi_{wkn}^{FE} = \left\{w_{k1}, w_{k2}, \ldots w_{ki}, w_{k(i+1)}, \ldots\right\}_n^T,$$

$$\psi_{\theta kn}^{FE} = \left\{\theta_{k1}, \theta_{k2}, \ldots \theta_{ki}, \theta_{k(i+1)}, \ldots\right\}_n^T$$

$k = 1$, 2, where subscript i is the left node of the i-th beam element and the right node of the $(i-1)$-th beam element, and superscript T stands for matrix transpose. When the number of beam elements is sufficient, the numerical modes and frequencies will be very close to the analytical modes and frequencies of the whole structure. Denote the i-th element shape function matrix of the k-th beam for w-displacement and u-displacement by $N_{wki}(\varepsilon)$ and $N_{uki}(\varepsilon)$ $(k = 1, 2)$. Then, one acquires an approximate expression of the n-th analytical mode of the blade-platform structure as

$$\psi_{w1n}(\varepsilon) = \begin{cases} N_{w11}(\varepsilon_1)\{u_{11}, \theta_{11}, u_{12}, \theta_{12}\}_n^T \ \varepsilon \subset \varepsilon_1 \\ N_{w12}(\varepsilon_2)\{u_{12}, \theta_{12}, u_{13}, \theta_{13}\}_n^T \ \varepsilon \subset \varepsilon_2 \\ \quad . \\ \quad . \\ \quad . \\ N_{w1i}(\varepsilon_i)\left\{u_{1i}, \theta_{1i}, u_{1(i+1)}, \theta_{1(i+1)}\right\}_n^T \ \varepsilon \subset \varepsilon_i \\ \quad . \\ \quad . \\ \quad . \end{cases} \tag{4}$$

$$\psi_{u1n}(\varepsilon) = \begin{cases} N_{u11}(\varepsilon_1)\{w_{11}, w_{12}\}_n^T \ \varepsilon \subset \varepsilon_1 \\ N_{u12}(\varepsilon_2)\{w_{12}, w_{13}\}_n^T \ \varepsilon \subset \varepsilon_2 \\ \quad . \\ \quad . \\ \quad . \\ N_{u1i}(\varepsilon_i)\left\{w_{1i}, w_{1(i+1)}\right\}_n^T \ \varepsilon \subset \varepsilon_i \\ \quad . \\ \quad . \\ \quad . \end{cases} \tag{5}$$

where ε_i is the local longitudinal coordinate within the i-th element of the blade. ψ_{u1n}^{FE} and ψ_{w1n}^{FE} are the numerical modes of the blade of the global coordinate axes X and Y, respectively. Because the local coordinate axes of the blade are perpendicular to the global coordinate axes, ψ_{u1n}^{FE} corresponds to the numerical modes of the blade in the lateral direction, while ψ_{w1n}^{FE} corresponds to the numerical modes of the blade in the axial direction.

$$\psi_{w2n}(\varepsilon) = \begin{cases} \mathbf{N}_{w21}(\varepsilon_1)\{w_{21}, \theta_{21}, w_{22}, \theta_{22}\}_n^{\mathrm{T}} \ \varepsilon \subset \varepsilon_1 \\ \mathbf{N}_{w22}(\varepsilon_2)\{w_{22}, \theta_{22}, w_{23}, \theta_{23}\}_n^{\mathrm{T}} \ \varepsilon \subset \varepsilon_2 \\ \quad\vdots \\ \quad\vdots \\ \mathbf{N}_{w2i}(\varepsilon_i)\left\{w_{2i}, \theta_{2i}, w_{2(i+1)}, \theta_{2(i+1)}\right\}_n^{\mathrm{T}} \ \varepsilon \subset \varepsilon_i \\ \quad\vdots \\ \quad\vdots \end{cases} \tag{6}$$

$$\psi_{u2n}(\varepsilon) = \begin{cases} \mathbf{N}_{u21}(\varepsilon_1)\{u_{21}, u_{22}\}_n^{\mathrm{T}} \ \varepsilon \subset \varepsilon_1 \\ \mathbf{N}_{u22}(\varepsilon_2)\{u_{22}, u_{23}\}_n^{\mathrm{T}} \ \varepsilon \subset \varepsilon_2 \\ \quad\vdots \\ \quad\vdots \\ \mathbf{N}_{u2i}(\varepsilon_i)\left\{u_{2i}, u_{2(i+1)}\right\}_n^{\mathrm{T}} \ \varepsilon \subset \varepsilon_i \\ \quad\vdots \\ \quad\vdots \end{cases} \tag{7}$$

where ε_i is the local longitudinal coordinate within the i-th element of the platform; $N_{wki}(\varepsilon_i) = \frac{1}{L_{ki}^3}[(L_{ki}^3 - 3L_{ki}\varepsilon_i^2 + 2\varepsilon_i^3), (L_{ki}^2\varepsilon_i - 2L_{ki}\varepsilon_i^2 + \varepsilon_i^3), (3L_{ki}\varepsilon_i^2 - 2\varepsilon_i^3), (\varepsilon_i^3 - L_{ki}\varepsilon_i^2)];$ and $N_{uki}(\varepsilon_i) = \left[1 - \frac{\varepsilon_i}{L_{ki}}, \frac{\varepsilon_i}{L_{ki}}\right];$ $(k = 1, 2);$ and L_{ki} is the length of the i-th element of the k-th beam. The reason for converting the FE modes to approximate analytical modes is to accommodate the tracking of horizontal positions of the friction dampers relative to the platform, which is a common formulation of moving-load problems [43]. Such a numerical-analytical combined formulation allows dynamic responses of complicated structures excited by moving loads to be determined conveniently [44].

The combined equations of motion of the blade-platform structure under external excitation can be written as

$$\mathbf{M}\begin{Bmatrix} \ddot{u}_1 \\ \ddot{u}_2 \\ \ddot{w}_1 \\ \ddot{w}_2 \end{Bmatrix} + \mathbf{K}\begin{Bmatrix} u_1 \\ u_2 \\ w_1 \\ w_2 \end{Bmatrix} = \mathbf{f} \tag{8}$$

$$\mathbf{M} = \begin{bmatrix} \rho_1 A_1 & 0 & 0 & 0 \\ 0 & \rho_2 A_2 & 0 & 0 \\ 0 & 0 & \rho_1 A_1 & 0 \\ 0 & 0 & 0 & \rho_2 A_2 \end{bmatrix} \quad \mathbf{K} = \begin{bmatrix} -E_1 A_1 \frac{\partial^2}{\partial \varepsilon^2} & 0 & 0 & 0 \\ 0 & -E_2 A_2 \frac{\partial^2}{\partial \varepsilon^2} & 0 & 0 \\ 0 & 0 & E_1 I_1 \frac{\partial^4}{\partial \varepsilon^4} & 0 \\ 0 & 0 & 0 & E_2 I_2 \frac{\partial^4}{\partial \varepsilon^4} \end{bmatrix}$$

$$\mathbf{f} = \begin{Bmatrix} 0 \\ f_1 \delta(\varepsilon - X_1) + f_2 \delta(\varepsilon - X_2) \\ F\delta(\varepsilon - Y_0) \\ N_1 \delta(\varepsilon - X_1) + N_2 \delta(\varepsilon - X_2) \end{Bmatrix}$$

where $\mathbf{M}$ and $\mathbf{K}$ are the "mass" and "stiffness" operator matrices of the blade-platform structure; and $\{u_1, u_2, w_1, w_2\}^{\mathrm{T}}$ and $\mathbf{f}$ are, respectively, the nodal displacement vector and force vector of the structure. The dot over a symbol represents differentiation with respect to time t, Y_0 is the location of the external excitation force F, X_1 and X_2 are the location of the contact points which are unknown a priori and vary with time. Please note that the first element of $\mathbf{f}$ is zero because there is no external force acting in the u_1 direction for beam 1 (the blade) while the second element denotes the friction forces at the two damper-platform interfaces.

The nodal displacement vector of the whole structure can be expressed as

$$
\left\{
\begin{array}{c}
\mathbf{u}_1(t) \\
\mathbf{u}_2(t) \\
\mathbf{w}_1(t) \\
\mathbf{w}_2(t) \\
\boldsymbol{\theta}_1(t) \\
\boldsymbol{\theta}_2(t)
\end{array}
\right\}
= \sum_{i=1}^{n}
\left\{
\begin{array}{c}
\boldsymbol{\psi}_{u1i}^{\mathrm{FE}} \\
\boldsymbol{\psi}_{u2i}^{\mathrm{FE}} \\
\boldsymbol{\psi}_{w1i}^{\mathrm{FE}} \\
\boldsymbol{\psi}_{w2i}^{\mathrm{FE}} \\
\boldsymbol{\psi}_{\theta1i}^{\mathrm{FE}} \\
\boldsymbol{\psi}_{\theta2i}^{\mathrm{FE}}
\end{array}
\right\}
q_i(t) =
\left\{
\begin{array}{c}
\boldsymbol{\Psi}_{u1}^{\mathrm{FE}} \\
\boldsymbol{\Psi}_{u2}^{\mathrm{FE}} \\
\boldsymbol{\Psi}_{w1}^{\mathrm{FE}} \\
\boldsymbol{\Psi}_{w2}^{\mathrm{FE}} \\
\boldsymbol{\Psi}_{\theta1}^{\mathrm{FE}} \\
\boldsymbol{\Psi}_{\theta2}^{\mathrm{FE}}
\end{array}
\right\}
\mathbf{q}(t)
\tag{9}
$$

where $q_i(t)$ is the i-th modal coordinate of the blade-platform structure.

$$
\mathbf{q}(t) = \left\{ q_1(t), q_2(t), q_3(t), \ldots q_n(t) \right\}^{\mathrm{T}}
$$

$$
\boldsymbol{\Psi}_{u1}^{\mathrm{FE}} = \left[\boldsymbol{\psi}_{u11}^{\mathrm{FE}}, \boldsymbol{\psi}_{u12}^{\mathrm{FE}}, \boldsymbol{\psi}_{u13}^{\mathrm{FE}}, \ldots \boldsymbol{\psi}_{u1n}^{\mathrm{FE}} \right], \boldsymbol{\Psi}_{u2}^{\mathrm{FE}} = \left[\boldsymbol{\psi}_{u21}^{\mathrm{FE}}, \boldsymbol{\psi}_{u22}^{\mathrm{FE}}, \boldsymbol{\psi}_{u23}^{\mathrm{FE}}, \ldots \left[\boldsymbol{\psi}_{u2n}^{\mathrm{FE}} \right] \right.
$$

$$
\boldsymbol{\Psi}_{w1}^{\mathrm{FE}} = \left[\boldsymbol{\psi}_{w11}^{\mathrm{FE}}, \boldsymbol{\psi}_{w12}^{\mathrm{FE}}, \boldsymbol{\psi}_{w13}^{\mathrm{FE}}, \ldots \boldsymbol{\psi}_{w1n}^{\mathrm{FE}} \right], \boldsymbol{\Psi}_{w2}^{\mathrm{FE}} = \left[\boldsymbol{\psi}_{w21}^{\mathrm{FE}}, \boldsymbol{\psi}_{w22}^{\mathrm{FE}}, \boldsymbol{\psi}_{w23}^{\mathrm{FE}}, \ldots \boldsymbol{\psi}_{w2n}^{\mathrm{FE}} \right]
$$

$$
\boldsymbol{\Psi}_{\theta1}^{\mathrm{FE}} = \left[\boldsymbol{\psi}_{\theta11}^{\mathrm{FE}}, \boldsymbol{\psi}_{\theta12}^{\mathrm{FE}}, \boldsymbol{\psi}_{\theta13}^{\mathrm{FE}}, \ldots \boldsymbol{\psi}_{\theta1n}^{\mathrm{FE}} \right], \boldsymbol{\Psi}_{\theta2}^{\mathrm{FE}} = \left[\boldsymbol{\psi}_{\theta21}^{\mathrm{FE}}, \boldsymbol{\psi}_{\theta22}^{\mathrm{FE}}, \boldsymbol{\psi}_{\theta23}^{\mathrm{FE}}, \ldots \boldsymbol{\psi}_{\theta2n}^{\mathrm{FE}} \right]
$$

$\boldsymbol{\psi}_{uki}^{\mathrm{FE}}$, $\boldsymbol{\psi}_{wki}^{\mathrm{FE}}$ and $\boldsymbol{\psi}_{\theta ki}^{\mathrm{FE}}$ ($k = 1$, 2) form the i-th mass-normalized FE mode of the k-th beam (in the order of ascending frequencies).

It follows from Equation (9) that the horizontal and the vertical displacements of the platform can be expressed as

$$
[u_2(\varepsilon, t), w_2(\varepsilon, t)] = \mathbf{q}^{\mathrm{T}}(t)[\boldsymbol{\psi}_{u2}(\varepsilon), \boldsymbol{\psi}_{w2}(\varepsilon)]
\tag{10}
$$

where $\boldsymbol{\psi}_{u2}(\varepsilon)$ and $\boldsymbol{\psi}_{w2}(\varepsilon)$ are approximate analytical u-component and w-component mode vectors converted from $\boldsymbol{\Psi}_{u2}^{\mathrm{FE}}$, $\boldsymbol{\Psi}_{w2}^{\mathrm{FE}}$ and $\boldsymbol{\Psi}_{\theta2}^{\mathrm{FE}}$ through element shape functions. Very similarly, the two displacements of the blade can also be expressed like Equation (10). Analytical modes (even though approximate), instead of numerical modes, are particular useful in dealing with moving loads [44].

Regardless of relative sticking or slipping of the dampers to the platform, friction forces f_1 and f_2 and normal forces N_1 and N_2 at the contact points always satisfy

$$
\frac{1}{2} m_1 \ddot{x}_1 + \frac{1}{2} k x_1 = f_1
\tag{11}
$$

$$
\frac{1}{2} m_2 \ddot{x}_2 + \frac{1}{2} k x_2 = f_2
\tag{12}
$$

$$
\frac{1}{2} F_1 - \frac{1}{2} m_1 \ddot{y}_1 = N_1
\tag{13}
$$

$$
\frac{1}{2} F_2 - \frac{1}{2} m_2 \ddot{y}_2 = N_2
\tag{14}
$$

where x_1 and x_2, and y_1 and y_2 are the local horizontal displacements and the vertical displacements of the two dampers, respectively; and the $\frac{1}{2}$ in the above equations is due to the equal share of a damper by two adjacent blades (see Figure 2). Platform horizontal displacements u_2 at the contact points can be found from Equation (10).

Denote $X_1 = u_2 - x_1 + x_{10}$, $X_2 = u_2 - x_2 + x_{20}$, $\frac{1}{2} F_1 = \frac{1}{2} F_2 = N_0$. N_0 is the preload, x_{10} is the coordinate of the X direction of the left damper coordinate system origin in the global coordinate system and x_{20} is the coordinate of the X direction of the right damper coordinate system origin in the global coordinate system. Figure 6 shows the un-deformed and deformed configuration of the blade-damper system.

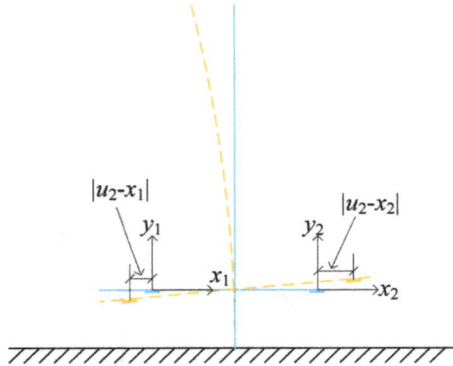

Figure 6. The un-deformed (in solid lines) and deformed (in dashed lines) configuration of the blade-damper system.

If the left damper slips relatively to the platform, the friction force is known as

$$f_1 = \mu_k N_1 \mathrm{sgn}(\dot{u}_2 - \dot{x}_1) = \mu_k N_1 \mathrm{sgn}(\dot{X}_1) \tag{15}$$

Similarly, if the right damper slips relatively to the platform, the friction force could be also written as

$$f_2 = \mu_k N_2 \mathrm{sgn}(\dot{u}_2 - \dot{x}_2) = \mu_k N_2 \mathrm{sgn}(\dot{X}_2) \tag{16}$$

If there is contact between the left damper and the platform, the relationship between the transverse displacement w_1 of the contact point of the platform and the vertical displacement y_1 of the damper is:

$$y_1(t) = w_2(X_1(t), t) = \mathbf{q}^T \boldsymbol{\psi}_{w2}(X_1) \tag{17}$$

Therefore,

$$\dot{y}_1(t) = \dot{\mathbf{q}}^T \boldsymbol{\psi}_{w2}(X_1) + \mathbf{q}^T \boldsymbol{\psi}'_{w2}(X_1)\dot{X}_1 \tag{18}$$

$$\ddot{y}_1(t) = \ddot{\mathbf{q}}^T \boldsymbol{\psi}_{w2}(X_1) + 2\dot{\mathbf{q}}^T \boldsymbol{\psi}'_{w2}(X_1)\dot{X}_1 + \mathbf{q}^T \boldsymbol{\psi}''_{w2}(X_1)\dot{X}_1^2 + \mathbf{q}^T \boldsymbol{\psi}'_{w2}(X_1)\ddot{X}_1 \tag{19}$$

where a dash denotes the differentiation with respect to the local spatial coordinate ε.

In the same way, if there is contact between the right damper and the platform, the relationship between the transverse displacement w_2 of the contact point of the platform and the vertical displacement y_2 of the damper is:

$$y_2(t) = w_2(X_2(t), t) = \mathbf{q}^T \boldsymbol{\psi}_{w2}(X_2) \tag{20}$$

and

$$\dot{y}_2(t) = \dot{\mathbf{q}}^T \boldsymbol{\psi}_{w2}(X_2) + \mathbf{q}^T \boldsymbol{\psi}'_{w2}(X_2)\dot{X}_2 \tag{21}$$

$$\ddot{y}_2(t) = \ddot{\mathbf{q}}^T \boldsymbol{\psi}_{w2}(X_2) + 2\dot{\mathbf{q}}^T \boldsymbol{\psi}'_{w2}(X_2)\dot{X}_2 + \mathbf{q}^T \boldsymbol{\psi}''_{w2}(X_2)\dot{X}_2^2 + \mathbf{q}^T \boldsymbol{\psi}'_{w2}(X_2)\ddot{X}_2 \tag{22}$$

By substituting Equations (19) and (22) into Equations (13) and (14), assuming $m_1 = m_2 = m$, normal contact forces N_1 and N_2 can be expressed as

$$N_1 = N_0 - \frac{1}{2}m[\ddot{\mathbf{q}}^T \boldsymbol{\psi}_{w2}(X_1) + 2\dot{\mathbf{q}}^T \boldsymbol{\psi}'_{w2}(X_1)\dot{X}_1 + \mathbf{q}^T \boldsymbol{\psi}''_{w2}(X_1)\dot{X}_1^2 + \mathbf{q}^T \boldsymbol{\psi}'_{w2}(X_1)\ddot{X}_1] \tag{23}$$

$$N_2 = N_0 - \frac{1}{2}m[\ddot{\mathbf{q}}^T \boldsymbol{\psi}_{w2}(X_2) + 2\dot{\mathbf{q}}^T \boldsymbol{\psi}'_{w2}(X_2)\dot{X}_2 + \mathbf{q}^T \boldsymbol{\psi}''_{w2}(X_2)\dot{X}_2^2 + \mathbf{q}^T \boldsymbol{\psi}'_{w2}(X_2)\ddot{X}_2] \tag{24}$$

Due to the orthogonality between modes of beam structures, the left-hand side of the equation of motion for the blade-platform structure, Equation (8), can be decoupled using the approximate analytical modes described in Equations (4)–(7) in the modal coordinator vector as

$$\ddot{\mathbf{q}}(t) + \text{diag}[\omega^2]\mathbf{q}(t) = F\psi_{w1}(Y_0) - f_1\psi_{u2}(X_1) - f_2\psi_{u2}(X_2) + N_1\psi_{w2}(X_1) + N_2\psi_{w2}(X_2) \tag{25}$$

If the two dampers are all in slip phase, by substituting Equations (15), (16), (23) and (24) into Equation (25), denoting $(X_1) = \psi_{w2}(X_1) - \mu_k \text{sgn}(\dot{X}_1)\psi_{u2}(X_1)$, $\mathbf{g}(X_2) = \psi_{w2}(X_2) - \mu_k \text{sgn}(\dot{X}_2)\psi_{u2}(X_2)$, one gets

$$\begin{aligned}
\Big[\mathbf{I} + \tfrac{1}{2}m\mathbf{g}&(X_1)\psi^{\mathrm{T}}_{w2}(X_1) + \tfrac{1}{2}m\mathbf{g}(X_2)\psi^{\mathrm{T}}_{w2}(X_2)\Big]\ddot{\mathbf{q}} \\
&= F\psi_{w1}(Y_0) + N_0[\mathbf{g}(X_1) + \mathbf{g}(X_2)] - \text{diag}[\omega^2]\mathbf{q} \\
&\quad - \tfrac{1}{2}m\mathbf{g}(X_1)[2\dot{\mathbf{q}}^{\mathrm{T}}\psi'_{w2}(X_1)\dot{X}_1 + \mathbf{q}^{\mathrm{T}}\psi''_{w2}(X_1)\dot{X}_1^2 + \mathbf{q}^{\mathrm{T}}\psi'_{w2}(X_1)\ddot{X}_1] \\
&\quad - \tfrac{1}{2}m\mathbf{g}(X_2)[2\dot{\mathbf{q}}^{\mathrm{T}}\psi'_{w2}(X_2)\dot{X}_2 + \mathbf{q}^{\mathrm{T}}\psi''_{w2}(X_2)\dot{X}_2^2 + \mathbf{q}^{\mathrm{T}}\psi'_{w2}(X_2)\ddot{X}_2]
\end{aligned} \tag{26}$$

If the two dampers are all in stick phase, the blade platform and the two dampers vibrate together and form one new system; $u_2 = \mathbf{q}^{\mathrm{T}}(t)\psi_{u2}(X_1)$; $u_2 - x_1 = c_1$; $u_2 - x_2 = c_2$; $\dot{X}_1 = 0$; $\ddot{X}_1 = 0$; $\dot{X}_2 = 0$; $\ddot{X}_2 = 0$; c_1 is the horizontal displacement difference between the left initial contact point and the left damper at the end of the previous slip phase and c_2 is the horizontal displacement difference between the right initial contact point and the right damper at the end of the previous slip phase. By substituting Equations (11), (12), (23) and (24) into Equation (25), then the equation of motion of the new system in the modal coordinator vector can be derived as

$$\begin{aligned}
\Big[\mathbf{I} + \tfrac{1}{2}m\psi_{u2}&(X_1)\psi^{\mathrm{T}}_{u2}(X_1) + \tfrac{1}{2}m\psi_{u2}(X_2)\psi^{\mathrm{T}}_{u2}(X_2) + \tfrac{1}{2}m\psi_{w2}(X_1)\psi^{\mathrm{T}}_{w2}(X_1) \\
&+ \tfrac{1}{2}m\psi_{w2}(X_2)\psi^{\mathrm{T}}_{w2}(X_2)\Big]\ddot{\mathbf{q}} \\
&= F\psi_{w1}(Y_0) + N_0[\psi_{w2}(X_1) + \psi_{w2}(X_2)] \\
&\quad - \tfrac{1}{2}k[\psi_{u2}(X_1)\psi^{\mathrm{T}}_{u2}(X_1) + \psi_{u2}(X_2)\psi^{\mathrm{T}}_{u2}(X_2)]\mathbf{q} - \text{diag}[\omega^2]\mathbf{q} \\
&\quad + \tfrac{1}{2}k[c_1\psi_{u2}(X_1) + c_2\psi_{u2}(X_2)]
\end{aligned} \tag{27}$$

If the left damper is in slip phase, and the right damper is in stick phase, one can get the equation of the new system in the modal coordinator as

$$\begin{aligned}
\Big[\mathbf{I} + \tfrac{1}{2}m\mathbf{g}&(X_1)\psi^{\mathrm{T}}_{w2}(X_1) + \tfrac{1}{2}m\psi_{u2}(X_2)\psi^{\mathrm{T}}_{u2}(X_2) + \tfrac{1}{2}m\psi_{w2}(X_2)\psi^{\mathrm{T}}_{w2}(X_2)\Big]\ddot{\mathbf{q}} \\
&= F\psi_{w1}(Y_0) + N_0[\mathbf{g}(X_1) + \psi_{w2}(X_2)] \\
&\quad - \tfrac{1}{2}m\mathbf{g}(X_1)[2\dot{\mathbf{q}}^{\mathrm{T}}\psi'_{w2}(X_1)\dot{X}_1 + \mathbf{q}^{\mathrm{T}}\psi''_{w2}(X_1)\dot{X}_1^2 + \mathbf{q}^{\mathrm{T}}\psi'_{w2}(X_1)\ddot{X}_1] \\
&\quad - \tfrac{1}{2}k\psi_{u2}(X_2)\psi^{\mathrm{T}}_{u2}(X_2)\mathbf{q} - \text{diag}[\omega^2]\mathbf{q} + \tfrac{1}{2}kc_2\psi_{u2}(X_2)
\end{aligned} \tag{28}$$

Similarly, if the left damper is in stick phase, and the right damper is in slip phase, one can get the equation of the new system in the modal coordinator as

$$\begin{aligned}
\Big[\mathbf{I} + \tfrac{1}{2}m\mathbf{g}&(X_2)\psi^{\mathrm{T}}_{w2}(X_2) + \tfrac{1}{2}m\psi_{u2}(X_1)\psi^{\mathrm{T}}_{u2}(X_1) + \tfrac{1}{2}m\psi_{w2}(X_1)\psi^{\mathrm{T}}_{w2}(X_1)\Big]\ddot{\mathbf{q}} \\
&= F\psi_{w1}(Y_0) + N_0[\mathbf{g}(X_2) + \psi_{w2}(X_1)] \\
&\quad - \tfrac{1}{2}m\mathbf{g}(X_2)[2\dot{\mathbf{q}}^{\mathrm{T}}\psi'_{w2}(X_2)\dot{X}_2 + \mathbf{q}^{\mathrm{T}}\psi''_{w2}(X_2)\dot{X}_2^2 + \mathbf{q}^{\mathrm{T}}\psi'_{w2}(X_2)\ddot{X}_2] \\
&\quad - \tfrac{1}{2}k\psi_{u2}(X_1)\psi^{\mathrm{T}}_{u2}(X_1)\mathbf{q} - \text{diag}[\omega^2]\mathbf{q} + \tfrac{1}{2}kc_1\psi_{u2}(X_1)
\end{aligned} \tag{29}$$

where $\mathbf{I}$ is identity matrix of appropriate dimension. Equations (11)–(14) and (26)–(29) must be solved simultaneously to obtain the dynamic responses of the two dampers and the blade-platform structure.

Please note that in Equations (26)–(29), X_1 and X_2 are unknown and vary with time t, so they represent a time-varying system. Additionally, X_1 and X_2 are nonsmooth because of the discontinuous

Coulomb's law of friction used, and the dynamic system is highly nonlinear. As a result, small time steps must be used. The precise time instants when stick regime switches to slip regime must be captured, and vice versa. These make numerical solutions of the system quite challenging.

A computational scheme for the numerical implementation in Matlab is shown in Figure 7.

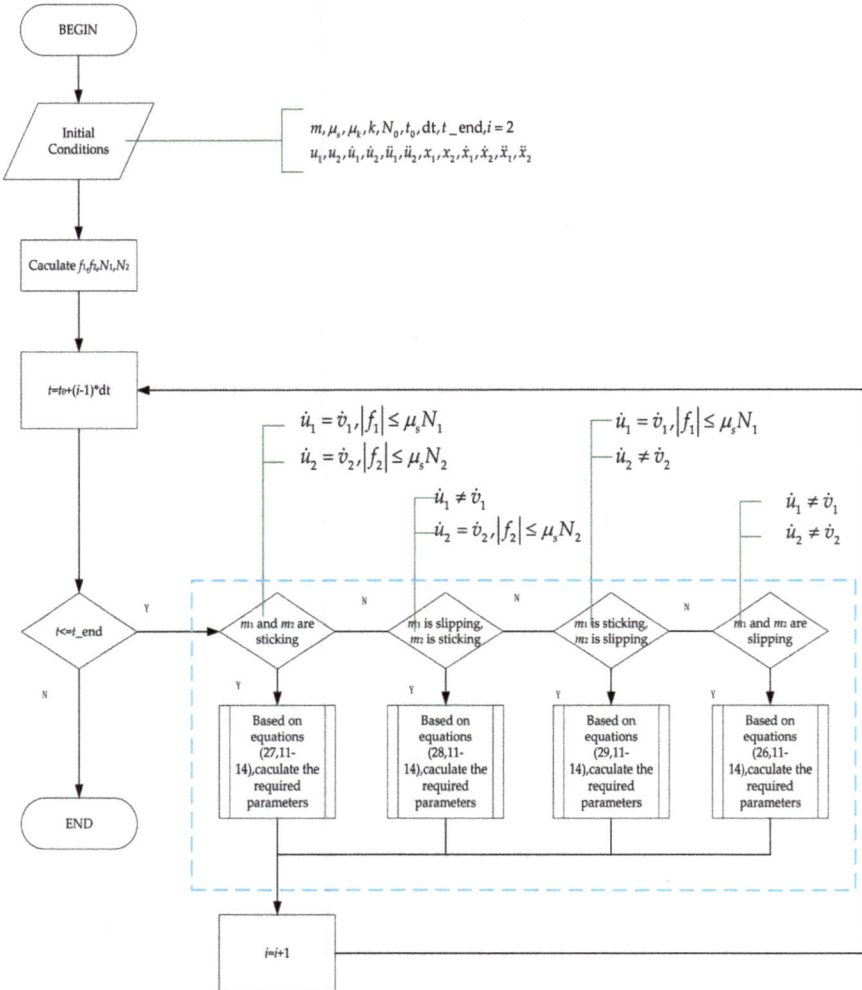

Figure 7. Flowchart for the numerical implementation in Matlab.

3. Numerical Simulation

As the state of the system switches between stick and slip phases, and the motions of the blade-platform structure and the two dampers are coupled, the dynamic behavior of the blade-damper system needs to be obtained by solving four different sets of governing equations of the blade-platform structure together with four governing equations of the two dampers at the same time, which brings about some difficulties in the numerical computations. A numerical integration scheme implementing Runge–Kutta algorithm appropriate for the second-order differential equations coded in MATLAB and capable of dealing with nonsmooth friction and contact behavior is developed to solve them in this paper. The states of the blade-platform structure and the two dampers during vibration, including values of the contact normal forces and tangential friction forces, and the forces on the spring and the two dampers, are monitored at each time step. If the results at the end of a time step do not satisfy the conditions for the system to stay in the same motion phase as at the start of the time step, then the bisection method is used to find the critical point where the dynamics switches from one phase to another phase. After getting the critical point, the current set of equations of motion changes to another set.

In this numerical example, the external excitation frequency is fixed as 200 rad/s and the amplitude is fixed at 20 N. The time step used in the numerical integrations is 0.0001 s. The simulation results have been visualized by means of time histories, frequency spectra and phase-plane plots. Due to the limited space, only some distinct and interesting results are presented. The basic parameter values used are: $k = 1.0 \times 10^4$ N/m; $\Omega = 50$ rad/s; $r = 0.46$; $E_1 = E_2 = 2.06 \times 10^{11}$ Pa; $\mu_s = 0.35$; $\mu_k = 0.3$; $\rho_1 = \rho_2 = 7800$ kg/m³; $A_1 = 6 \times 10^{-5}$ m²; $A_2 = 6 \times 10^{-5}$ m²; $I_1 = 2 \times 10^{-11}$ m⁴; $I_2 = 2 \times 10^{-11}$ m⁴; $L_1 = 0.2$ m; $L_2 = 0.04$ m; L_1 and L_2 are the length of the blade and the platform, respectively. The first natural frequency of the blade-platform structure alone is 32 Hz, while the first natural frequency of the blade platform structure and the two dampers with their cantilevers together as one system is 45.7 Hz.

3.1. Gross Slip Regime

Figures 8–11 show the results when the preload N_0 is 1 N. The steady-state response and frequency spectrum of the blade tip are illustrated in Figure 8, and the normal contact forces and friction forces versus time t curves and frequency spectrums of the normal contact forces and friction forces are given in Figure 9, and the results for the motion and frequency spectrums of the dampers with dry friction are given in Figure 10. Phase plane plots of the relative velocity of the contact points and displacements of the two dampers are shown in Figure 11. It can be seen that the forced vibration of the blade tip is periodic and has the same frequency as the excitation force (Figure 8). This is not unexpected since during gross sliding, the amplitudes of the normal contact forces and the friction forces are fairly small in comparison with the amplitude of excitation, and therefore their effect on blade tip vibration is small so that the blade tip vibration only contains one frequency component (the driving frequency). Figure 9a,b demonstrates that the normal contact forces fluctuate around the preload, and the first-order and higher-order harmonics of the driving frequency appear (Figure 9c,d), so the normal contact forces are periodic. Figure 9e,f shows that friction force amplitudes f_1 and f_2 are equal to $\mu_k N_1$ and $\mu_k N_2$, respectively, and they are also periodic (Figure 9g,h). Figure 10 shows that the motions of the two dampers are not harmonic. Effectively, the friction forces and the normal contact forces act as excitations of horizontal and vertical directions to the two dampers, respectively. Equations (11)–(14) suggest that the horizontal and vertical motions of the dampers should contain the same frequency components as the friction forces and the normal contact forces. Thus it can be concluded that the vibrations of the dampers are periodic whose period are the period of the excitation. From Figure 11, it is clear that the phase plane maps of the dampers are smooth closed curves.

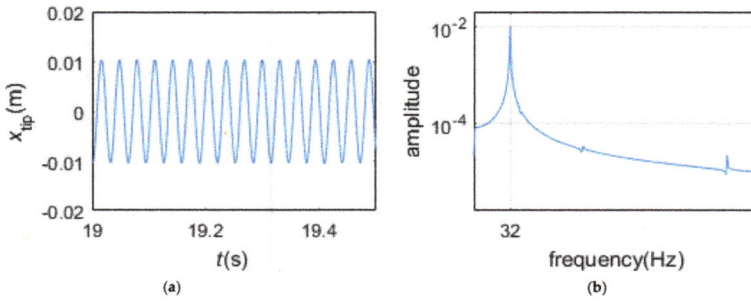

Figure 8. (**a**) The steady-state response of the blade tip; (**b**) the frequency spectrum of the blade tip.

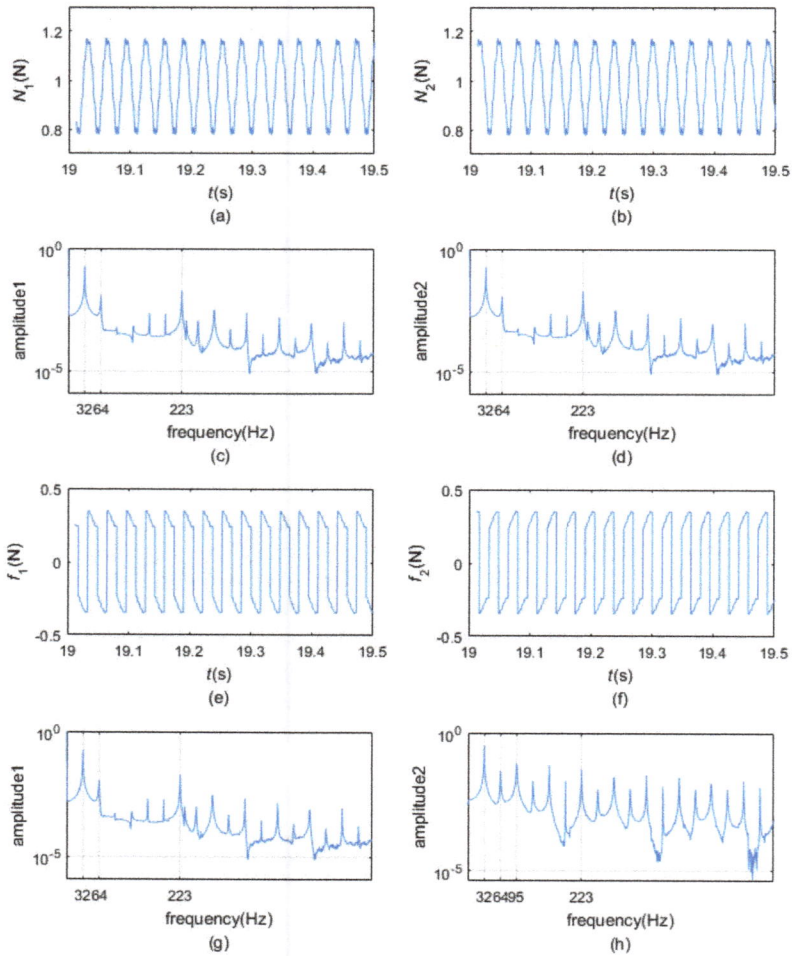

Figure 9. The normal contact forces and friction forces versus time *t* curves (**a,b,e,f**); and frequency spectrums of the normal contact forces and friction forces (**c,d,g,h**).

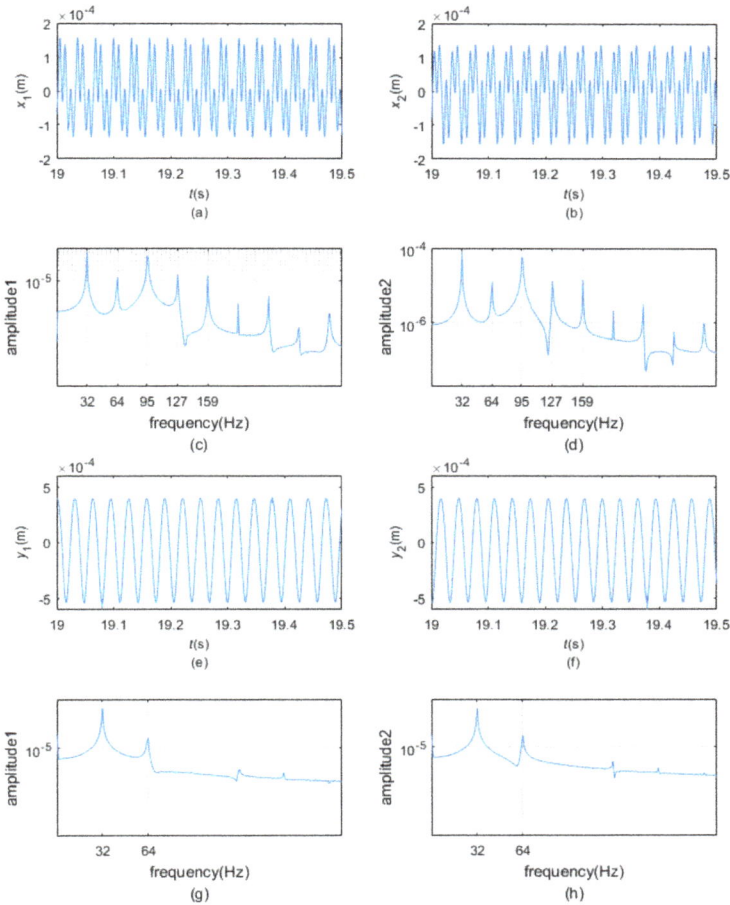

Figure 10. The motion and frequency spectrums of the dampers in the x direction (**a**, **b**, **c**, **d**); the motion and frequency spectrums of the dampers in the y direction (**e**, **f**, **g**, **h**).

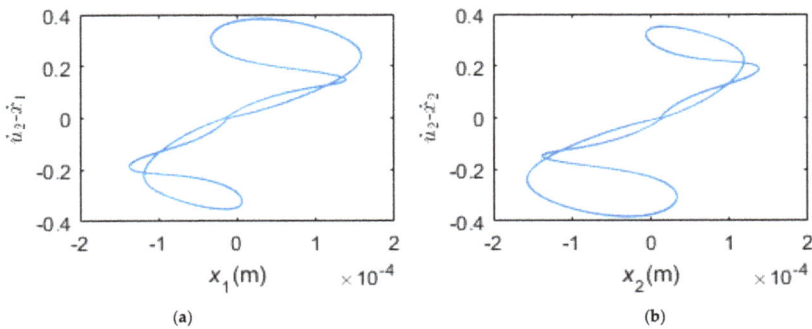

Figure 11. The relative velocity of the contact points versus displacements of the dampers, (**a**) left damper; (**b**) right damper.

3.2. Stick-Slip Regime

Figures 12–15 show the results when the preload N_0 is 12 N. Figure 12 shows the steady-state response and frequency spectrum of the blade tip, and the normal contact forces and friction forces versus time t curves and frequency spectrums of the normal contact forces are provided in Figure 13, and the results for the motion and frequency spectrums of the dampers with dry friction are given in Figure 14. Phase plane plots of the relative velocity of the contact points and displacements of the two dampers are shown in Figure 15. As Figure 12 shows, the forced vibration of the blade tip remains periodic but is no longer harmonic, and it has the same frequency as the excitation force as in the gross slip regime, but the difference here is the appearance of the higher order harmonics of the driving frequency. This is because the increase of the normal contact forces leads to horizontal stick-slip vibration of the dampers. As a result, an increase in friction forces brings about a greater contribution of its influence on blade tip vibration. As illustrated in Figure 13a,b, the normal contact forces continue to fluctuate around the preload and are periodic, and the first-order and higher-order harmonics of the driving frequency appear (see Figure 13c,d). However, the friction forces are now very interesting. Friction forces f_1 and f_2 fluctuate between $\pm\mu_s N_1$ and $\pm\mu_s N_2$, respectively, and are periodic. At times, f_1 equals $\pm\mu_k N_1$ and f_2 equals $\pm\mu_k N_2$, which is when the dampers slip to the platform. At any other time, the dampers stick relatively to the platform. Therefore, the dampers are sometimes slipping and sometimes sticking relatively to the blade platform. When the dampers stick to the platform, the relative velocity between the dampers and the blade platform is zero; on the other hand, while the dampers slip relatively to the platform, the relative velocity between the dampers and the blade platform is non-zero. These lead to the phase plane plots of Figure 15. The motions of the dampers in stick-slip regime are similar to the motions of the dampers in gross slip regime, as shown in Figure 14.

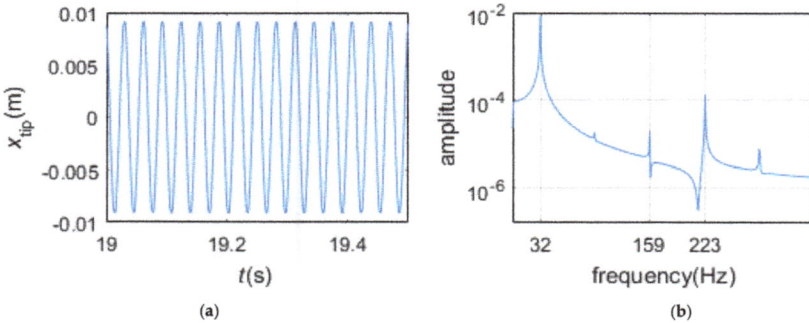

Figure 12. (a) The steady-state response of the blade tip; (b) the frequency spectrum of the blade tip.

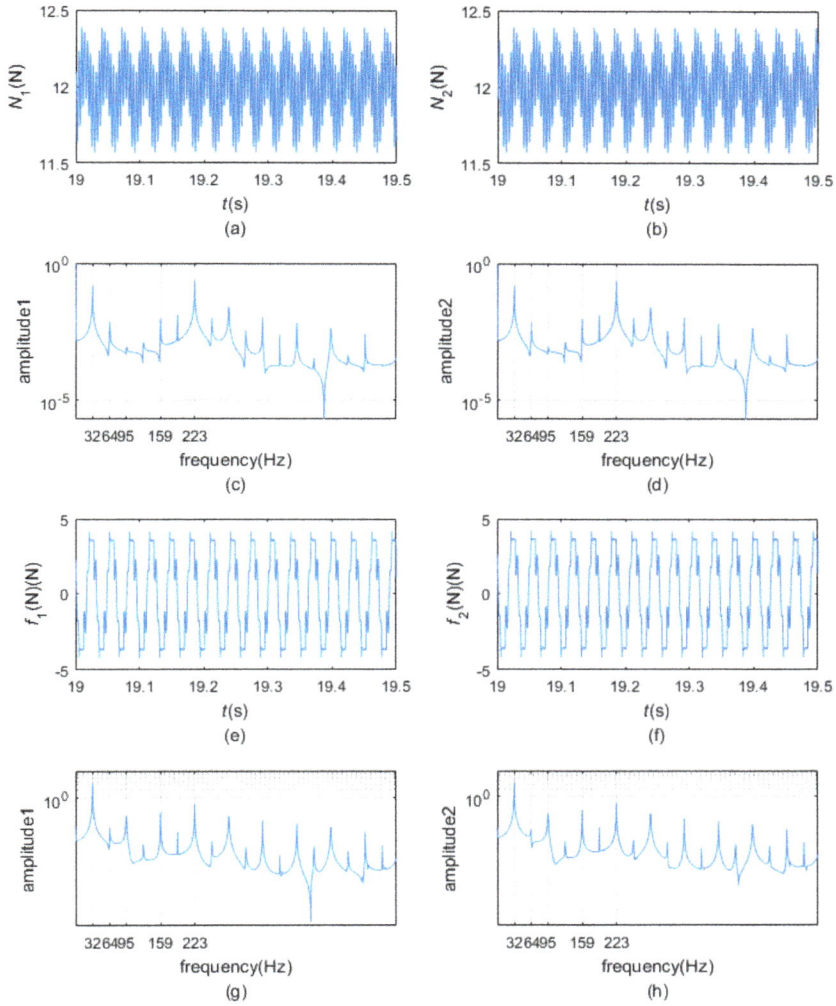

Figure 13. The normal contact forces and friction forces versus time *t* curves (**a**,**b**,**e**,**f**); and frequency spectrums of the normal contact forces and friction forces (**c**,**d**,**g**,**h**).

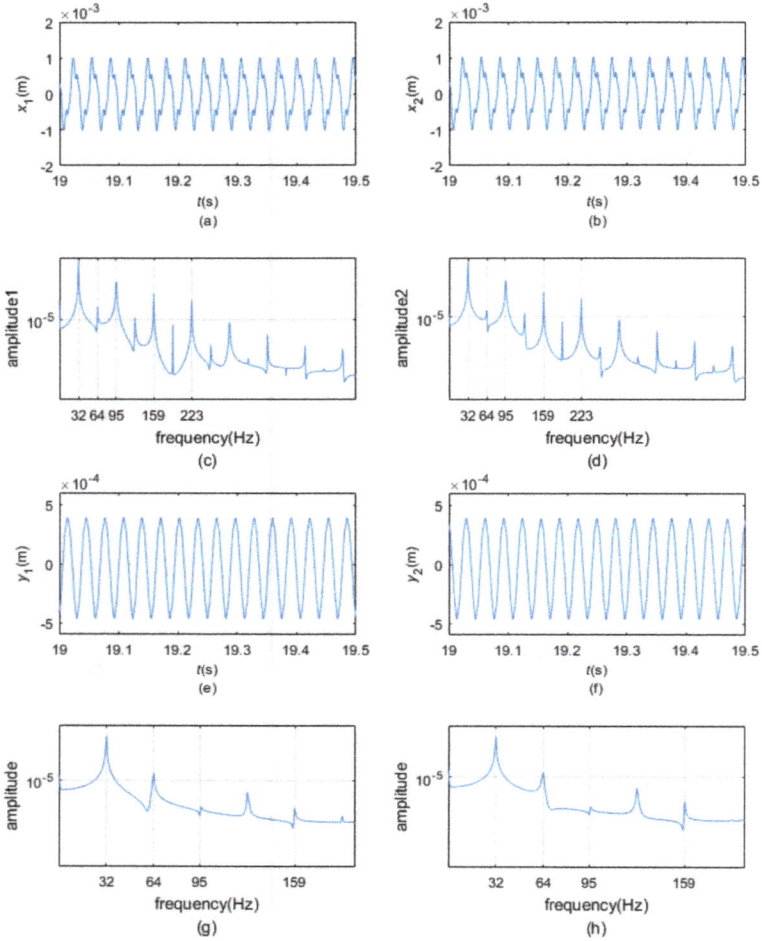

Figure 14. The motion and frequency spectrums of the dampers in the x direction (**a, b, c, d**); the motion and frequency spectrums of the dampers in the y direction (**e, f, g, h**).

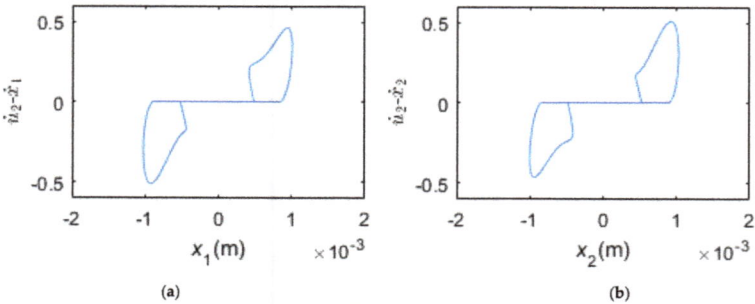

Figure 15. The relative velocity of the contact points versus displacements of the dampers, (**a**) left damper; (**b**) right damper.

3.3. Complete Stick Regime

Figures 16–19 show the results when the preload N_0 is 26 N. Figure 16 displays the steady-state response and frequency spectrum of the blade tip, the normal contact forces and friction forces versus time t curves and frequency spectrums of the normal contact forces are given in Figure 17, and the results for the motion and frequency spectrums of the dampers with dry friction are illustrated in Figure 18. Phase plane plots of the relative velocity of the contact points and displacements of the two dampers are presented in Figure 19. Figure 16 shows that the forced vibration of the blade tip is now quasi-periodic and it has three kinds of harmonic components (excitation frequency and the first-order and fifth-order bending frequency of the blade-platform structure with the two dampers and the short cantilever beams as a whole). Figure 17 indicates that the normal contact forces and the friction forces exhibit a quasi-periodic variation. Figure 19 shows that the relative velocity of the contact points of the dampers and the blade platform is always zero, hence, the dampers are always sticking to the horizontal beam. As a result, the blade-platform structure and the dampers form one new system. As shown in Figure 18, the motions of the dampers are non-periodic. There are three peaks in the frequency spectrum plots: the frequency of the first peak is the same as the excitation frequency, but the frequencies of the second peak and the third peak are now the first-order bending frequency and the fifth-order bending frequency of the whole new system.

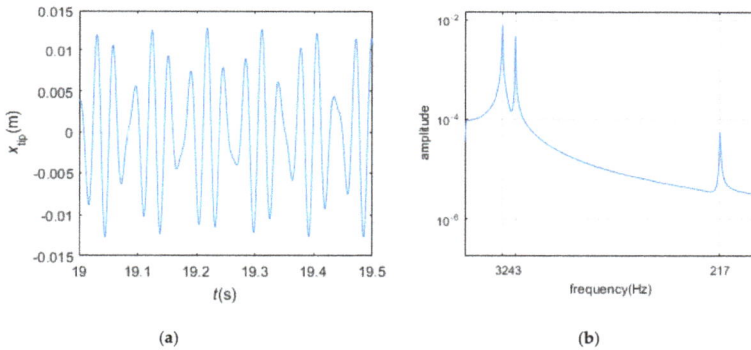

Figure 16. (a) The steady-state response of the blade tip; (b) the frequency spectrum of the blade tip.

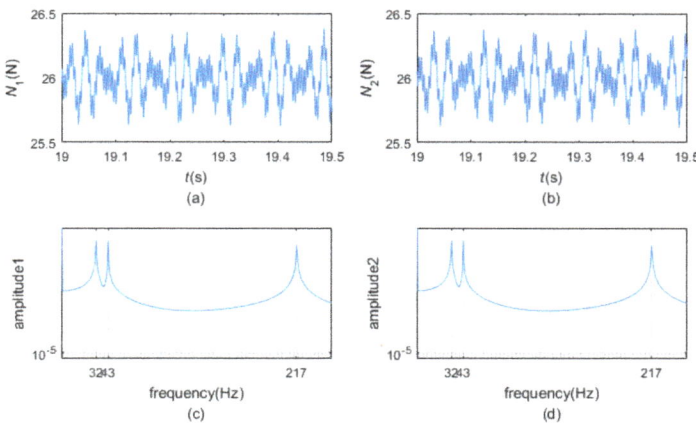

Figure 17. *Cont.*

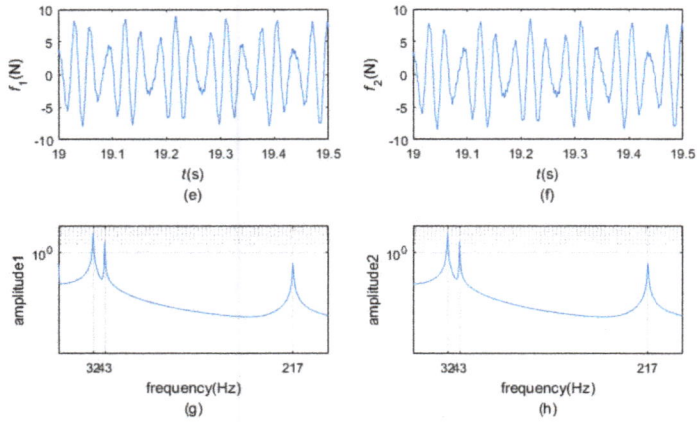

Figure 17. The normal contact forces and friction forces versus time *t* curves (**a,b,e,f**); and frequency spectrums of the normal contact forces (**c,d,g,h**).

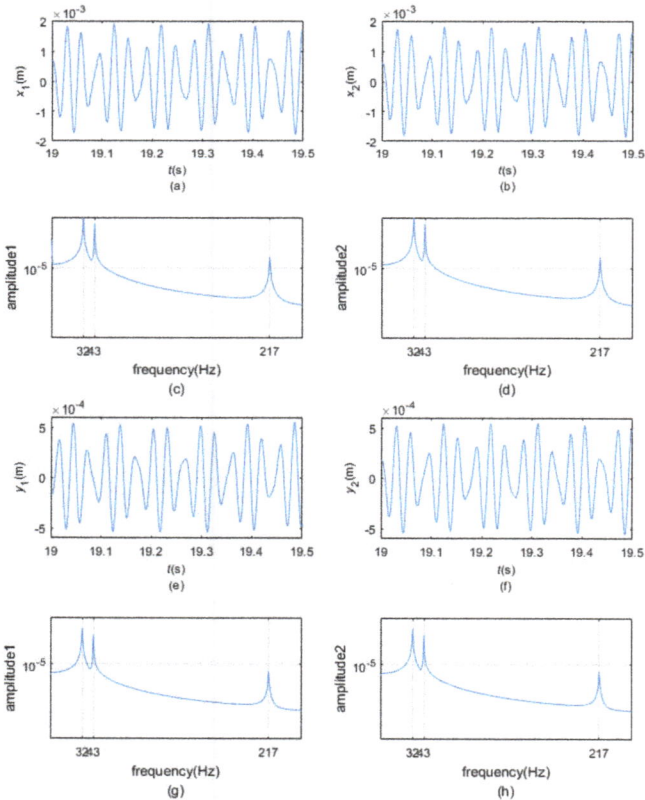

Figure 18. The motion and frequency spectrums of the dampers in the *x* direction (**a, b, c, d**); the motion and frequency spectrums of the dampers in the *y* direction (**e, f, g, h**).

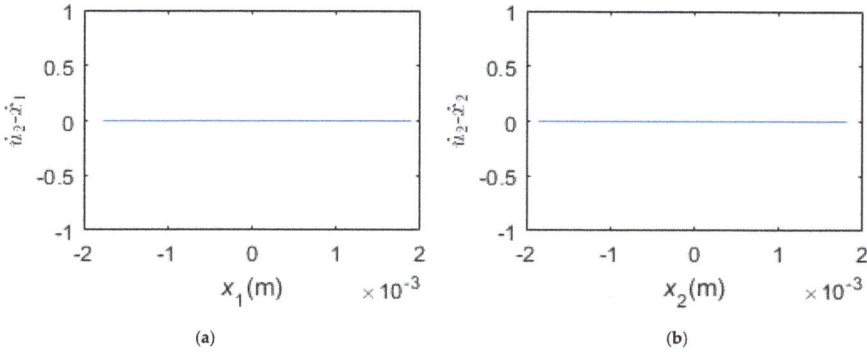

Figure 19. The relative velocity of the contact points versus displacements of the dampers, (**a**) left damper; (**b**) right damper.

3.4. The Effect of the Preload N_0 for Vibration Reduction

In order to predict the effect of the preload N_0 on the blade tip vibration reduction, a normalized energy density (E_ρ) is defined below as a measure of vibration response so that the vibration reduction effect of the damper in various conditions can be assessed. The smallest E_ρ is considered to indicate the best vibration reduction. $E_\rho = \left(\int_{t_1}^{t_2} x_{\text{tip}}(t)^2 dt \right) / (t_2 - t_1)$ ($t_1 = 19$ s; $t_2 = 19.5$ s).

As illustrated in Figure 20, with the increase of the pre-load (N_0), the normalised energy density (E_ρ) decreases continuously at first, then increases and finally stays the same. It is found that there is an optimal pre-load which can make the biggest vibration reduction. Hence, the damper mass can be optimized in order to reduce the vibration and thus dynamic stresses of a blade to the maximum extent. This could provide a useful guideline for the design of the damper. Other parameters can also affect vibration reduction but are not reported here.

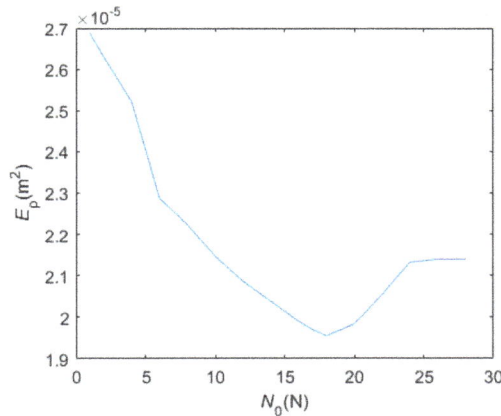

Figure 20. The pre-load versus Normalized energy density curve.

3.5. Further Discussion

As shown in the previous sub-sections, depending on the values of the preload, the dry friction dampers can display distinct dynamic behavior, and a suitable value of the preload can

Appl. Sci. **2017**, *7*, 228

be determined that would allow best energy dissipation of the dynamic system so that blade vibration can be contained.

The difference between μ_s and μ_k influences the degree of nonsmoothness of the dynamic process and is also worth studying. In this paper, a simple classical friction law is used. In reality, friction is often more complicated than what the Coulomb's law of friction can cover. More sophisticated friction laws, even state-dependent dynamic friction laws, may be needed. A comprehensive review of friction, in particular in dynamic context, can be found in [45].

Spring constant k obviously affects the horizontal motion of the damper. If it is too big, the damper would slide most time; on the other hand, if it is too small, the damper would stick most time. Thus, its value regulates the time durations of stick and slip and thus can be used as another design parameter.

The study of the last two aspects (the friction laws and the stiffness of the short vertical beams) is beyond the scope of this paper and will be conducted in near future.

4. Conclusions

This paper presents a study on the forced vibration response of a simplified turbine blade model with a new kind of under-platform dry friction dampers. It establishes a numerical approach which uses numerical modes of the structure and is capable of accommodating the discontinuous classical Coulomb's law of friction and dealing with moving loads efficiently.

In this study, both the normal contact forces and the resultant friction forces at the damper-platform interfaces are modeled as moving loads, and the normal contact force variation as a result of platform vibration is also considered. Parametric analysis demonstrates that the dampers will experience three motion regimes: gross slip regime, stick-slip regime, and pure stick regime, as the preloads at the contact points between the dampers and blade platform increase. The following conclusions can be drawn:

(1) When the pre-load N_0 is very small, the two dampers are all in gross slip regime in the steady state. The motion of the blade tip is harmonic. The normal contact forces fluctuate around the preload and are periodic, the friction forces f_1 and f_2 undulate between $\pm\mu_k N_1$ and $\pm\mu_k N_2$, respectively, and are also periodic. The motions of the two dampers are not harmonic but periodic whose periods are the period of the excitation. The phase plane maps of the dampers are smooth closed curves.

(2) When the pre-load N_0 is bigger and appropriate, the two dampers could undergo stick-slip vibration in the steady state. The motion of the blade tip is periodic rather than harmonic. The normal contact forces continue to fluctuate around the preload and are periodic as in gross slip regime. Friction forces f_1 and f_2 fluctuate between $\pm\mu_s N_1$ and $\pm\mu_s N_2$, respectively, and are periodic. At times, f_1 equals $\pm\mu_k N_1$ and f_2 equals $\pm\mu_k N_2$; at any other time, friction forces f_1 and f_2 could reach $\pm\mu_s N_1$ and $\pm\mu_s N_2$, separately. The motions of the dampers are similar to those in the gross slip regime. The phase plane maps of the dampers are closed curves but not smooth.

(3) When the pre-load N_0 is high enough, the two dampers are all in pure stick regime in the steady state. During pure stick scheme, the blade-platform structure and the dampers form one new system. The normal contact forces and the friction forces display a quasi-periodic variation, and the motion of the blade tip changes to quasi-periodic vibration as well. The motions of the dampers are non-periodic but quasi-periodic. There are three peaks in the frequency spectrum plots: the frequency of the first peak is the same as the excitation frequency, but the frequencies of the second peak and the third peak are now the first-order bending frequency and the fifth-order bending frequency of the whole new system.

(4) There is an optimal pre-load which can make the biggest vibration reduction.

Acknowledgments: The authors are grateful to the support by the National Natural Science Foundation of China (Grants No. 51405452 and No. 11672052) and "the Fundamental Research Funds for the Central Universities" (DUT16RC(3)027). Part of this work was carried out during the first author's visit to the University of Liverpool.

Author Contributions: Bingbing He and Huajiang Ouyang conceived the study. Bingbing He wrote the paper and clarified the key methods and results. Huajiang Ouyang edited the manuscript. Shangwen He and Xingmin Ren helped in simulation and data analysis.

Conflicts of Interest: The authors declare no conflict of interest.

References

1. Xie, Y.; Meng, Q. Numerical model for steam turbine blade fatigue life. *J. Xi'an Jiaotong Univ.* **2002**, *36*, 912–915.
2. Griffin, J. Friction damping of resonant stresses in gas turbine engine airfoils. *J. Eng. Gas Turbines Power* **1980**, *102*, 329–333. [CrossRef]
3. Menq, C.-H.; Bielak, J.; Griffin, J. The influence of microslip on vibratory response, part I: A new microslip model. *J. Sound Vib.* **1986**, *107*, 279–293. [CrossRef]
4. Csaba, G. *Microslip Friction Damping: With Special Reference to Turbine Blade Vibrations*; Linköping University: Linköping, Sweden, 1995.
5. Sanliturk, K.; Ewins, D. Modelling two-dimensional friction contact and its application using harmonic balance method. *J. Sound Vib.* **1996**, *193*, 511–523. [CrossRef]
6. Xia, F. Modelling of a two-dimensional Coulomb friction oscillator. *J. Sound Vib.* **2003**, *265*, 1063–1074. [CrossRef]
7. Yang, B.; Chu, M.; Menq, C. Stick–slip–separation analysis and non-linear stiffness and damping characterization of friction contacts having variable normal load. *J. Sound Vib.* **1998**, *210*, 461–481. [CrossRef]
8. Cigeroglu, E.; Lu, W.; Menq, C.-H. One-dimensional dynamic microslip friction model. *J. Sound Vib.* **2006**, *292*, 881–898. [CrossRef]
9. Cigeroglu, E.; An, N.; Menq, C.-H. A microslip friction model with normal load variation induced by normal motion. *Nonlinear Dyn.* **2007**, *50*, 609–626. [CrossRef]
10. Cigeroglu, E.; An, N.; Menq, C.-H. Wedge damper modeling and forced response prediction of frictionally constrained blades. In *ASME Turbo Expo 2007: Power for Land, Sea, and Air*; American Society of Mechanical Engineers: New York, NY, USA, 2007; pp. 519–528.
11. Allara, M. A model for the characterization of friction contacts in turbine blades. *J. Sound Vib.* **2009**, *320*, 527–544. [CrossRef]
12. Panning, L.; Popp, K.; Sextro, W.; Götting, F.; Kayser, A.; Wolter, I. Asymmetrical underplatform dampers in gas turbine bladings: Theory and application. In *ASME Turbo Expo 2004: Power for Land, Sea, and Air*; American Society of Mechanical Engineers: New York, NY, USA, 2004; pp. 269–280.
13. Michelis, S. Linear and Nonlinear Dynamics of a Turbine Blade in Presence of an Underplatform Damper with Friction. Ph.D. Thesis, University of Illinois at Chicago, Chicago, IL, USA, 2014.
14. Petrov, E. Explicit finite element models of friction dampers in forced response analysis of bladed disks. *J. Eng. Gas Turbines Power* **2008**, *130*, 022502. [CrossRef]
15. Sever, I.A.; Petrov, E.P.; Ewins, D.J. Experimental and numerical investigation of rotating bladed disk forced response using underplatform friction dampers. *J. Eng. Gas Turbines Power* **2008**, *130*, 042503. [CrossRef]
16. Gastaldi, C.; Gola, M.M. On the relevance of a microslip contact model for under-platform dampers. *Int. J. Mech. Sci.* **2016**, *115*, 145–156. [CrossRef]
17. Ostachowicz, W. The harmonic balance method for determining the vibration parameters in damped dynamic systems. *J. Sound Vib.* **1989**, *131*, 465–473. [CrossRef]
18. Guillen, J.; Pierre, C. An Efficient, Hybrid, Frequency-Time Domain Method for The Dynamics of Large-Scale Dry-Friction Damped Structural Systems. In *IUTAM Symposium on Unilateral Multibody Contacts*; Springer: Berlin/Heidelberg, Germany, 1999; pp. 169–178.
19. Firrone, C.M.; Zucca, S. Underplatform dampers for turbine blades: The effect of damper static balance on the blade dynamics. *Mech. Res. Commun.* **2009**, *36*, 515–522. [CrossRef]
20. Půst, L.; Pešek, L.; Radolfová, A. Various types of dry friction characteristics for vibration damping. *Eng. Mech.* **2011**, *18*, 203–224.
21. Schwingshackl, C.; Petrov, E.; Ewins, D. Measured and estimated friction interface parameters in a nonlinear dynamic analysis. *Mech. Syst. Signal Process.* **2012**, *28*, 574–584. [CrossRef]

22. Gola, M.M.; dos Santos, M.B.; Liu, T. Measurement of the scatter of underplatform damper hysteresis cycle: Experimental approach. In Proceedings of the ASME 2012 International Design Engineering Technical Conferences and Computers and Information in Engineering Conference, Chicago, IL, USA, 12–15 August 2012; American Society of Mechanical Engineers: New York, NY, USA, 2012; pp. 359–369.

23. Pesaresi, L.; Salles, L.; Jones, A.; Green, J.; Schwingshackl, C. Modelling the nonlinear behaviour of an underplatform damper test rig for turbine applications. *Mech. Syst. Signal Process.* **2017**, *85*, 662–679. [CrossRef]

24. Di Maio, D. SLDV Technology for Measurement of Mistuned Bladed Disc Vibration. Ph.D. Thesis, Imperial College London, London, UK, 2008.

25. Zucca, S.; Di Maio, D.; Ewins, D. Measuring the performance of underplatform dampers for turbine blades by rotating laser Doppler vibrometer. *Mech. Syst. Signal Process.* **2012**, *32*, 269–281. [CrossRef]

26. Nikhamkin, M.S.; Sazhenkov, N.; Semenova, I.; Semenov, S. The Basic Mechanisms of Turbine Dummy-Blades Assembly and Dry-Friction Dampers Interaction Experimental Investigation. In *Applied Mechanics and Materials*; Trans Tech Publications: Zurich, Switzerland, 2015; pp. 346–350.

27. Bessone, A.; Toso, F.; Berruti, T. Investigation on the Dynamic Response of Blades with Asymmetric Under Platform Dampers. In *ASME Turbo Expo 2015: Turbine Technical Conference and Exposition*; American Society of Mechanical Engineers: New York, NY, USA, 2015; p. V07BT33A003.

28. Gola, M.; Liu, T.; Dos Santos, M.B. Investigation of under-platform damper kinematics and its interaction with contact parameters (nominal friction coefficient). In Proceedings of the WTC 2013, 5th World Tribology Congress, Turin, Italy, 8–13 September 2013.

29. Rastogi, V.; Kumar, V.; Bhagi, L.K. Dynamic Modeling of Underplateform Damper used in Turbomachinery. *Int. Sch. Sci. Res. Innov.* **2012**, *6*, 460–469.

30. Bessone, A.; Traversone, L. Simplified Method to Evaluate the "Under Platform" Damper Effects on Turbine Blade Eigenfrequencies Supported by Experimental Test. In *ASME Turbo Expo 2014: Turbine Technical Conference and Exposition*; American Society of Mechanical Engineers: New York, NY, USA, 2014; p. V07BT33A005.

31. Pesaresi, L.; Salles, L.; Elliott, R.; Jones, A.; Green, J.S.; Schwingshackl, C.W. Numerical and Experimental Investigation of an Underplatform Damper Test Rig. *Appl. Mech. Mater.* **2016**, *849*, 1–12. [CrossRef]

32. Lu, X.X.; Huang, S.H.; Liu, Z.Q.; Li, L.P.; Deng, X.H. Study on contact-impact damping characteristics of shrouded blades based on harmonic balance method. *J. Chin. Soc. Power Eng.* **2010**, *30*, 578–583.

33. Ramirez, J.; Tirca, L. Numerical Simulation and Design of Friction-Damped Steel Frame Structures damped. In Proceedings of 15th World Conference in Earthquake Engineering, Lisbon, Portugal, 24–28 September 2012.

34. Kang, D.W.; Jung, S.W.; Nho, G.H.; Ok, J.K.; Yoo, W.S. Application of Bouc-Wen model to frequency-dependent nonlinear hysteretic friction damper. *J. Mech. Sci. Technol.* **2010**, *24*, 1311–1317. [CrossRef]

35. Chandiramani, N.K.; Srinivasan, K.; Nagendra, J. Experimental study of stick-slip dynamics in a friction wedge damper. *J. Sound Vib.* **2006**, *291*, 1–18. [CrossRef]

36. Popp, K.; Panning, L.; Sextro, W. Vibration damping by friction forces: Theory and applications. *J. Vib. Control* **2003**, *9*, 419–448. [CrossRef]

37. Pascal, M. Sticking and nonsticking orbits for a two-degree-of-freedom oscillator excited by dry friction and harmonic loading. *Nonlinear Dyn.* **2014**, *77*, 267–276. [CrossRef]

38. Li, Z.; Ouyang, H.; Guan, Z. Friction-induced vibration of an elastic disc and a moving slider with separation and reattachment. *Nonlinear Dyn.* **2017**, *87*, 1045–1067. [CrossRef]

39. Olsson, H.; Åström, K.J.; De Wit, C.C.; Gäfvert, M.; Lischinsky, P. Friction models and friction compensation. *Eur. J. Control* **1998**, *4*, 176–195. [CrossRef]

40. Rao, S.S. *Vibration of Continuous Systems*; John Wiley & Sons: Hoboken, NJ, USA, 2007.

41. Han, S.M.; Benaroya, H.; Wei, T. Dynamics of transversely vibrating beams using four engineering theories. *J. Sound Vib.* **1999**, *225*, 935–988. [CrossRef]

42. Boeraeve, P. *Introduction to the Finite Element Method (FEM)*; Institut Gramme: Liege, Belgium, 2010; pp. 2–68.

43. Ouyang, H. Moving-load dynamic problems: A tutorial (with a brief overview). *Mech. Syst. Signal Process.* **2011**, *25*, 2039–2060. [CrossRef]

44. Baeza, L.; Ouyang, H. Vibration of a truss structure excited by a moving oscillator. *J. Sound Vib.* **2009**, *321*, 721–734. [CrossRef]

45. Berger, E. Friction modeling for dynamic system simulation. *Appl. Mech. Rev.* **2002**, *55*, 535–577.

applied
sciences

MDPI

Article

Preliminary Study on the Damping Effect of a Lateral Damping Buffer under a Debris Flow Load

Zheng Lu [1,2], Yuling Yang [2], Xilin Lu [1,2] and Chengqing Liu [3,*]

[1] State Key Laboratory of Disaster Reduction in Civil Engineering, Tongji University, Shanghai 200092, China; luzheng111@tongji.edu.cn (Z.L.); lxlst@tongji.edu.cn (X.L.)
[2] Research Institute of Structural Engineering and Disaster Reduction, Tongji University, Shanghai 200092, China; yangyl_tj@163.com
[3] School of civil engineering, Southwest JiaoTong University, Chengdu 610031, China
* Correspondence: lcqtj@163.com; Tel.: +86-21-6598-6186

Academic Editors: Gangbing Song, Steve C.S. Cai and Hong-Nan Li
Received: 29 December 2016; Accepted: 13 February 2017; Published: 20 February 2017

Abstract: Simulating the impact of debris flows on structures and exploring the feasibility of applying energy dissipation devices or shock isolators to reduce the damage caused by debris flows can make great contribution to the design of disaster prevention structures. In this paper, we propose a new type of device, a lateral damping buffer, to reduce the vulnerability of building structures to debris flows. This lateral damping buffer has two mechanisms of damage mitigation: when debris flows impact on a building, it acts as a buffer, and when the structure vibrates due to the impact, it acts as a shock absorber, which can reduce the maximum acceleration response and subsequent vibration respectively. To study the effectiveness of such a lateral damping buffer, an impact test is conducted, which mainly involves a lateral damping buffer attached to a two-degree-of-freedom structure under a simulated debris flow load. To enable the numerical study, the equation of motion of the structure along with the lateral damping buffer is derived. A subsequent parametric study is performed to optimize the lateral damping buffer. Finally, a practical design procedure is also provided.

Keywords: debris flow; shock absorbers; buffer; impact test; numerical simulation; parametric study

1. Introduction

Debris flows can cause serious damage to infrastructures and threaten lives. There have been a considerable number of debris flows involving the destruction of mountain villages and small towns, such as the disasters which occurred in Venezuela in December 1999 [1] and in Zhouqu, China, in August 2010 [2]. Mountainous areas with low vegetation cover and up-and-down landform are extremely vulnerable to debris flows. Therefore, buildings in these mountainous areas are prone to damage by debris flows. Actually, buildings play an important role in resisting debris flows. Hence, drastic measures are urgent to ensure buildings withstand such kinds of disasters.

Studies on buildings against debris flows specifically are relatively deficient; however, there is still some research involving mitigation measures to reduce the damage by debris flows and the interaction between debris flows and buildings [3,4]. Debris flow mitigation measures can be categorized into two classifications, namely engineering measures and non-engineering measures [5]. Meanwhile, engineering measures include drainage systems [6,7], check dams [8–10], and flexible barriers [11], etc., while non-engineering measures involve disaster warning and evacuation systems, retrofitting and reinforcement of buildings [12], etc. In this paper, the main goal is to propose some specific structural measures to reduce the damage to buildings caused by debris flows from the perspective of building retrofitting.

It is known that debris flow is characterized by high velocity, strong striking force and severe destruction [13]. Therefore, if prevention measures are based on reinforcement, the construction cost would be drastically high, while the safety of buildings still cannot be guaranteed for the randomness of debris flows. Therefore, people should seek a method to retrofit buildings with some devices or equipment of a lower cost. On the other hand, the application of structural vibration control in aseismic engineering field provides a new solution to this problem. In civil engineering, the systematic knowledge of structural vibration control was first introduced by Yao [14], and structural control plays an important role in engineering nowadays [15–17]. It is a typical method to attach devices [18] or to study the failure modes [19] to prevent structures from being destroyed under earthquake. Moreover, in the field of vehicle control, the semi-active suspension [20–23] and active suspension [24] play an important role in guaranteeing both safety and comfort of cars. In this way, devices or equipment suitable for different structures or applicable to different scenarios (e.g., [25–30]) are put forward to reduce the structure's response under different dynamic loads.

Enlightened by various structural control methods [25–30] and semi-active suspensions [20–23] as well as active suspensions [24], a lateral damping buffer is proposed to retrofit buildings in debris flow-prone areas. Specifically, it is a combination of traditional buffer and shock absorber. By attaching such devices to existing buildings, the structures' debris flow resistance can be enhanced a lot, especially when the loading level is high; the maximum acceleration response can be reduced by nearly half. In addition, the construction is fairly simple because the buffer can be directly attached to the existing structure without large-scale reconstruction. To examine the performance of the lateral damping buffer, an impact test of such a device attached to a two-degree-of-freedom (DOF) structure under a simulated debris flow load is completed. In China, there are mainly three types of domestic structures in mountain regions of the western area, namely, reinforced framed structures, masonry structures, and reinforced masonry structures [31]. In this paper, we take a frame structure as an example to study the effect of the buffer. Then, a numerical simulation is performed to validate its rationality. Finally, based on parametric study, a practical design procedure is provided.

2. Schematic of a Lateral Damping Buffer

A schematic of the lateral damping buffer is shown in Figure 1. Two pieces of board are connected by springs, acting together as a cushion to the impact. In addition, eight boxes with eight particles uniformly adhere to one of the boards, whose main function is to enlarge the damping as energy-dissipating dampers. Meanwhile, the diameter of the particles is 12 mm, and each particle weighs 7 g. The size of the boxes is 25 mm × 25 mm × 18 mm, and each one weighs 6 g.

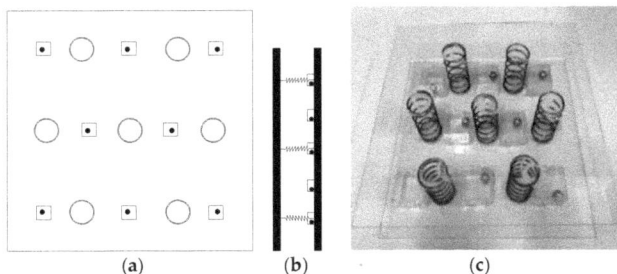

Figure 1. A lateral damping buffer: (**a**) Front view; (**b**) Side view; (**c**) Model picture.

As mentioned above, a lateral damping buffer takes effective roles in two ways. Firstly, it is installed in the impact point, and works as a cushion. In this way, the lateral damping buffer can reduce the peak intensity of applied stresses effectively at the beginning of impact. Then, the particles collide with the boxes to enlarge the damping coefficient of the system, so that the subsequent vibration of structures can be further reduced. On the whole, the lateral damping buffer reduces the response of structures subjected to debris flow impacts by decreasing both the peak value and the mean value.

3. Experiment Validation

3.1. Experiment Design

In the experiment, responses of an uncontrolled structure and a structure with the lateral damping buffer under simulated debris flows are measured to make a detailed comparison. For the buffer case, the lateral damping buffer is installed at the potential impact point, as is shown in Figure 2. In this way, the effectiveness of the lateral damping buffer can be examined. In most mountainous areas of China, the buildings to be hit are typically reinforced framed structures, masonry structures or reinforced masonry structures. In this paper, a two-degree-of-freedom frame structure is used as a primary structure, whose mass is 0.9 kg, with added mass weighing 6 kg evenly placed on the two layers. The first and second order frequencies of the structure are 3.56 Hz and 14.07 Hz, and its damping ratio is 4.0% [32].

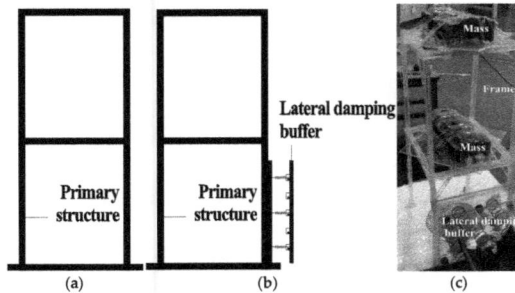

Figure 2. Schematic diagram of the test: (**a**) Uncontrolled structure; (**b**) Structure with the lateral damping buffer; (**c**) Experimental model picture.

A test site is designed according to the situation when debris flows break through and the test operability, shown in Figure 3. The loading device includes a slope with an angle of 30° on the horizontal plane and a fixed mount. Referring to other experts' experience, the load of debris flows is simulated by an impact force caused by a ball rolling down from a slope [33–35]. There is one accelerometer placed at the top of the test model to record the structural response.

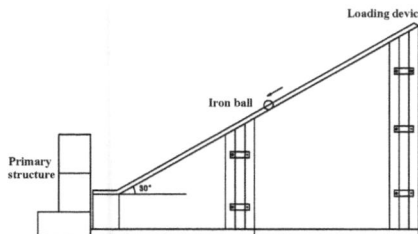

Figure 3. Experimental device.

In the test, the impact load is exerted on the primary structure when the steel ball weighing 0.51 kg rolls down from different heights. The loading procedures are divided into ten grades. The relative height of the first loading is 0.05 m, and the height difference between two loading grades is 0.05 m. Response of the structure is measured by accelerometers.

3.2. Responses in Time Domain

Figure 4 shows a typical acceleration time history curve of the structure with and without a buffer. It can be found that the response of the structure with the lateral damping buffer is smaller than that of the uncontrolled structure, in which the peak value of the acceleration is reduced evidently. The maximal acceleration response appears at the beginning of the time history curves. As is mentioned above, the lateral damping buffer can be regarded as a combination of a shock absorber and a buffer. Given that energy dissipation devices need time to work, when the maximum value of acceleration response appears, the shock absorbers in the device have not functioned yet. Therefore, the buffer of the device makes a major contribution to the reduction of the maximal acceleration response of the primary structure.

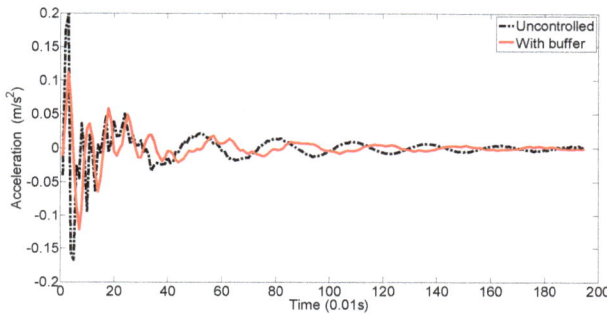

Figure 4. Time history curves of structure with/without a lateral damping buffer (grade 7).

To study the effectiveness of the damping buffer in reducing the subsequent vibration of the primary structure after impact, root mean square (r.m.s) response is taken as a measurement index. Furthermore, the influence of the buffer can be eliminated by removing the wavelengths where the peak value was.

The response reduction effects of both the maximum acceleration and the root mean square acceleration are shown in Figure 5. It can be seen that the responses of both the maximum acceleration and the root mean square acceleration of the structure with the lateral damping buffer are reduced largely. In addition, the response reduction effect is relatively small when the loading grade is low, while the response reduction effects of both the maximum acceleration and the root mean square acceleration increase step by step as the loading grade increases. That is, the larger the impact force, the better the effectiveness of the device, which is obviously beneficial to the practical engineering application.

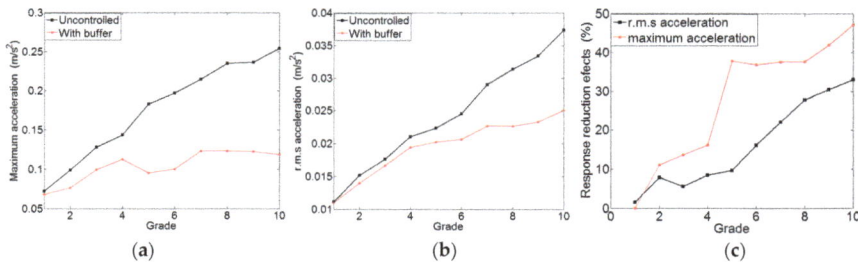

Figure 5. Acceleration response and the reduction efficiency: (**a**) Maximum acceleration; (**b**) Root mean square (r.m.s) acceleration; (**c**) Response reduction effects.

3.3. Responses in Frequency Domain

Figure 6 shows the responses of the uncontrolled structure and the structure with buffer in the frequency domain under the loading cases of grade 4 and grade 6.

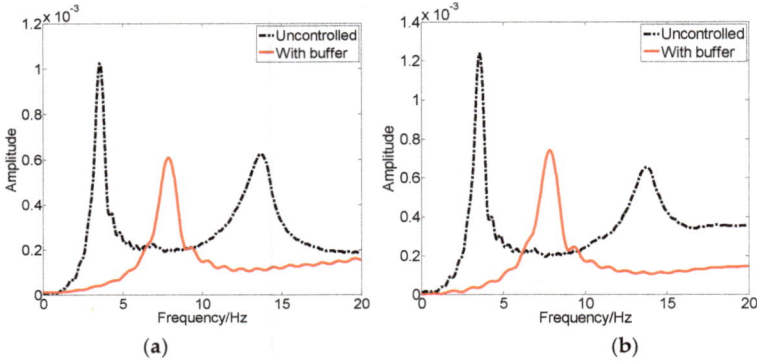

(a)   (b)

Figure 6. The frequency domain response: (a) Grade 4; (b) Grade 6.

From the figure, it can be seen that:

(1) The Fourier spectrum of the acceleration time history curve of the uncontrolled structure has two obvious peak values in the vicinity of 3.6 Hz and 14 Hz respectively, which correspond to the measured natural frequencies. However, the Fourier spectrum of the structure with buffer has just one peak value in the vicinity of 8 Hz, since the attachment of the buffer has strengthened the integrity of the structure.

(2) As for the amplitude, the response of the structure with a buffer is clearly reduced. That is, the vibration of the structure is under control and the lateral damping buffer has good effects.

(3) The Fourier spectral lines show the distribution of the vibration power of the primary structure in the frequency domain. The area under the Fourier spectral line of the structure with the buffer is smaller than that of the uncontrolled one, which shows the lateral damping buffer can greatly decrease the vibration energy of structures.

(4) Compared the response under fourth loading grade and sixth loading grade, the latter is evidently smaller than the former, indicating that with increasing impact, the buffering and vibration controlling effects of the device will increase.

3.4. Equivalent Damping Ratio

According to the test, the equivalent damping ratios are calculated based on its definition [32] (shown in Table 1). The equivalent damping ratio of the uncontrolled structure is about 4.0%, while the equivalent damping ratio of the structure with lateral damping buffer is 5.2%–7.5%, with improvement being 30.0%–87.5%. The lateral damping buffer has control effectiveness on account of the subsequent structural vibration.

Table 1. Equivalent damping ratio (%).

Loading Grades	1	2	3	4	5	6	7	8	9	10
Uncontrolled	4.0	4.0	4.0	4.0	4.0	4.0	4.0	4.0	4.0	4.0
With isolators	5.2	6.2	6.1	6.2	5.4	6.1	6.7	7.5	7.0	6.3
Improvement	30.0	55.0	52.5	55.0	35.0	52.5	67.5	87.5	75.0	57.5

4. Numerical Simulation

As introduced in Section 2, a lateral damping buffer can be regarded as a combination of a buffer and a shock absorber. Meanwhile, the element of the buffer functions in a remarkably short period of time, mainly in the vicinity of the moment of impact. Specifically, the peak value of the acceleration of the controlled structure is reduced greatly compared to the uncontrolled structure. In this limited timespan, the shock absorber has not come into action yet because the increase of the damping coefficient by the collision of the boxes and particle takes time. Then, after impact, the portion of the shock absorber functions step by step and reduces the subsequent vibration of the primary structure gradually. In summary, the two roles of the lateral damping buffer take effect in different periods of time: the buffer portion works in a short time in the vicinity of the impact, while the shock absorber portion works mainly in the process of the structural vibration after impact. Consequently, the effectiveness of the two parts can be considered respectively.

To simulate the whole process of the structural vibration reasonably, the process of the response of the structure is divided into two parts: the maximum acceleration and the root mean square of the subsequent response.

4.1. The Cushion Phase

In this section, the buffer action of the device is the main consideration and the calculation of peak acceleration is the major subject. The structure is simulated as an entirety with infinite rigidity of the beam assumption and the steel ball is simplified to a particle. In addition, the buffer is simulated by a contact spring. The mechanical model is shown in Figure 7. The equations of motion can be established on the basis of the *D'alembert principle*.

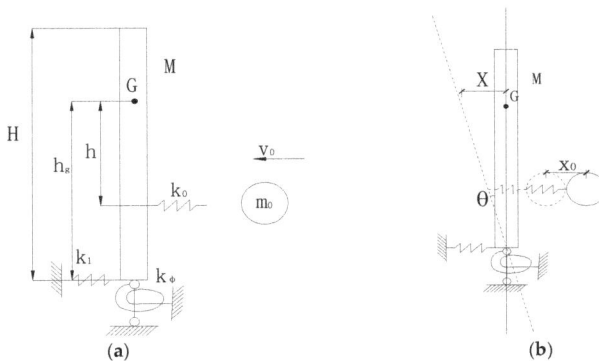

Figure 7. Mechanical model: (**a**) Geometric parameters; (**b**) Motion parameters.

In the system shown in Figure 7, m_0 and M represent the impact ball and the primary structure respectively, k_0, k_1, and k_ϕ denote the stiffness of the contact spring, the lateral bracing rigidity and the anti-rotation stiffness of the structure respectively, H, h_g, h stand for the total height of the structure, the height of the center of gravity from the bottom of the structure and the difference of height between the center of gravity and the contact spring respectively, and J means inertia rotation.

In the process of the impact, the displacement of the steel ball m_0 is represented by x_0, the displacement of the primary structure is represented by X, the rotational angle at the bottom of the structure is represented by θ, and the bottom lateral translation of the structure is small enough to be neglected. Meanwhile, the influence of the damping is not considered because the peak value appears at the very beginning of the process of the response and the damping has not come into play to reduce the vibration. The equation of motion is given as:

$$m_0\ddot{x}_0 + k_0[x_0 - (x_1 - h\theta)] = 0$$
$$M\ddot{X} + k_0[x_0 - (x_1 - h\theta)] = 0 \tag{1}$$
$$J\ddot{\theta} + k_\varphi \varphi + k_0 h[x_0 - (x_1 - h\theta)] = 0$$

The velocities v_0 of the steel ball in different loading grades can be calculated by the theorem of kinetic energy (shown in Table 2). Therefore, the equation of motion can be solved and the simulated peak values of the acceleration response of the structure in different loading grades can be obtained.

Table 2. Impact velocities of the iron ball.

Loading Grades	1	2	3	4	5	6	7	8	9	10
Height (m)	0.05	0.1	0.15	0.2	0.25	0.3	0.35	0.4	0.45	0.5
Velocity (m/s)	0.77	1.08	1.33	1.53	1.71	1.88	2.03	2.17	2.3	2.42

4.2. The energy Dissipation Phase

To study the vibration damping performance of the shock absorber portion, the root mean square response of the subsequent waves is calculated. The structure with the lateral damping buffer is simplified as a 3-DOF model (the primary structure is 2-DOF and the lateral damping buffer is the third DOF), and the equation of motion is established. The waveband where the peak value is located has been obtained in Section 4.1. Therefore, the velocity and the acceleration at the end of the mentioned waveband can be regarded as initial conditions to solve the equation.

The equation of the motion is given as:

$$[K]\{x\} + [C]\{\dot{x}\} + [M]\{\ddot{x}\} = 0 \tag{2}$$

where stiffness matrix $[K] = \begin{bmatrix} k_1 + k_2 + k_3 & -k_2 & -k_3 \\ -k_2 & k_2 & 0 \\ -k_3 & 0 & k_3 \end{bmatrix}$; mass matrix $[M] = \begin{bmatrix} m_1 & 0 & 0 \\ 0 & m_2 & 0 \\ 0 & 0 & m_3 \end{bmatrix}$;

damping matrix $[C]$ is determined according to the Rayleigh damping assumption. m_1, m_2, m_3 represent the mass of the first storey and the second storey of the primary structure, as well as the mass of the lateral damping buffer, respectively; k_1, k_2, k_3 represent the stiffness of the first storey and the second storey of the primary structure, as well as the stiffness of the lateral damping buffer, respectively, and $\{x\}$, $\{\dot{x}\}$, $\{\ddot{x}\}$ represent the displacements, velocities and accelerations of both the primary structure and the lateral damping buffer given by the impact.

4.3. Calculation Parameters

In the test, the primary structure with added mass weighs 6.9 kg totally, so $m_1 = m_2 = 3.45$ kg, and the mass of the lateral damping buffer $m_3 = 0.33$ kg. The first and second order frequencies of the structure are $f_1 = 3.56$ Hz and $f_2 = 14.07$ Hz. The stiffness of the structure is $k_1 = k_2 = 10085$ N/m. Moreover, the stiffness of the buffer is obtained by measurement, $k_3 = 2321.4$ N/m.

The damping matrix [C] is determined according to the Rayleigh damping assumption

$$[C] = \begin{bmatrix} 35.25 & -9.14 & -1.65 \\ -9.14 & 24.46 & 0 \\ -1.65 & 0 & 3.11 \end{bmatrix} \text{N·s·m}^{-1}$$

4.4. Calculation Results

In this paper, the numerical simulation is accomplished. The experimental maximum acceleration and the calculated maximum acceleration are shown in Figure 8. The curves of experimental and calculated acceleration time histories are shown in Figure 9. It can be seen that the calculated values agree well with the experimental results. Extraordinarily, the calculated peak value shows high degree of fit. Considering the debris flow characteristics of high velocity, strong striking force and severe destruction, the study aimed at peak value of the response is more meaningful.

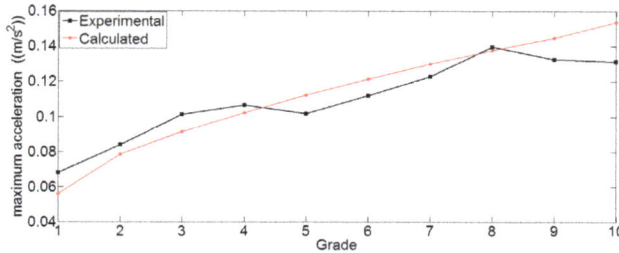

Figure 8. The comparison of the experimental and calculated maximum acceleration.

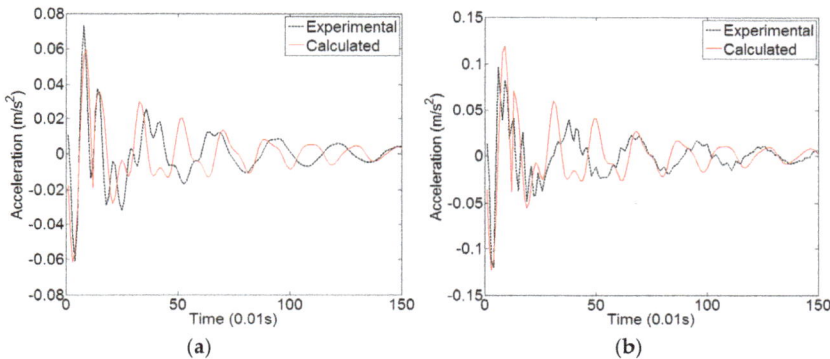

Figure 9. The comparison of the experimental and calculated time history curves: (**a**) Grade 2; (**b**) Grade 8.

4.5. Parametric Study

To study the influence of different parameters and obtain reasonable design of the lateral damping buffer, the effects of its contact stiffness, damping ratio and the mass ratio are discussed.

4.5.1. Contact Stiffness

To study the influence of contact stiffness on the effectiveness of the lateral damping buffer, the response of the primary structure can be calculated by the methods above with different contact stiffness. The initial velocity is taken as the velocity in grade 5, and the contact stiffness ranges from 1000 N/m to 3000 N/m. Then, the peak values of the structural response (as shown in Figure 10) are calculated by the methods mentioned in Section 4.1.

From Figure 10, it can be seen that the maximum acceleration increases with the enlargement of the contact stiffness. In other words, the decrease of the contact stiffness can improve the reduction performance, but there is a deformation limit for springs in reality. Therefore, although the relative "soft" spring is preferred, the choice of contact stiffness will also be constrained by practical considerations. That is, the relative "soft" spring is preferred according to the parametric study. However, the stiffness of the spring cannot be infinitely small. There is a restriction, which is the limit of spring deformation. At the very least, it should be ensured that the spring can operate normally under the impact load of debris flow, and this is related to site investigation of the designed target area.

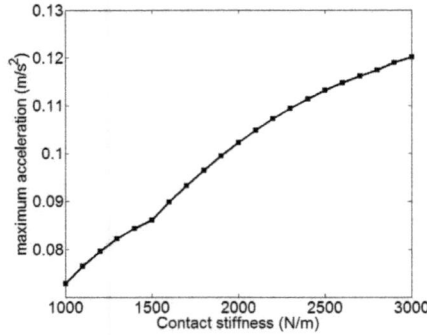

Figure 10. The influence of the contact stiffness.

4.5.2. Damping Ratio

To study the influence of the damping ratio, the response of the primary structure can be calculated by methods illustrated in Section 4.2 with different damping ratios. The initial velocity and initial acceleration are taken as the velocity and acceleration in grade 5, and the damping ratio is taken as $\xi = 0.04, 0.05, 0.06, 0.07, 0.08$, and 0.09. Then, the root mean square accelerations (as shown in Figure 11) are calculated by the methods in Section 4.2.

From Figure 11, it can be seen that the root mean square accelerations decrease with the enlargement of the damping ratios. In other words, the increase of the damping ratio can improve the reduction performance. Hence, by optimizing the parameters of the boxes and the particles, the structural damping can be increased, thereby reducing the subsequent structural vibration.

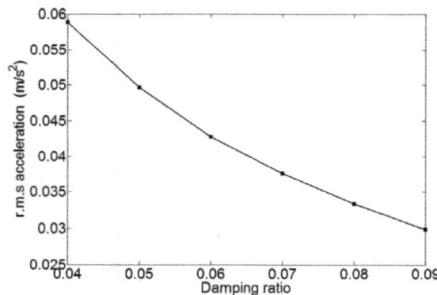

Figure 11. The influence of the damping ratio.

4.5.3. Mass Ratio

To study the influence of the mass ratio (the ratio of total auxiliary mass, including the boards, springs, particles and boxes, to the mass of test model), we can calculate the response of the structure by the methods in Sections 4.1 and 4.2 with different mass ratios. The mass ratio is different from the previous two parameters in that it affects both the structural response in two phases. Therefore, the analysis of this parameter can be considered from the two aspects of peak acceleration and root mean square acceleration, respectively. The initial velocity and initial acceleration are taken as the velocity and acceleration in grade 5, and the mass ratio ranges from 1.0% to 4.0%. Then, the response reduction effects of the peak acceleration and the root mean square acceleration (as shown in Figure 12) are calculated by the methods in Sections 4.1 and 4.2.

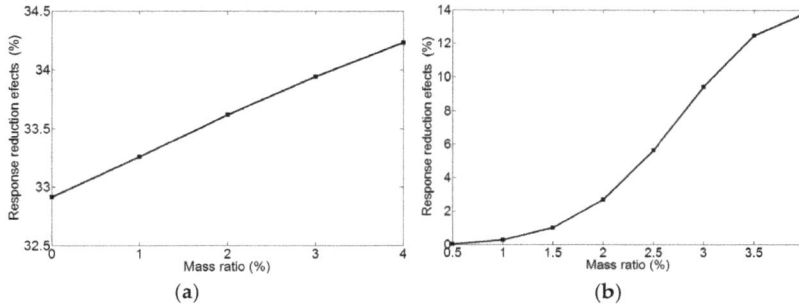

Figure 12. The influence of the mass ratio: (**a**) Maximum acceleration; (**b**) r.m.s acceleration.

From Figure 12a, it can be seen that with the increase of mass ratio, the decrease rate of the peak response of the structure increases monotonously, but the increase range is small, indicating that the increase of the mass ratio has a certain effect on reducing the peak response of the structure, but the effect is not prominent. In Figure 12b, the variation of the root mean square response reduction rate is the same as in Figure 12a in general. However, the rate of increase of the root mean square reduction rate is obviously faster when the mass ratio is changed from 2.0% to 3.5%. Moreover, the influence of the mass ratio on the root mean square response reduction is obviously greater than that on the peak response. The design of the device can be considered in terms of its impact on both of the root mean square response and the peak response. However, in practice, it is clear that the peak response of the structure is more meaningful considering the characteristics of the debris flow load.

5. Design Procedure

According to the mechanisms and parametric study of the lateral damping buffer, the design process of such a buffer in resisting the debris flows is proposed, as shown in Figure 13.

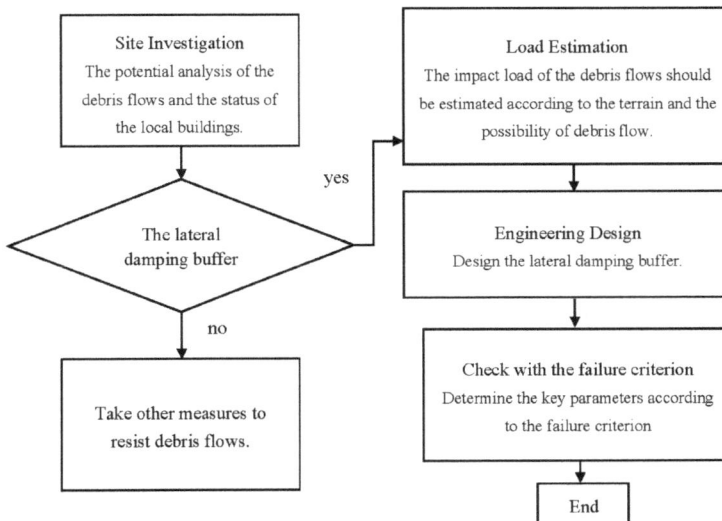

Figure 13. The design process of the lateral damping buffer in resisting debris flows.

The buffer is mainly made of boards, springs, particles and boxes. The specific steps to determine the necessary parameters for engineering design are described below.

(1) The size of the boards should be determined based on the specific engineering information of the target area, especially the statistical characterizations of the vulnerable debris flows. The length of the boards should be equal to the length of target buildings facing debris flows, and the height should be determined by the impact height of debris flows. For example, the height of the board can be twice of the height of the impact height;

(2) The stiffness of the springs should be determined by numerical simulation and the actual circumstance, considering the debris flow load, spring deformation limit and disaster reduction effect synthetically. Generally, a relative "soft" spring is preferred according to the parametric study. However, its stiffness is also constrained by the workability under the impact load of debris flows;

(3) The mass of particles and boxes should be determined by the mass ratio, which is the mass of particles and boxes to the mass of the primary structure. For the determination of sizes, refer to the previous study on filling ratio in particle damping technology [36,37].

In addition, the failure criterion is related to the regulations in the code. For example, in the "Code for Seismic Design of Buildings" [38] in China, the maximum inter-story drift under major earthquake is regulated as $1/50$, and this value can be used to be a kind of failure criterion.

6. Conclusions

In this paper, we propose and study a new type of device, named a lateral damping buffer, to reduce the vulnerability of building structures to debris flows. The lateral damping buffer can be regarded as a combination of a buffer and a shock absorber, which can reduce the maximum acceleration response and the subsequent vibration respectively. To examine the effectiveness of the lateral damping buffer, an impact test of the lateral damping buffer attached to a two-degree-of-freedom structure under the simulated debris flows load is completed. After the test, a corresponding numerical simulation is performed to validate its rationality. Then, a following parametric study is performed to optimize the lateral damping buffer. Finally, an engineering design procedure is put forward.

The impact test shows that the lateral damping buffer has good effectiveness in reducing both the maximum acceleration response and the subsequent vibration. Particularly for the tenth loading grade, the maximum acceleration response can be reduced by nearly half and the root mean square acceleration can be reduced by over 30%. In addition, the simulated values fit well with the experimental results. Moreover, the effectiveness of the lateral damping buffer increases with the increase in the loading grade. This characteristic is beneficial to the application in practical engineering.

The lateral damping buffer can also reduce the subsequent vibration of the primary structure by energy dissipation. However, considering the characteristics of the debris flows, the attachment of the cushion devices at the impact point is more effective.

Acknowledgments: Financial support from the National Key Technology R&D Program through grant 2014BAL05B01 is highly appreciated.

Author Contributions: Zheng Lu proposed the new device, conceived the experiments and wrote the paper; Yuling Yang performed the experiments and analyzed the data; Xilin Lu provided valuable discussions and revised the paper; Chengqing Liu proposed the new device, conceived and designed the experiments and revised the paper.

Conflicts of Interest: The authors declare no conflict of interest.

References

1. Osti, R.; Egashira, S. Method to improve the mitigative effectiveness of a series of check dams against debris flows. *Hydrol. Processes* **2008**, *22*, 4986–4996. [CrossRef]

2. Hu, K.H.; Zhang, J.Q. Characteristics of damage to buildings by debris flows on 7 August 2010 in Zhouqu, Western China. *Nat. Hazards Earth Syst. Sci.* **2012**, *12*, 2209–2217. [CrossRef]

3. Zanuttigh, B.; Lamberti, A. Experimental analysis of the impact of dry avalanches on structures and implications for debris flows. *J. Hydraul. Res.* **2006**, *44*, 522–534. [CrossRef]

4. Chen, J.G.; Chen, X.Q.; Li, Y.; Wang, F. An experimental study of dilute debris flow characteristics in a drainage channel with an energy dissipation structure. *Eng. Geol.* **2015**, *193*, 224–230. [CrossRef]

5. Chen, X.Q.; Cui, P.; You, Y.; Chen, J.G.; Li, D.G. Engineering measures for debris flow hazard mitigation in the Wenchuan earthquake area. *Eng. Geol.* **2014**, *194*, 73–85. [CrossRef]

6. You, Y.; Pan, H.L.; Liu, J.F.; Ou, G.Q. The optimal cross-section design of the "Trapezoid-V" shaped drainage channel of viscous debris flow. *J. Mt. Sci.* **2011**, *8*, 103–107. [CrossRef]

7. Armanini, A.; Larcher, M. Rational criterion for designing opening of slit-check dam. *J. Hydraul. Eng.* **2001**, *127*, 94–104. [CrossRef]

8. Chanson, H. Sabo check dams-mountain protection systems in Japan. *Int. J. River Basin Manag.* **2004**, *2*, 301–307. [CrossRef]

9. Hassanli, A.M.; Nameghi, A.E.; Beecham, S. Evaluation of the effect of porous check dam location on fine sediment retention (a case study). *Environ. Monit. Assess.* **2009**, *152*, 319–326. [CrossRef] [PubMed]

10. Canelli, L.; Ferrero, A.M.; Migliazza, M.; Segalini, A. Debris flow risk mitigation by the means of rigid and flexible barriers-experimental tests and impact analysis. *Nat. Hazards Earth Syst. Sci.* **2012**, *12*, 1693–1699. [CrossRef]

11. Navratil, O.; Liébault, F.; Bellot, H.; Travaglini, E.; Theule, J.; Chambon, G.; Laigle, D. High-frequency monitoring of debris-flow propagation along the Réal Torrent, Southern French Prealps. *Geomorphology* **2013**, *201*, 157–171. [CrossRef]

12. Okano, K.; Suwab, H.; Kanno, T. Characterization of debris flows by rainstorm condition at a torrent on the Mount Yakedake volcano, Japan. *Geomorphology* **2012**, *136*, 88–94. [CrossRef]

13. Chengdu Institute of Mountain Hazards and Environment, Chinese Academy of Sciences and Ministry of Water Resources. *Debris Flows in China*, 1st ed. The Commercial Press: Beijing, China, 2000. (In Chinese)

14. Yao, J.T.P. Concept of structural control. *J. Struct. Div. ASCE* **1972**, *98*, 1567–1574.

15. Lu, Z.; Lu, X.L.; Lu, W.S.; Masri, S.F. Shaking table test of the effects of multi-unit particle dampers attached to an MDOF system under earthquake excitation. *Earthq. Eng. Struct. Dyn.* **2012**, *41*, 987–1000. [CrossRef]

16. Zhou, Y.; Zhang, C.Q.; Lu, X.L. Seismic performance of a damping outrigger system for tall buildings. *Struct. Control. Health Monit.* **2016**, *24*. [CrossRef]

17. Lu, Z.; Lu, X.L.; Jiang, H.J.; Masri, S.F. Discrete element method simulation and experimental validation of particle damper system. *Eng. Comput.* **2014**, *31*, 810–823. [CrossRef]

18. Zhang, P.; Song, G.B.; Lin, Y. Seismic Control of Power Transmission Tower Using Pounding TMD. *J. Eng. Mech.* **2013**, *139*, 1395–1406. [CrossRef]

19. Lu, Z.; Chen, X.Y.; Lu, X.L.; Yang, Z. Shaking table test and numerical simulation of an RC frame-core tube structure for earthquake-induced collapse. *Earthq. Eng. Struct. Dyn.* **2016**, *45*, 1537–1556. [CrossRef]

20. Poussot-Vassal, C.; Spelta, C.; Sename, O.; Savaresi, S.M.; Dugard, L. Survey and performance evaluation on some automotive semi-active suspension control methods: A comparative study on a single-corner mode. *Annu. Rev. Control* **2012**, *36*, 148–160. [CrossRef]

21. Poussot-Vassal, C.; Sename, O.; Dugard, L.; Gaspar, P.; Szabo, Z.; Bokor, J. A new semi-active suspension control strategy through LPV technique. *Control Eng. Pract.* **2008**, *16*, 1519–1534. [CrossRef]

22. Lozoya-Santos, J.D.J.; Hernandez-Alcantara, D.; Morales-Menendez, R.; Ramirez-Mendoza, R.A. Modeling of dampers guided by their characteristic diagrams. *Rev. Iberoam. Autom. Inf. Ind.* **2015**, *12*, 282–291. [CrossRef]

23. Lozoya-Santos, J.D.J.; Morales-Menendez, R.; Ramirez-Mendoza, R.A. Evaluation of on-off semi-active vehicle suspension systems by using the hardware-in-the-loop approach and the software-in-the-loop approach. *Proc. Inst. Mech. Eng. D J. Autom. Eng.* **2015**, *229*, 52–69. [CrossRef]

24. Poussot-Vassal, C.; Sename, O.; Dugard, L.; Gaspar, P.; Szabo, Z.; Bokor, J. Attitude and handling improvements through gain-scheduled suspensions and brakes control. *Control Eng. Pract.* **2011**, *19*, 252–263. [CrossRef]

25. Li, H.N.; Zhang, P.; Song, G.B.; Li, L.; Patil, D.; Mo, Y.L. Robustness study of the pounding tuned mass damper for vibration control of subsea jumpers. *Smart Mater. Struct.* **2015**, *24*, 1–12. [CrossRef]

26. Zhang, P.; Li, L.; Patil, D.; Singla, M.; Li, H.N.; Mo, Y.L.; Song, G.B. Parametric study of pounding tuned mass damper for subsea jumpers. *Smart Mater. Struct.* **2016**, *25*, 1–7. [CrossRef]

27. Lu, Z.; Wang, D.C.; Masri, S.F.; Lu, X.L. An experimental study of vibration control of wind-excited high-rise buildings using particle tuned mass dampers. *Smart Struct. Syst.* **2016**, *25*, 1–7. [CrossRef]

28. Lu, Z.; Chen, X.Y.; Zhang, D.C.; Dai, K.S.; Masri, S.F.; Lu, X.L. Experimental and analytical study on the performance of particle tuned mass dampers under seismic excitation. *Earthq. Eng. Struct. D* **2016**. [CrossRef]

29. Dai, K.S.; Wang, J.Z.; Mao, R.F.; Lu, Z.; Chen, S.E. Experimental investigation on dynamic characterization and seismic control performance of a TLPD system. *Struct. Des. Tall Spec. Build.* **2016**. [CrossRef]

30. Gong, S.M.; Zhou, Y.; Zhang, C.Q.; Lu, X.L. Experimental study and numerical simulation on a new type of viscoelastic damper with strong nonlinear characteristics. *Struct. Control Health* **2016**. [CrossRef]

31. Zhang, Y.; Wei, F.Q.; Wang, Q. Dynamic response of buildings struck by debris flows. Debris-Flow Hazards Mitigation: Mechanics, Prediction, and Assessment. In Proceedings of the 4th International Conference on Debris-Flow Hazards Mitigation—Mechanics, Prediction, and Assessment, Chengdu, China, 10–13 September 2007.

32. Clough, R.W.; Penzien, J. Analysis of free vibration. In *Dynamics of Structures*, 3rd ed.; Computers & Structures, Inc.: Berkeley, CA, USA, 2003.

33. Zhang, Y.; Wei, F.Q.; Jia, S.W.; Liu, B. Experimental research of unreinforced masonry wall under dynamic impact of debris flow. *J. Mt. Sci.* **2006**, *24*, 340–345. (In Chinese)

34. Kozo, O.; Hiroki, T.; Hendro, S. Shock-absorbing capability of lightweight concrete utilizing volcanic pumice aggregate. *Constr. Build. Mater.* **2015**, *83*, 261–274.

35. Japan Society of Civil Engineers. *Practical Methods for Impact Test and Analysis*; Maruzen: Tokyo, Japan, 2004. (In Japanese)

36. Lu, Z.; Masri, S.F.; Lu, X.L. Parametric studies of the performance of particle dampers under harmonic excitation. *Struct. Control Health* **2011**, *18*, 79–98. [CrossRef]

37. Lu, Z.; Wang, D.C.; Zhou, Y. Experimental parametric study on wind-induced vibration control of particle tuned mass damper on a benchmark high-rise building. *Struct. Des. Tall Spec. Build.* **2017**. [CrossRef]

38. PRC Ministry of Housing and Urban-Rural Development; General Administration of Quality Supervision, Inspection and Quarantine of the People's Republic of China. *Code for Seismic Design of Buildings (GB50011-2010)*; China Architecture & Building Press: Beijing, China, 2010. (In Chinese)

Article

Self-Tuning Fuzzy Control for Seismic Protection of Smart Base-Isolated Buildings Subjected to Pulse-Type Near-Fault Earthquakes

Dahai Zhao [1],*, Yang Liu [1] and Hongnan Li [2]

[1] School of Civil Engineering and Mechanics, Yanshan University, Qinhuangdao 066004, China; zdhly0227@126.com
[2] School of Civil Engineering, Dalian University of Technology, Dalian 116024, China; hnli@dlut.edu.cn
* Correspondence: zhaodahai@126.com; Tel.: +86-335-806-7926

Academic Editor: Steve C.S. Cai
Received: 10 December 2016; Accepted: 9 February 2017; Published: 16 February 2017

Abstract: Pulse-type near-fault earthquakes have obvious long-duration pulses, so they can cause large deformation in a base-isolated system in contrast to non-pulse-type near-fault and far-field earthquakes. This paper proposes a novel self-tuning fuzzy logic control strategy for seismic protection of a base-isolated system, which can operate the control force of the piezoelectric friction damper against different types of earthquakes. This control strategy employs a hierarchic control algorithm, in which a higher-level supervisory fuzzy controller is implemented to adjust the input normalization factors and output scaling factor, while a sub-level fuzzy controller effectively determines the command voltage of the piezoelectric friction damper according to current level of earthquakes. The efficiency of the proposed control strategy is also compared with uncontrolled and maximum passive cases. Numerical results reveal that the novel fuzzy logic control strategy can effectively reduce the isolation system deformations without the loss of potential advantages of base-isolated system.

Keywords: pulse-type near-fault earthquake; base-isolation; fuzzy logic control (FLC); piezoelectric friction damper; vibration control

1. Introduction

Over the past decades, base isolation has been demonstrated to be an effective means to protect crucial structures and their contents from the destructive effects of dynamic excitations. The performance of base-isolated structures against near-fault earthquakes characterized by long-duration pulses with peak velocities has been investigated by several researchers [1,2]. Recent research results have shown that near-fault earthquakes characterized by long-duration pulses with peak velocities will result in significant base displacements and inter-story drifts of a seismically isolated structure [3]. Kalkan and Kunnath [4] showed that the fling-step effect would mainly excite the first order modal response of the structures with middle and long natural vibration period, and cause the maximum deformation of the structure at the bottom, which would result in the failure mode of the structure. Providakis [5] pointed out that isolation bearings will have a large deformation when the base-isolated structures with long natural vibration period are subjected to pulse-type near-fault earthquakes. Yang [6] examined the performance of a base-isolated structure under the action of near-fault earthquakes with fling-step effect. He found that the base displacement of the base-isolated structure exceeded the permissible displacement of the isolation bearings. The base-isolated structure may cause lateral instability, and the fling-step effect is more harmful than the rupture forward directivity for the base-isolated structure.

Due to the large isolator displacements, the isolated structure will suffer some serious problems, such as large permanent base displacement, failure of the isolation bearing and overturning of the superstructure [7]. In addition, as a result of large isolator displacements, the size of the isolation device will significantly increase, which may require much larger seismic gaps between buildings or much larger bridge expansion joints. Therefore, these requirements increase the cost of the construction, which contradicts the primary goal of seismic isolation, which is to design structures more efficiently and economically [8]. In order to enhance the performance of the base-isolated structures subjected to pulse-type near-fault excitations, many kinds of control devices have been proposed [9–11]. Passive supplemental damping devices can reduce the deformations of the isolation bearings during strong ground motions. However, they may also induce large damper forces and result in increased floor accelerations [12]. Furthermore, when the structure is subjected to weak or moderate earthquakes, passive devices may adversely affect the isolation system since the desired isolation characteristics may be different for these earthquakes.

Semi-active control is an effective control alternative that regulates the output power of dampers according to current structural responses. Therefore, semi-active control has better adaptability than traditional passive control. Furthermore, semi-active control requires less power compared to active control, which uses powerful actuators to achieve ideal effects. For these reasons, semi-active control devices have been receiving considerable attentions in recent years. Nagarajaiah and Sahasrabudhe [13] proposed a variable stiffness device that is used in a sliding isolation system to reduce the seismic response of a base-isolated structure, and the effectiveness of the semi-active device was proved by numerical analytical and experimental studies. Shook et al. [14] proposed a super elastic semi-active damping device in a base-isolated structure and its damping performance was thoroughly analyzed and evaluated. Madhekar and Jangid [15] analyzed and evaluated the dynamic response of a seismically isolated benchmark bridge equipped with viscous and variable dampers, and assessed the performance of these dampers.

The goal of utilizing a semi-active device in a base-isolated structure is to mitigate the displacement response of the isolation system according to the current stage of excitation without an obvious increase in the superstructure response. In order to utilize the full capabilities of the semi-active devices employed in a semi-active isolation system, an effective control algorithm is essential. Due to the uncertainties in the nature of earthquakes and the characteristics of base-isolated systems, the task of developing an optimal controller is a challenging task. For instance, a controller designed for near-fault earthquakes that cause significant deformations in the isolation system might develop large damper forces during a far field earthquake of generally moderate excitation. As a result, the isolation system may not perform as expected, and a significant increase in the acceleration response of the superstructure may be observed. Alternatively, if the controller is designed for an earthquake with far-field characteristics, the damper force may not be large enough to effectively reduce the structural responses during the pulse-type ground motion. Thus, semi-active dampers need an adaptive control strategy for both far-field and near-field earthquakes.

Recently, fuzzy logic control has been shown to be a promising control algorithm and has been widely used by many researchers for the control of semi-active devices due to its superiority and effectiveness in dealing with complex, uncertain, and nonlinear systems [16]. Das et al. [17] studied the fuzzy control for seismic protection of civil structures using MR dampers. Kim and Roschke [18] adopted the supervisory fuzzy control technique in a similar manner to MR dampers in order to reduce seismic response of a base-isolated benchmark building. Ozbulut and Stefan [19] developed a supervisory fuzzy logic controller and a GA-based self-organizing fuzzy logic controller for piezoelectric friction dampers that are employed as semi-active devices in the base-isolated structure. An adaptive fuzzy neural control strategy was also developed by Ozbulut et al. [20] to adjust the contact force of the variable friction dampers (VFDs) that are used in a smart isolation system. In addition, Zhao and Li [21] proposed a new fuzzy logic controller that is designed for seismic protection of base-isolated structures utilizing piezoelectric friction damper against near-fault earthquakes for different ground sites.

In this study, an improved self-tuning fuzzy logic control strategy is proposed in order to adjust the contact force of the piezoelectric friction damper (PFD) that is used in a smart isolation system. The control strategy employs an intelligent upper-level supervisory controller and a sub-level knowledge-based fuzzy controller. In this control strategy, the supervisory controller provides a mechanism to identify the nature of the earthquake, while the sub-level fuzzy controller specifies the command voltage for the damper by using isolation displacement and velocity as the two input variables. In particular, the supervisory controller tunes normalization factors of the sub-level controller inputs and the scaling factor of the sub-level controller outputs in order to improve the performance of the controller under different types of earthquakes. Taking into account that the pulse-type near-fault earthquake contains great acceleration and velocity pulses, seismic acceleration and seismic velocity are selected as two inputs of the supervisory controller. In order to verify the adaptability of the proposed control strategy under earthquakes with different intensities, and pulse-type near-fault earthquakes, non-pulse-type near-fault earthquakes and far-field earthquakes are employed as external excitations in the numerical simulations. For comparison purposes, maximum passive operation of the PFD and uncontrolled case are also considered in the simulations. A series of numerical simulations for the base-isolated building is performed to assess the performance of the control strategy.

2. Pulse-Type near Fault Ground Motion

In recent decades, many destructive strong earthquakes occurred throughout the world. Some researchers have obtained a large amount of data records for near-fault strong earthquakes. These earthquakes produced very serious damage to engineering structures, which aroused the attention to the seismological community and the engineering field [22].

A large number of previous studies indicated that some near-fault earthquakes have unique characteristics compared with far-field earthquakes, such as permanent ground displacement, significant rupture forward directivity and fling-step effect. Among these characteristics, pulse effects, such as rupture forward directivity and fling-step effect, can cause remarkable velocity pulses. Thus, compared with non-pulse-type near-fault earthquakes and far-field earthquakes, there exist significant differences in the amplitude, frequency and duration in pulse-type near-fault earthquakes, which aggravates the damage to the engineering structure. In general, for pulse-type near-fault earthquakes, the amplitudes of displacement, velocity and acceleration are all large. The low frequency component is abundant and the characteristic period is prolonged. Therefore, the near-fault earthquakes have high energy in the initial stage, which will result in large structural damage [23].

Previous research results indicate that, for pulse-type near-fault earthquakes, the ratio of peak ground velocity (PGV) to peak ground acceleration (PGA) is large. Some scholars have pointed out that PGV/PGA is mainly utilized to identify the pulse-effect of near-fault earthquakes. When PGV/PGA is more than 0.2, the pulse-effect is very obvious. Therefore, PGV/PGA of pulse-type near-fault earthquakes is more than 0.2 in this paper, and the fault distances are less than 20 km. In addition, for the selection of pulse-type near-fault earthquakes, the following criteria are used to distinguish the velocity pulse: the velocity time history curve of pulse-type near-fault earthquake contains a sharp bulge, and the duration for velocity pulse is above 0.5 s. In order to compare the control effectiveness and adaptability of the proposed control strategy under different types of earthquakes, non-pulse-type near-fault earthquakes and far-field earthquakes are selected. Table 1 shows the selected seismic records and the characteristics of the data. The data source is from the Pacific Earthquake Engineering Research (PEER) Center strong earthquake database: http://peer.berkeley.edu. In this table, code A represents near-fault earthquakes with fling-step effect, code B represents near-fault earthquakes with rupture forward directivity, code C represents non-pulse-type near-fault earthquakes, and code D represents far-field earthquakes.

Table 1. Earthquake records and their characteristic parameters.

Type	Code	Earthquake	Station and Direction	Magnitude (M_w)	PGA (g)	PGV (cm/s)	PGV/ PGA	Fault Distance (km)
Fling-step effect near-fault earthquakes	A1	Chi-Chi	TCU052-NS	7.6	0.411	95.7	0.233	0.06
	A2	Chi-Chi	TCU052-EW	7.6	0.341	159.0	0.466	0.06
	A3	Chi-Chi	TCU068-NS	7.6	0.452	263.0	0.581	0.50
	A4	Chi-Chi	TCU068-EW	7.6	0.555	176.5	0.318	0.50
Rupture forward directivity near-fault earthquakes	B1	Northridge	WPI-316	7.1	0.319	67.4	0.211	7.10
	B2	Northridge	WPI-046	7.1	0.446	92.7	0.208	7.10
	B3	Chi-Chi	TCU102-EW	7.6	0.308	87.2	0.290	1.19
	B4	Chi-Chi	TCU120-EW	7.6	0.233	62.6	0.280	9.87
Non-pulse-type near-fault earthquakes	C1	Chi-Chi	TCU071-EW	7.6	0.530	69.83	0.080	4.88
	C2	Chi-Chi	TCU072-EW	7.6	0.480	85.51	0.180	7.87
	C3	Chi-Chi	TCU078-EW	7.6	0.442	42.14	0.100	8.27
	C4	Chi-Chi	TCU079-EW	7.6	0.590	64.49	0.110	10.95
Far-field earthquakes	D1	Taft	No.095-S69E	7.7	0.172	17.2	0.099	43.0
	D2	Taft	No.095-N21E	7.7	0.153	17.8	0.116	43.0
	D3	El Centro	No.117-S90W	7.1	0.214	36.9	0.172	40.0
	D4	El Centro	No.117-S00E	7.1	0.349	33.5	0.096	40.0

The acceleration and velocity time history curves of the selected four types of earthquakes are shown in Figure 1. It can be seen that the energy of pulse-type near-fault earthquakes accumulates rapidly. This will result in an impact-type ground motion with immense amplitude, obvious pulse waveform and short seismic duration. In addition, from the time history curves, it can be shown that velocity pulse induced by fling-step effect has the characteristic of a single pulse, and the velocity pulse induced by rupture forward directivity has the characteristic of bidirectional velocity pulses. In addition, the velocity time history curve with rupture forward directivity possesses multiple-pulse characteristics. The average elastic acceleration and velocity response spectra for the selected earthquakes with 5% damped ratio are shown in Figure 2. It can be seen from the figure that, with respect to non-pulse-type near-fault earthquake and far-field earthquake, the spectra of acceleration and velocity for pulse-type near-fault earthquake in long period section significantly increase. For a base-isolated structure, the basic period is generally large. Therefore, it is very sensitive to pulse-type near-fault earthquakes.

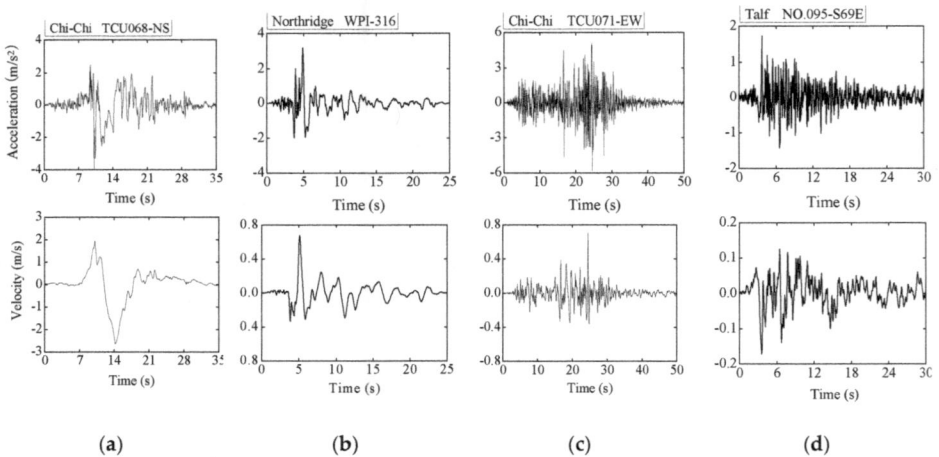

Figure 1. Acceleration and velocity time history curves for different types of earthquakes: (**a**) fling-step effect; (**b**) rupture forward directivity; (**c**) non-pulse-type; and (**d**) far-field.

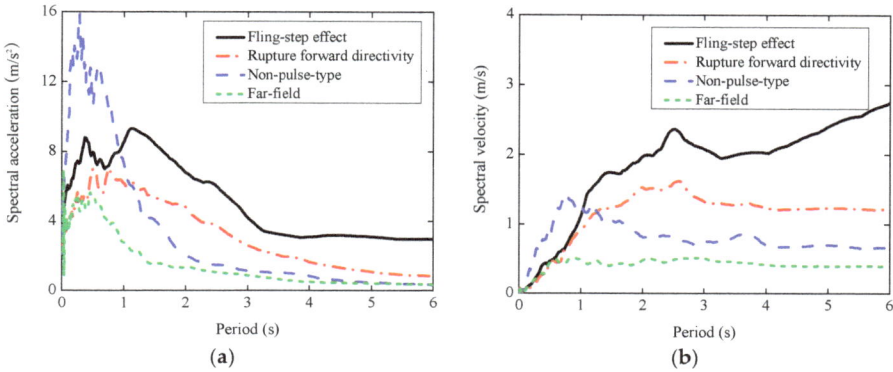

Figure 2. Average acceleration and velocity response spectra for different types of earthquakes: (a) spectra acceleration; and (b) spectra velocity.

3. Modeling of Base-Isolated Structure with Piezoelectric Friction Dampers

In this section, a five-story base-isolated structure is selected to investigate the performance of the following proposed control strategy. This base-isolated structure is equivalent to a lumped-mass model with one degree of freedom per floor. It is assumed that the superstructure of the base-isolated building remains elastic during the seismic excitation, as is usually the case for a base-isolated building. The schematic diagram of the semi-active base-isolated structure with PFD is shown in Figure 3.

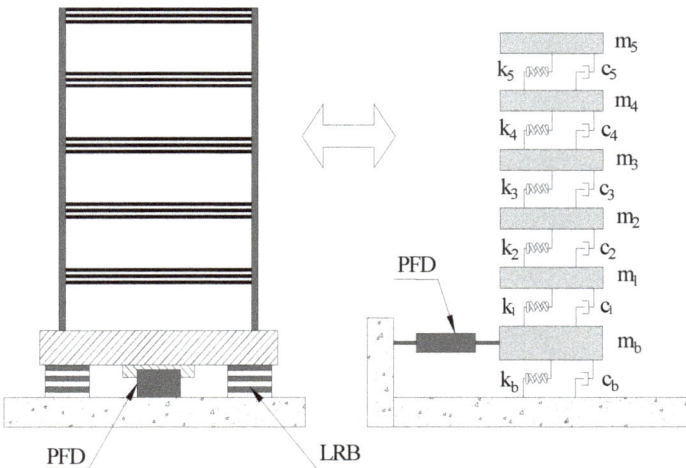

Figure 3. Model of the base-isolated structure.

The fundamental period of the five-story base-fixed structure equals 0.3 s, with a damping of 2% in the first mode. The isolation system is composed of lead rubber bearing (LRB) and assumed to have nonlinear force–deformation behavior with viscous damping. The relevant parameters of the superstructure and the isolation system are listed in Table 2. To improve the performance of the base-isolated structure under different types of earthquakes, PFD is installed on the base of the base-isolated structure.

Table 2. Parameters for the base-isolated structure.

Floor	Mass	Stiffness (KN/m)	Damping Coefficient (KN·s/m)
Base floor	6800 kg	2315	7.45
1	5897 kg	33,732	67
2	5897 kg	29,093	58
3	5897 kg	28,621	57
4	5897 kg	24,954	50
5	5897 kg	19,059	38

3.1. Equation of Motion for a Base-Isolated Structure

Consider an n-degree of freedom base-isolated structure with piezoelectric friction dampers at the isolation floor subject to seismic acceleration $\ddot{x}_g$. The equation governing the dynamic response of the structural system is given by

$$\mathbf{M}\ddot{\mathbf{x}}(t) + \mathbf{C}\dot{\mathbf{x}}(t) + \mathbf{K}\mathbf{x}(t) = -\mathbf{M}\mathbf{I}\ddot{x}_g(t) + \mathbf{D}F(t) \tag{1}$$

where $\mathbf{M}$, $\mathbf{C}$, $\mathbf{K}$ represent the $n \times n$ mass, damping, and stiffness matrices, respectively; $\mathbf{x}(t)$, $\dot{\mathbf{x}}(t)$, and $\ddot{\mathbf{x}}(t)$ are $n \times 1$ displacement, velocity, and acceleration vectors, respectively; $\mathbf{I}$ is an n-dimensional identity matrix; $\mathbf{D}$ is an $n \times 1$ damper location vector; and $F(t)$ is the damping force of damper and LRB.

Rewriting Equation (1) in state-space form:

$$\dot{\mathbf{z}}(t) = \mathbf{A}\mathbf{z}(t) + \mathbf{B}F(t) + \mathbf{H}\ddot{x}_g(t) \tag{2}$$

where

$$\mathbf{z}(t) = \begin{bmatrix} \mathbf{x}(t) \\ \dot{\mathbf{x}}(t) \end{bmatrix}, \ \mathbf{A} = \begin{bmatrix} \mathbf{0} & \mathbf{I} \\ -\mathbf{M}^{-1}\mathbf{K} & \mathbf{M}^{-1}\mathbf{C} \end{bmatrix}, \ \mathbf{B} = \begin{bmatrix} \mathbf{0} \\ -\mathbf{M}^{-1}\mathbf{D} \end{bmatrix}, \ \mathbf{H} = \begin{bmatrix} \mathbf{0} \\ -\mathbf{I} \end{bmatrix}. \tag{3}$$

3.2. Model of Piezoelectric Friction Damper

As a novel semi-active control device, piezoelectric friction dampers have been widely investigated by many researchers [24,25]. Recently, the authors also proposed a piezoelectric friction damper and investigated in their performances theoretically and experimentally [26]. Piezoelectric friction dampers utilize piezoelectric stacks to regulate the damping force and provide a satisfied friction force. The controllable friction force is favorable to ensure energy dissipation for various levels of earthquakes.

The contact force of the damper is expressed as

$$N(t) = N_{pre} + C_{pz}V(t) \tag{4}$$

where $N(t)$ denotes the total contact force, N_{pre} denotes the constant preload of the piezoelectric friction damper, C_{pz} denotes the piezoelectric coefficient of the piezoelectric actuator, and $V(t)$ denotes the input voltage of piezoelectric stack actuator.

During the movement of the base-isolated structure, a friction damper has two possible motion states: sticking and slipping phases. The friction force of the piezoelectric friction damper is expressed as

$$f(t) = -\mu N(t)\mathrm{sgn}(\dot{x}), \ \dot{x} \neq 0 \tag{5}$$

$$-\mu N(t) \leq f(t) \leq \mu N(t), \ \dot{x} = 0 \tag{6}$$

where μ denotes the friction coefficient of the damper, sgn( ) denotes the sign function related to the slip rate of the damper, and $f(t)$ denotes the damping force of the piezoelectric friction damper.

In the sticking phase, the absolute value of the friction force, f_s, can be approximately expressed as follows:

$$f_s = -\mu N(t)\mathrm{sgn}(\dot{x}) \text{ when } f_s = |f_i + f_r|, \ |f(t)| \geq f_s$$

$$f_i = m_t \ddot{x}_g = \left(m_b + \sum_{i=1}^{n} m_i \right) \ddot{x}_g, \ f_r = k_b x_b \tag{7}$$

where f_i is the inertial force applied to the mass; f_r is the restoring force provided by the isolation bearing; m_i is the mass of the superstructure; m_b and k_b are the mass and stiffness of the base isolator, respectively; and x_b is the displacement of the base isolator. The parameters N_{pre}, C_{pz}, μ and V_{max} of the piezoelectric friction damper are set to 2500 N, 2.5 N/V, 0.4 and 1000 V, respectively.

3.3. Modeling of Lead Rubber Bearing

In a lead rubber bearing, a central lead-core is used to provide an additional means of energy dissipation and initial rigidity against minor earthquakes and winds. The LRBs also provide an additional hysteretic damping through the yielding of the lead-core. The force–deformation behavior of the LRB is generally represented by non-linear characteristics. For the present study, Wen's model is used to characterize the hysteretic behavior of the LRBs. The restoring force developed in the isolation bearing is given by

$$f_b(t) = \alpha k_b x_b + (1 - \alpha) F_y Z \tag{8}$$

where α is an index which represents the ratio of post- to pre-yielding stiffness; k_b is the initial stiffness of the bearing; x_b is the displacement of the isolation layer; F_y is the yield strength of the bearing; and Z is the non-dimensional hysteretic displacement component satisfying the following non-linear first order differential equation:

$$q\dot{Z} = A\dot{x}_b - \beta |\dot{x}_b| Z |Z|^{n-1} - \tau \dot{x}_b |Z|^n \tag{9}$$

where q is the yield displacement; and dimensionless parameters A, β, τ and n are selected to determine the shape of the nonlinear hysteresis loops. The parameter n is an integer constant, which controls the smoothness of transition from elastic to plastic response. The values of $q = 0.025$m, $A = 1$, $\beta = 0.5$, $\tau = 0.5$ and $n = 2$ have been selected in this study to characterize the hysteretic behavior of the LRB.

4. Self-Tuning Fuzzy Control Strategy

In this section, self-tuning control strategy is designed to adjust the contact force of PFDs by controlling the real-time voltage of PFD according to the level and type of current external excitation.

4.1. Process Architecture of Self-Tuning Control Strategy

The self-tuning control strategy adopts a hierarchic structure that consists of a supervisory controller and a sub-level controller. Taking into account that the pulse-type near-fault earthquake has large acceleration amplitudes and velocity pulses, the supervisory controller is employed to evaluate the nature of the earthquake according to the current acceleration and velocity level. A sub-level controller is designed to determine the command voltage of the PFD. Base displacement and velocity are employed as two input variables, and the command voltage is defined as a single output for the sub-level controller. Since the amplitudes of base displacement and velocity differ greatly from near-fault and far-field earthquakes, the decision on the normalization factor and the scaling factor is evaluated by the supervisory controller. The block diagram of the proposed self-tuning fuzzy control strategy is shown in Figure 4.

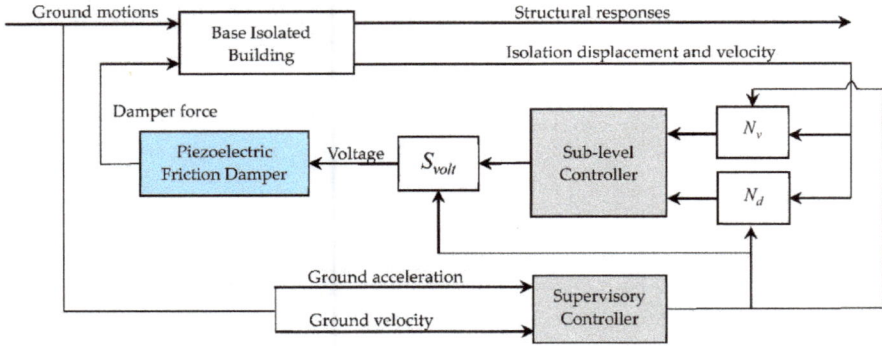

Figure 4. Block diagram of the self-tuning fuzzy control strategy.

4.2. Self-Tuning Hierarchic Fuzzy Logic Control (SHFLC) Strategy

Because of the uncertainties and nonlinearities in the nature of the earthquake and the isolation system, as well as the discontinuous characteristics of the friction force of PFD, the development of semi-active control methods is a challenging task. Fuzzy logic approach is an effective method to deal with complex nonlinear systems. It can describe relationships between inputs and outputs of a controller using simple verbose statements instead of complicated mathematical terms. Due to its inherent robustness and simplicity, fuzzy logic theory has been widely used to develop controllers for semi-active devices [17–19].

In the self-tuning hierarchic fuzzy logic control strategy, fuzzy logic controllers (FLC) are used in the design of the sub-level controller and the supervisory controller. In the sub-level FLC, seven Gaussian membership functions were defined for each input variable as shown in Figure 5. The fuzzy sets for the input variables are defined as follows: NL: negative large, NM: negative medium, NS: negative small, ZE: zero, PS: positive small, PM: positive medium, and PL: positive large. Furthermore, the universe of discourse for each input variable ranges from −1 to 1. Five Gaussian membership functions were defined to cover the universe of discourse of the output variable voltage that varies from 0 to 1000 V. The fuzzy sets for the output variables are defined as: VL: very large, L: large, M: medium, S: small, and ZE: zero.

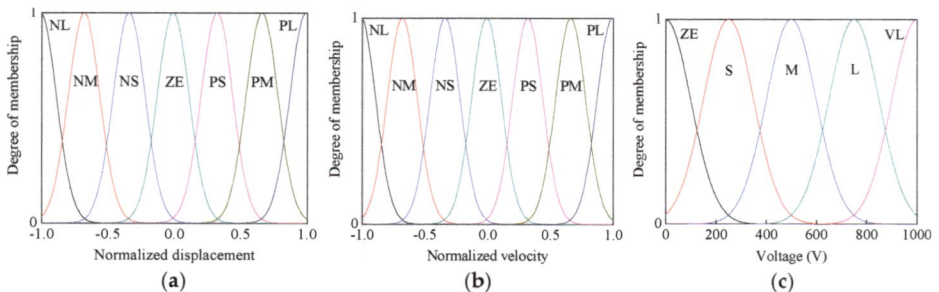

Figure 5. Input and output membership functions for sub-level fuzzy controller: (**a**) input normalized displacement; (**b**) input normalized velocity; and (**c**) output voltage.

After the fuzzification of the input and output variables, fuzzy control rules were established for the sub-level FLC, as shown in Table 3. The control rules are in the form of if-then rules and map the links between the input and output membership functions. The control principles are as follows: (1) if the base displacement and velocity have opposite signs (i.e., the isolation system returns to its original position), then the output voltage becomes small in order to ensure the PFD output small damping force; and (2) if the base displacement and velocity have the same sign, then the output voltage becomes large. At the same time, the magnitude of the output is linearly proportional to the magnitude of the input variables. When the displacement and velocity are almost zero or small, the command voltage is about zero, which means that PFD acts as a passive Coulomb damper. The defuzzification of fuzzy control employs the method of centroid to get a crisp output value. The control surface for the sub-level FLC is shown in Figure 6.

Table 3. Fuzzy rule base for sub-level FLC.

Isolation Velocity	Isolation Displacement						
	NL	**NM**	**NS**	**ZE**	**PS**	**PM**	**PL**
NL	VL	VL	L	L	M	S	ZE
NM	VL	L	L	M	S	ZE	S
NS	L	L	M	S	ZE	S	M
ZE	L	M	S	ZE	S	M	L
PS	M	S	ZE	S	M	L	L
PM	S	ZE	S	M	L	L	VL
PL	ZE	S	M	L	L	VL	VL

Note: NL: negative large, NM: negative medium, NS: negative small, ZE: zero, PS: positive small, PM: positive medium, and PL: positive large.

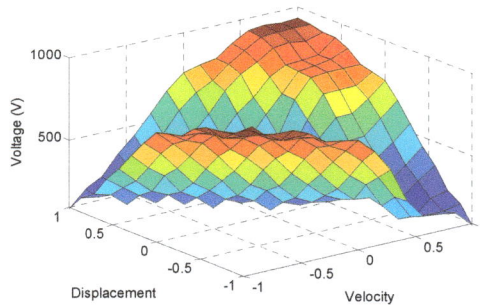

Figure 6. Fuzzy control surface for sub-level FLC.

In order to enhance the adaptability of the proposed control strategy under different types of earthquakes, the type and intensity of earthquakes should be distinguished during the design of the supervisory controller. As discussed above, pulse-type near-fault earthquakes possess large acceleration amplitudes and large long duration velocity impulses. Therefore, seismic acceleration $\ddot{u}_g$ and velocity $\dot{u}_g$ are selected as the input variables of the supervisory FLC in order to determine the characteristics of the earthquake. The universe of discourse for input seismic velocity is chosen from 0 to 1.8 m/s. Furthermore, the universe of discourse for input seismic acceleration is chosen from 0 to 4 m/s^2. The outputs of the supervisory FLC are normalization factors N_d and N_v for the displacement and velocity of the base isolation, and normalization factor S_{volt} for the voltage of the PFD. It should be noted that normalization factors are used to keep the input variables of the sub-level FLC in the range of the universe of discourse. For each input variable, a reasonable normalization factor should be determined. If the factors are too small, the input value of the sub-level FLC will be converted to a small value in the universe of discourse, and the output voltage will be too low. Thus, the performance of the PFD cannot be fully utilized. On the other hand, if the factors are too large, the input value will be converted to a larger value in the universe of discourse. This will lead to a large damping force and limit the effectiveness of the controller. The supervisory FLC can automatically adjust the value of the normalization factors to different types and intensities of earthquakes. To also ensure that the PFD will operate in an effective manner during a non-pulse-type near-fault earthquake or a far-field earthquake, a lower command voltage would be necessary for the damper otherwise, the damping force will be too large. If that were the case, the damper would not dissipate much energy since it would not slide as expected. On the other hand, to ensure that the damping force is not reduced too much, therefore not providing enough energy dissipation under strong earthquakes, such as pulse-type near-fault earthquakes, the output scaling factor with a range from 0 to 1.3 was used to scale the command voltage of the piezoelectric friction damper.

In the preliminary simulations, the universe of discourse for each normalization factor was selected according to the maximum structural response of the base-isolated building subjected to different types and intensity of earthquakes. Gaussian-type membership functions were selected for both input and output variables, as shown in Figure 7. The membership functions of the input variables are defined as S: small, M: medium, and L: large, and the output membership functions are defined as low: LOW, medium: MED and high: HIG.

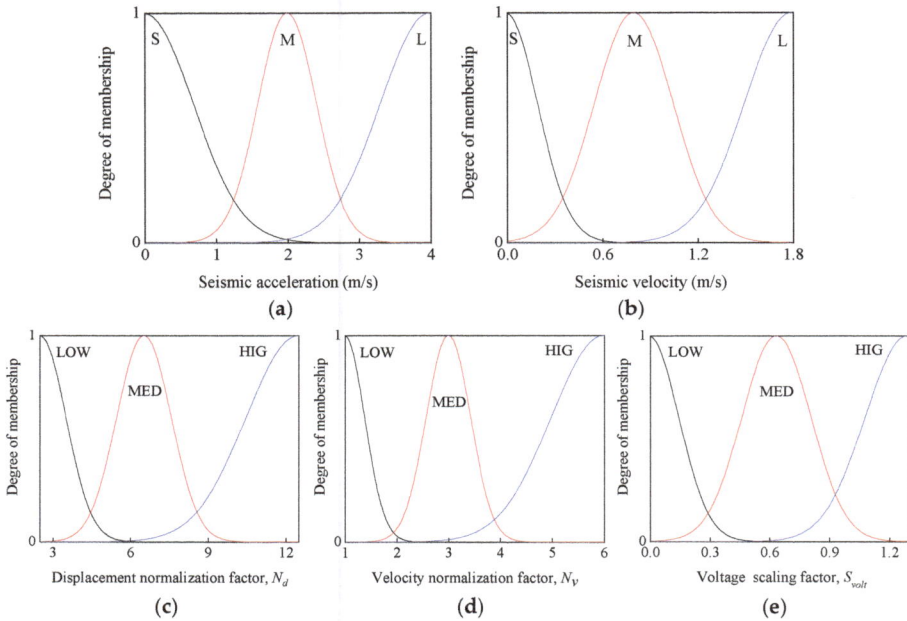

Figure 7. Input and output membership functions for supervisory FLC: (**a**) input seismic acceleration; (**b**) input seismic velocity; (**c**) output displacement normalization factor; (**d**) output velocity normalization factor; and (**e**) output voltage scaling factor.

The control principles for the supervisory FLC are described as follows. If the input values are large (L), which indicate the earthquake intensity is strong and the structural response is large, then the values of normalization factors should be set to low. This will prevent the structural response from over exceeding the input universe of discourse of the sub-level FLC, and avoid the damper to be set in a full power condition for a long time. At the same time, to make the friction damper output a large damping force, the scaling factor S_{volt} should be increased to enhance the output voltage of the sub-level fuzzy controller. On the contrary, if the input value is small (S) or medium (M), which indicates the earthquake intensity is low and structure response is small, then the normalization factors of the sub-level FLC should be appropriately increased. To make the damper output an appropriate damping force, the scaling factor S_{volt} should be decreased to reduce the output voltage of the sub-level FLC. Furthermore, the smaller damping force can be output by the friction damper. The control rules of the supervisory fuzzy controller are shown in Tables 4 and 5, respectively.

Table 4. Fuzzy rule base for N_d and N_v of supervisory FLC.

Input ($\ddot{u}_g$)	Input ($\dot{u}_g$)		
	S	M	L
S	HIG	HIG	MED
M	HIG	MED	LOW
L	MED	LOW	LOW

Table 5. Fuzzy rule base for S_{volt} of supervisory FLC.

Input ($\ddot{u}_g$)	Input ($\dot{u}_g$)		
	S	M	L
S	LOW	LOW	MED
M	LOW	MED	HIG
L	MED	HIG	HIG

5. Numerical Simulations

A total of sixteen earthquake records, shown in Table 1, were selected to evaluate the performance of the proposed self-tuning hierarchic fuzzy control strategy. The peak accelerations of the earthquakes are scaled to 0.4 g. A series of time history analyses for the base-isolated structure were performed with MATLAB/Simulink under the mentioned earthquakes. To evaluate the performance of the base-isolated structure for different cases, a total of nine performance indices are employed in this paper, as shown in Table 6 [27]. Here, the performance indices J_1 through J_5 denote peak base shear, peak structural shear, peak base displacement, peak inter-story drift, and peak floor acceleration in the controlled structure normalized by the corresponding values in the uncontrolled structure, respectively; J_7 and J_8 denote root mean square (RMS) base displacement and floor accelerations of the controlled structure that are normalized likewise, respectively; J_6 denotes the peak force generated by piezoelectric friction dampers normalized by the peak base shear in the controlled structure; and J_9 denotes energy dissipated by the dampers normalized by the earthquake input energy to the controlled structure.

Table 6. Performance indices.

Peak Base Shear	Peak Inter-Story Drift	RMS Base Displacement
$J_1 = \dfrac{\max_t\|V_0(t)\|}{\max_t\|\hat{V}_0(t)\|}$	$J_4 = \dfrac{\max_{t,f}\|d_f(t)\|}{\max_{t,f}\|\hat{d}_f(t)\|}$	$J_7 = \dfrac{\max_i\|\sigma_d(t)\|}{\max_i\|\hat{\sigma}_d(t)\|}$
Peak Structural Shear	**Peak Floor Acceleration**	**RMS Floor Acceleration**
$J_2 = \dfrac{\max_t\|V_1(t)\|}{\max_t\|\hat{V}_1(t)\|}$	$J_5 = \dfrac{\max_{t,f}\|a_f(t)\|}{\max_{t,f}\|\hat{a}_f(t)\|}$	$J_8 = \dfrac{\max_f\|\sigma_a(t)\|}{\max_f\|\hat{\sigma}_a(t)\|}$
Peak Base Displacement	**Peak Control Force**	**Energy Dissipated by PFD**
$J_3 = \dfrac{\max_t\|x_b(t)\|}{\max_t\|\hat{x}_b(t)\|}$	$J_6 = \dfrac{\max_t\|f_d(t)\|}{\max_t\|\hat{f}_d(t)\|}$	$J_9 = \dfrac{\int_0^T f_d(t)\dot{x}_b(t)dt}{\int_0^T V_0(t)\dot{U}_g(t)dt}$

In Table 6, $V_0(t)$ and $\hat{V}_0(t)$ denote the base shear for the controlled and uncontrolled structures, respectively; $V_1(t)$ and $\hat{V}_1(t)$ denote the structure shear for the controlled and uncontrolled structures, respectively; $x_b(t)$ and $\hat{x}_b(t)$ denote the base displacement for the controlled and uncontrolled structures, respectively; $d_f(t)$ and $\hat{d}_f(t)$ denote the inter-story drift for the controlled and uncontrolled structures, respectively; $a_f(t)$ and $\hat{a}_f(t)$ denote the absolute floor acceleration for the controlled and uncontrolled structures, respectively; $f_d(t)$ and $\hat{f}_d(t)$ denote the force generated by the PFD and the base shear in the controlled structures, respectively; $\sigma_d(t)$ and $\hat{\sigma}_d(t)$ denote the RMS base displacement for the controlled and uncontrolled structures, respectively; $\sigma_a(t)$ and $\hat{\sigma}_a(t)$ the RMS absolute floor acceleration for the controlled and uncontrolled structures, respectively; and $U_g(t)$ is vector of the ground accelerations.

To evaluate the influence of the pulse effects of the pulse-type near-fault earthquakes (namely, fling-step effect and rupture forward directivity effect) on the base-isolated structure controlled by the SHFLC, the peak inter-story drift, peak floor acceleration and peak base displacement of the SHFLC isolated structure are firstly evaluated. Figure 8 shows the comparison of the average peak inter-story drift and floor acceleration of the SHFLC isolated structure under the four types of earthquake. It can be seen from the figure that, compared to non-pulse-type near-fault earthquakes (type C) and far-field earthquakes(type D), pulse-type near-fault earthquakes (types A and B) can cause much larger inter-story drifts and floor accelerations at the bottom and center of SHFLC isolated structure, and this will decrease gradually with the height of the floors. It is also noted that, for the type C near-fault earthquakes, the peak inter-story drifts and floor accelerations at the top of the isolated superstructure exceed the corresponding values of the pulse-type near-fault earthquakes. For the pulse-type near-fault earthquakes, the earthquakes containing fling-step effect (type A) can induce much larger responses of the SHFLC isolated superstructure excepting at the bottom floor. In addition, far-field earthquakes (type D) have a minimal influence on the inter-story drifts of SHFLC isolated superstructure.

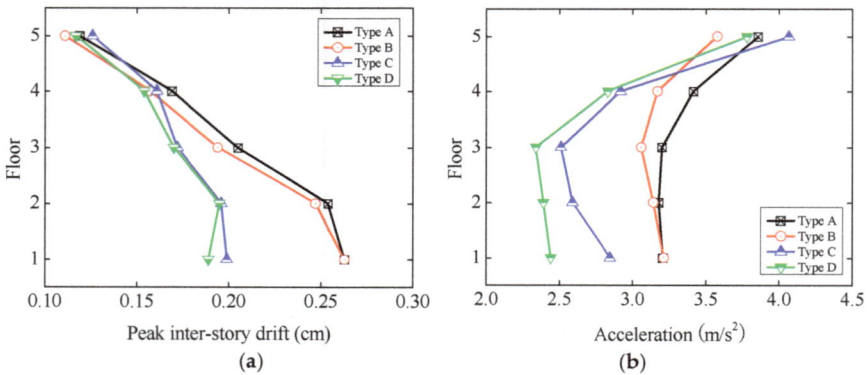

Figure 8. Average peak inter-story drift and floor acceleration of controlled base-isolated structure subjected to earthquakes: (**a**) average peak inter-story drift; and (**b**) average peak floor acceleration.

To explore the influence of pulse-type near-fault earthquakes on base displacements, the peak base displacements of SHFLC isolated structure under four types of earthquakes are shown in Figure 9. It can be seen from this figure that the peak base displacements of SHFLC isolated structure are less than 0.2 m for most of the pulse-type near-fault earthquakes. For the two types of near-fault earthquakes with the fling-step effect and the rupture forward directivity, the average peak base displacement is 0.167 m and 0.187 m, respectively. However, the base displacement decreased was evident when subjected to non-pulse-type near-fault and far field earthquakes, with the average peak base displacements equal to 0.065 m and 0.046 m, respectively. It can be concluded that pulse-type near-fault earthquakes can lead to much larger responses for SHFLC isolated structure as compared to non-pulse-type near fault earthquake and far-field earthquake, especially in the isolation layer and the bottom of the superstructure.

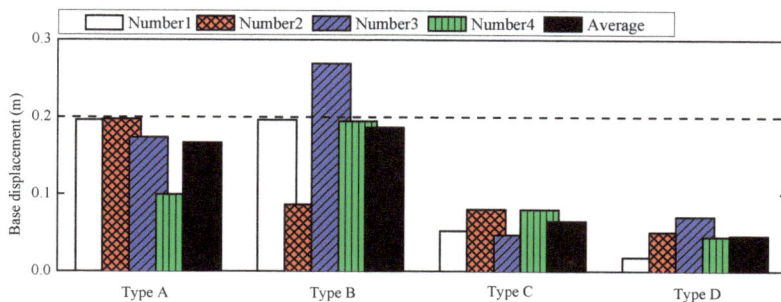

Figure 9. Peak base displacement of controlled base-isolated structure subjected to earthquakes.

The performance indices for the pulse-type near-fault earthquakes and the earthquakes without pulse-effect are listed in Tables 7 and 8, respectively. In these tables, the self-tuning hierarchic fuzzy logic control strategy is named SHFLC. Moreover, for comparison, the results for the maximum passive operation of PFDs are also listed in these tables. Numbers in bold font indicate the best results among the two control cases for the given excitations. Base displacement is an important index to evaluate the effectiveness of the proposed control strategy. For a base-isolated structure subjected to pulse-type near-fault earthquakes, one of the primary purposes is to prevent permanent damage on the isolation bearing or collision with adjacent structures. In these tables, it can be seen that although the maximum passive operation of the PFDs can successfully reduce the performance indices associated with the peak and RMS base displacements (J_3 and J_7), it will lead to significant amplification in peak and RMS floor acceleration (J_5 and J_8) for most of the selected earthquakes due to the large constant damper force. For pulse-type near-fault earthquakes, such as A1, A4, B1, B2, and B4, the maximum passive operation of the PFD increases the peak floor acceleration by 40%, 74%, 31%, 52% and 33%, respectively, compared to the uncontrolled structure. However, for the earthquake without pulse-effect, the maximum passive operation of the PFD caused an increase in the responses compared to the uncontrolled structure, except for the C4 earthquake. The increase in peak acceleration (J_5) for C1, C2, C3, D1, D2, D3, and D4 are 48%, 29%, 42%, 17%, 37%, 22% and 39%, respectively.

Table 7. Performance indices for various control cases under pulse-type near-fault earthquakes.

Earthquake	Control Case	J_1	J_2	J_3	J_4	J_5	J_6	J_7	J_8	J_9
A1	Passive-max	**0.88**	0.92	**0.61**	1.00	1.40	**0.63**	**0.70**	1.29	0.85
	SHFLC	0.94	**0.90**	0.70	**0.96**	**1.26**	0.69	0.83	**1.10**	0.79
A2	Passive-max	**0.95**	1.00	0.75	1.08	**0.89**	0.65	0.83	**0.79**	0.85
	SHFLC	0.98	**0.94**	**0.72**	**0.94**	0.98	**0.64**	0.86	0.92	0.85
A3	Passive-max	**0.83**	**0.93**	**0.41**	1.29	0.99	0.66	**0.46**	0.92	0.84
	SHFLC	0.88	0.95	0.62	**0.95**	**0.93**	0.66	0.65	**0.81**	0.89
A4	Passive-max	**0.61**	**0.68**	**0.35**	1.14	1.74	0.57	**0.54**	1.46	0.76
	SHFLC	0.79	0.76	0.47	**1.09**	**1.30**	**0.39**	0.66	**1.21**	0.77
B1	Passive-max	0.82	0.85	**0.67**	1.14	1.31	0.89	**0.72**	1.13	**0.89**
	SHFLC	**0.72**	**0.75**	0.81	**1.07**	**1.16**	0.89	0.81	**0.95**	0.81
B2	Passive-max	**0.94**	0.94	**0.54**	0.95	1.52	0.81	**0.63**	1.36	0.60
	SHFLC	0.95	**0.95**	0.54	0.95	**1.02**	0.68	0.66	**0.91**	0.64
B3	Passive-max	0.98	1.01	0.63	**1.14**	0.97	**0.67**	**0.70**	0.81	**0.82**
	SHFLC	**0.89**	**0.89**	**0.62**	0.94	**0.81**	0.78	0.83	**0.66**	0.78
B4	Passive-max	**0.96**	**0.97**	**0.52**	1.00	1.33	0.85	0.66	1.12	0.45
	SHFLC	0.99	0.99	**0.52**	1.07	**1.20**	**0.53**	**0.62**	**1.01**	0.59

Table 8. Performance indices for various control cases under earthquakes without pulse-effect.

Earthquake	Control Case	J_1	J_2	J_3	J_4	J_5	J_6	J_7	J_8	J_9
C1	Passive-max	0.92	0.92	0.66	0.94	1.48	0.43	0.74	1.34	0.33
	SHFLC	0.74	0.74	0.60	0.75	1.40	0.33	0.71	1.20	0.28
C2	Passive-max	0.94	0.98	0.81	1.12	1.29	0.40	0.81	0.72	0.45
	SHFLC	0.93	0.95	0.74	0.96	1.03	0.42	0.77	0.57	0.37
C3	Passive-max	0.85	0.93	0.62	0.93	1.42	0.51	0.67	1.19	0.41
	SHFLC	0.81	0.97	0.83	0.97	1.17	0.33	0.89	0.97	0.35
C4	Passive-max	0.90	0.89	1.06	1.00	0.95	0.45	1.00	0.85	0.39
	SHFLC	0.98	0.89	1.21	1.09	1.07	0.40	1.15	0.84	0.32
D1	Passive-max	1.09	1.17	0.69	1.46	1.17	0.87	0.76	1.10	0.19
	SHFLC	0.89	0.91	0.90	0.94	0.89	0.52	0.86	0.77	0.23
D2	Passive-max	1.29	1.32	0.39	1.60	1.37	0.97	0.42	1.22	0.41
	SHFLC	0.96	0.99	0.51	1.00	1.14	0.72	0.43	1.03	0.27
D3	Passive-max	0.99	0.93	0.65	1.18	1.22	0.70	0.59	0.96	0.36
	SHFLC	1.03	0.80	0.75	1.27	1.12	0.55	0.69	0.81	0.27
D4	Passive-max	0.78	0.83	0.52	0.83	1.39	0.91	0.55	1.23	0.48
	SHFLC	0.89	0.64	0.70	0.90	1.30	0.57	0.67	1.10	0.30

Although the SHFLC also increases the peak floor acceleration with respect to the uncontrolled structure under most of the earthquakes, the SHFLC can successfully limit the increase in peak floor accelerations for the selected earthquakes. For example, for A4, B2, C2, and D2 earthquakes, maximum passive operation of PFD increases peak floor acceleration (J_5) by 74%, 52%, 29%, and 37%, respectively, while SHFLC increases these responses by 30%, 2%, 3%, and 14%. Additionally, the SHFLC case can lead to obvious decreases in RMS floor acceleration (J_8) for most of the selected earthquakes. In addition, it can be seen that the SHFLC case can reduce peak and RMS base displacements for most of the selected earthquakes, which indicates the effectiveness of the proposed control strategy subjected to different types of earthquakes.

According to the definition of performance indices, J_3, J_5, J_7 and J_8 for the uncontrolled base-isolated structure are equal to 1. In order to graphically compare the control effectiveness of maximum passive control case and the developed SHFLC case with the uncontrolled case for peak and RMS base displacements and floor accelerations, Figures 10 and 11 show the difference between the performance indices J_3, J_5, J_7 and J_8 for control cases and the performance index 1 for uncontrolled case. The vertical coordinate of these figures is the difference between the performance indices and 1, a negative value indicates that the structural response is reduced relative to the uncontrolled structure, and a positive value illustrates that the structural response is amplified relative to the uncontrolled structure. The absolute value of the vertical coordinate is the proportion of the reduction or amplification. In Figure 10, it can be seen that, compared to the uncontrolled case, the peak and RMS displacements of the base for the two control cases are significantly reduced. The control effect of the maximum passive control for the base displacement is much larger. For the maximum passive control case and the SHFLC case, the average reduction ratios of the peak base displacement under types A and B in Table 1 are 38.3% and 29.8%, respectively. For the passive control case and SHFLC case, the average reduction ratios of the RMS base displacement are 32.6% and 25.1%, respectively. However, it can be seen in Figure 11 that the max passive control can significantly amplify the peak and RMS floor acceleration for most of the selected earthquakes. Similarly, for the SHFLC case, the peak and RMS floor accelerations will be amplified for some of the selected earthquakes. For the maximum passive control case and SHFLC case, the average amplification ratios of peak floor acceleration are 27.8% and 3.7%, respectively. For the maximum passive control case, the average amplification ratio of RMS floor acceleration is 28.3%, while the average reduction ratio of the RMS floor acceleration for the SHFLC case is only 9.2%. In summary, as compared to maximum passive control case for all the excitation cases, the SHFLC case produces an average of 24.1% and 19.1% reductions in peak and RMS floor acceleration at an expense of an average of 8.5% and 7.6% increase in peak and RMS base displacements. In Figures 10 and 11, it can be seen that, compared to the maximum passive control, for most of the excitation cases, the SHFLC case considerably improves the performance of the base-isolated structure in peak and RMS floor accelerations at the cost of slight deterioration of peak and RMS base displacements. Therefore, semi-active operation of PFDs with the SHFLC is more favorable than the passive operation of the dampers in simultaneously suppressing the peak base displacement and peak floor acceleration.

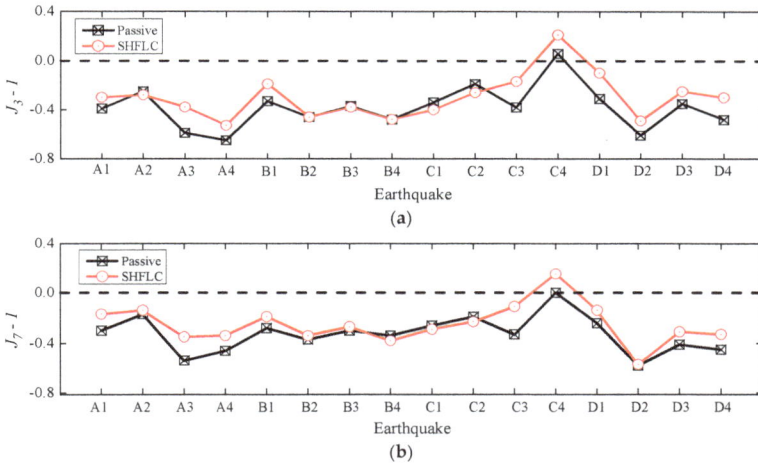

Figure 10. Performance indices J_3 and J_7 for different control cases under various types of earthquakes: (**a**) Performance index J_3 ; and (**b**) Performance index J_7 .

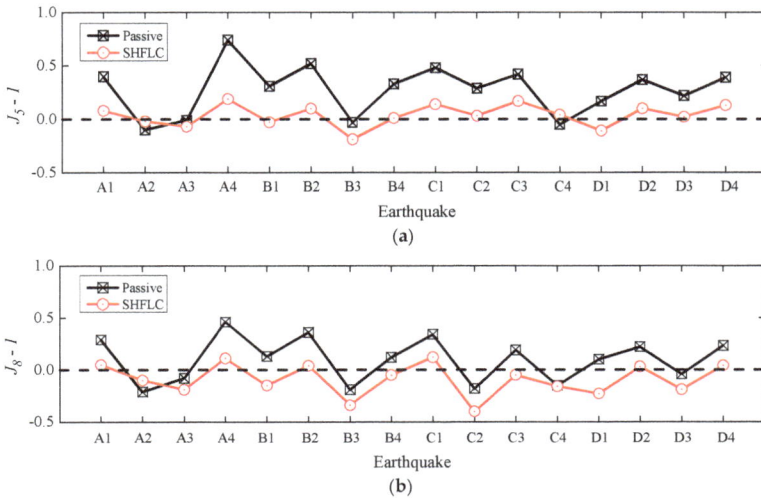

Figure 11. Performance indices J_5 and J_8 for different control cases under various types of earthquakes: (**a**) Performance index J_5; and (**b**) Performance index J_8.

Figure 12 illustrates the profiles of average inter-story drifts for uncontrolled, maximum passive control and SHFLC cases. In the figure, it can be seen that the SHFLC case is better than the maximum passive control in inter-story drift reductions of base-isolated structure. Compared with uncontrolled base-isolated structure, there is no obvious amplification in inter-story drifts in the SHFLC case. Furthermore, the inter-story drifts of the SHFLC case at the bottom of the superstructure have an appropriate reduction, which can be illustrated by the performance indices J_4 shown in Tables 7 and 8, respectively. The values of J_4 for the SHFLC case subjected to most earthquakes is less than 1, which demonstrates that the peak inter-story drift of the SHFLC case is reduced relative to uncontrolled structure.

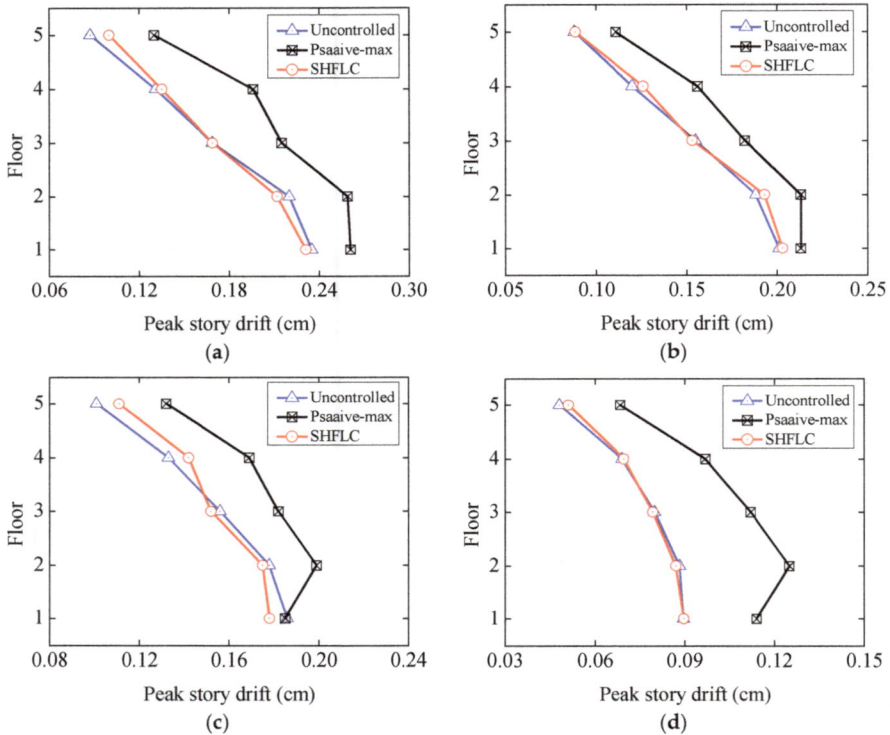

Figure 12. Average inter-story drifts with four types of earthquakes: (**a**) near-fault earthquakes with fling effect; (**b**) near-fault earthquakes with rupture forward directivity; (**c**) near-fault earthquakes without pulse-effect; and (**d**) far-field earthquakes.

In order to further evaluate the adaptability of the proposed SHFLC control strategy under different types of earthquakes, the command voltage time history, the damping force time history and the force–displacement curve of the PFD with the SHFLC control strategy are shown in Figure 13, for different earthquake records. As seen in Figure 13, under the pulse-type near-fault earthquakes, such as the TCU068-NS record of the Chi-Chi earthquake with fling effect and the WPI-316 record of the Northridge earthquake with rupture forward directivity, the SHFLC method can output the full amplitude voltage in real time when subjected to large pulse earthquakes. In these cases, large structural responses will be produced and the PFD will output large damping forces to control the base displacement. Therefore, the large output command voltage of the controller will lead to many irregular serrations in the force–displacement curve of the PFD.

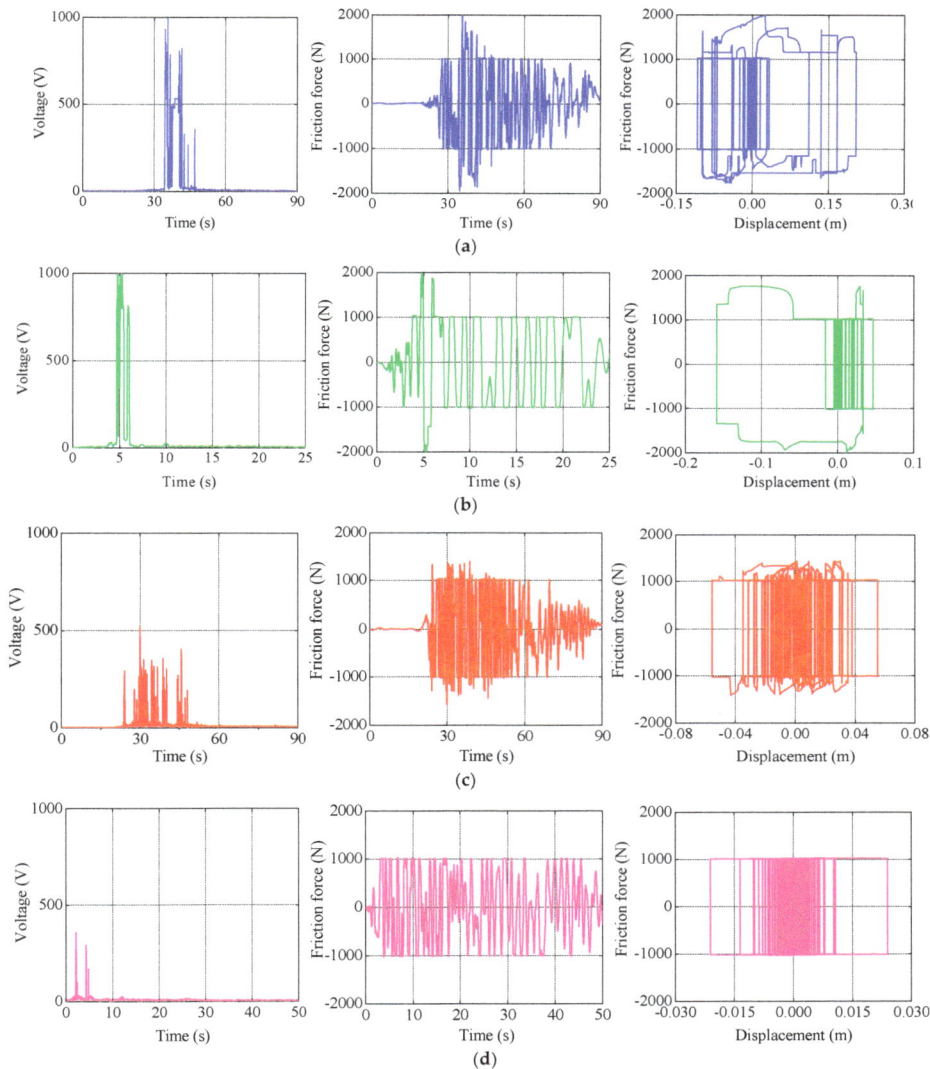

Figure 13. Time history of command voltage, damper force and force–displacement diagram of PFD under four earthquakes: (**a**) TCU068-NS record for Chi-Chi earthquake; (**b**) WPI-316 record for Northridge earthquake; (**c**) TCU071-EW record for Chi-Chi earthquake; and (**d**) No.095-S69E record for Taft earthquake.

However, for non-pulse-type near-fault earthquakes and far-field earthquakes, such as the TCU071-EW record of the Chi-Chi earthquake and the No.095-S69E record of the Taft earthquake, the output command voltage of the SHFLC controller is much smaller. In fact, for the No.095-S69E record of the Taft earthquake, the SHFLC controller operates the PFD as a passive Coulomb damper with nearly zero voltage. Therefore, the PFD outputs a small damping force produced by pre-pressure and the force–displacement curve of the PFD is much smoother and similar to that of the passive friction damper. In addition, when comparing the control effort (performance index J_6) of the SHFLC controller to that of with the maximum passive controller, it can be seen from Tables 7 and 8 that the SHFLC case requires a much smaller control force to be produced by the PFD for most earthquakes.

In order to further evaluate the performance of the developed control strategy, the energy balance equations of a base-isolated structure are established. For a base-isolated structure with a PFD, the energy equations for the base-isolated structure are expressed as

$$E_k + E_\xi + E_S + E_H = E_I \tag{10}$$

$$E_k = \frac{1}{2}\dot{\mathbf{x}}_t^T \mathbf{M}\dot{\mathbf{x}}_t \tag{11}$$

$$E_\xi = \int_0^t \dot{\mathbf{x}}^T \mathbf{C} d\mathbf{x} = \int_0^t \dot{\mathbf{x}}^T \mathbf{C}\dot{\mathbf{x}} dt \tag{12}$$

$$E_S = \int_0^t \mathbf{x}^T \mathbf{K} d\mathbf{x} \tag{13}$$

$$E_H = -\int_0^t F \mathbf{D}^T d\mathbf{x} \tag{14}$$

$$E_I = -\int_0^t \ddot{x}_g \mathbf{M} \mathbf{I} d\mathbf{x} \tag{15}$$

where E_k, E_ξ, E_S, E_H, and E_I denote the absolute kinetic energy, the damping energy, the elastic strain energy, the hysteretic energy provided by the PFD, and the absolute input energy, respectively.

Figure 14 shows the time histories of the input energy E_I, the viscous damped energy of the structure, lead rubber bearings E_ξ, and the energy dissipated by the PFD E_H for the SHFLC base-isolated building subjected to the A3, B1, C1, and D1 earthquakes. It can be seen from the figures that the PFD can successfully dissipate most of the input energy of the pulse-type near-fault earthquakes, such as the TCU068-NS record of the Chi-Chi earthquake and the WPI-316 record of the Northridge earthquake. In Table 7, it can be seen that the energy performance indices J_9 of the base-isolated structure with SHFLC under A3 and B1 earthquakes equal 0.89 and 0.81, respectively. It can also be seen that the energy performance index J_9 for the SHFLC case for other pulse-type near-fault earthquakes is generally large. Compared to the maximum passive control case, the energy performance index J_9 for the SHFLC case can dissipate a lot more energy for most of the pulse-type near fault earthquakes. However, the energy consumption proportion of the PFD controlled with SHFLC will decrease under the earthquakes without pulse effect (types C and D). In Table 8, it can be seen that the energy performance index J_9 for the SHFLC case is generally less than 0.4. For example, the performance index J_9 for the SHFLC case subjected to the TCU071-EW record of the Chi-Chi earthquake and the No.095-S69E record of the Taft earthquake are 0.28 and 0.23, respectively. This means that the energy dissipation proportion of the PFD for the SHFLC case is 28% and 23%, respectively. From the energy time history curves for the SHFLC case under C1 and D1 earthquakes in Figure 14, it can also be seen that the energy dissipation proportion of the isolation building will increase, which means that most of the seismic input energy will be dissipated. The above results reflect the adaptability of the proposed SHFLC control strategy for different types of earthquakes.

Appl. Sci. **2017**, *7*, 185

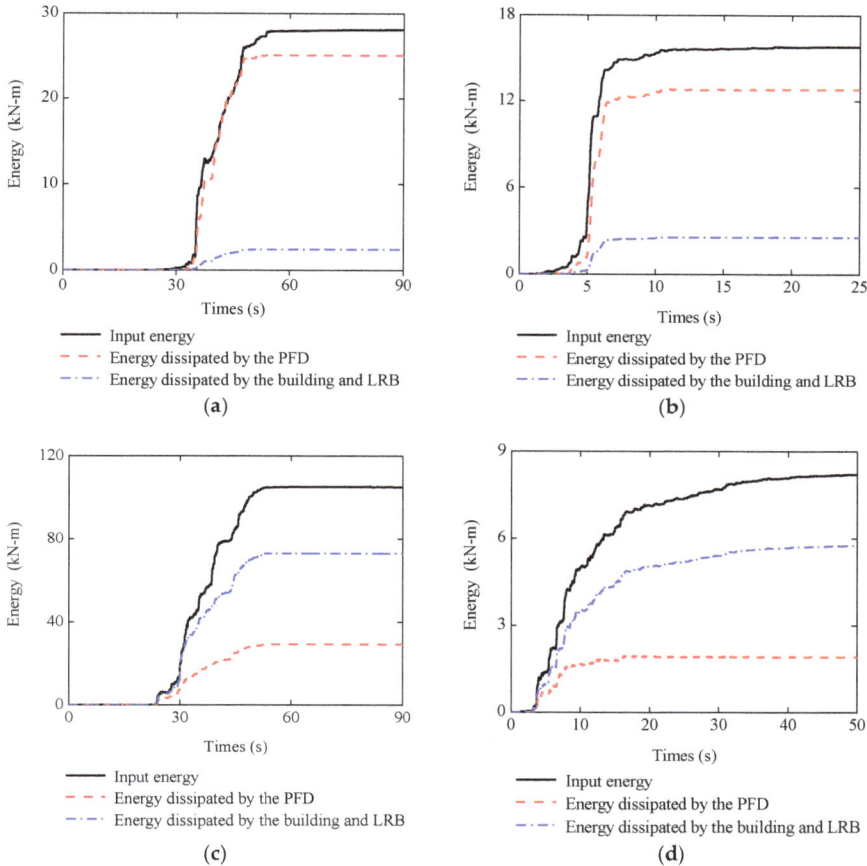

Figure 14. Energy time histories for a base-isolated structure controlled by SHFLC subjected to four types of earthquakes: (**a**) TCU068-NS record of Chi-Chi earthquake; (**b**) WPI-316 record of Northridge earthquake; (**c**) TCU071-EW record of Chi-Chi earthquake; and (**d**) No.095-S69E record of Taft earthquake.

6. Conclusions

In this study, to simultaneously reduce the responses of isolation system and superstructure for a base-isolated building under various types and intensities of earthquakes, a self-tuning hierarchic fuzzy logic control strategy is proposed and implemented. This controlled method can be mainly divided into two parts. The first part of the developed control strategy, named a supervisory fuzzy logic controller, can adjust two input normalization factors and one output scaling factor of the sub-level fuzzy logic controller in order to identify the type of earthquake. The second controller, namely the sub-level controller in the control strategy, is used to determine the command voltage of the PFD according to the seismic responses of the base-isolated system. Uncontrolled case and maximum passive operation case are compared to the proposed SHFLC strategy to evaluate the performance of the fuzzy controllers. The simulation results reveal that PFDs controlled by the developed strategies can effectively improve the responses of the base-isolated structure against different types of earthquakes including both far-field and near-fault earthquakes. Furthermore, it can be concluded that the proposed SHFLC strategy shows good adaptability to different types of earthquakes.

Acknowledgments: This research is supported by the National Natural Science Foundation of China (51308487) and Hebei Provincial Natural Science Foundation of China (E2014203055). These supports for this research are greatly appreciated.

Author Contributions: Dahai Zhao conceived the idea, Yang Liu set up the numerical model and analyzed the data, Dahai Zhao and Yang Liu wrote the paper, and Hongnan Li modified the grammar in this paper.

Conflicts of Interest: The authors declare no conflict of interest.

References

1. Jangid, R.S.; Kelly, J.M. Base isolation for near-fault motions. *Earthq. Eng. Struct. Dyn.* **2001**, *30*, 691–707. [CrossRef]

2. Shen, J.; Tsai, M.H.; Chang, K.C.; Lee, G.C. Performance of a seismically isolated bridge under near-fault earthquake ground motions. *J. Struct. Eng.* **2004**, *130*, 861–868. [CrossRef]

3. Mazza, F.; Vulcano, A. Nonlinear response of RC framed buildings with isolation and supplemental damping at the base subjected to near-fault earthquakes. *J. Earthq. Eng.* **2009**, *13*, 690–715. [CrossRef]

4. Kalkan, E.; Kunnath, S.K. Effects of fling-step and forward directivity on seismic response of buildings. *Earthq. Spectra* **2006**, *22*, 367–390. [CrossRef]

5. Providakis, C.P. Effect of LRB isolators and supplemental viscous dampers on seismic isolated buildings under near-fault excitations. *Eng. Struct.* **2008**, *30*, 1187–1198. [CrossRef]

6. Yang, D.; Zhao, Y. Effects of rupture forward directivity and fling-step of near-fault ground motions on seismic performance of base-isolated building structure. *Acta Seismol. Sin.* **2010**, *32*, 579–587.

7. Li, S.; Xie, L. Progress and trend on near-field problems in civil engineering. *Acta Seismol. Sin.* **2007**, *29*, 102–111. [CrossRef]

8. Ismail, M. Influence of isolation Elimination of torsion and pounding of isolated asymmetric structures under near-fault ground motions. *Struct. Control Health Monit.* **2015**, *22*, 1295–1324. [CrossRef]

9. Barbat, A.H.; Rodellar, J.; Ryan, E.P.; Molinares, N. Active control of nonlinear base-isolated buildings. *J. Eng. Mech.* **1995**, *121*, 676–685. [CrossRef]

10. Pardo-Varela, J.; de-la-Llera, J.C. A semi-active piezoelectric friction damper. *Earthq. Eng. Struct. Dyn.* **2015**, *44*, 333–354. [CrossRef]

11. Ozbulut, O.E.; Hurlebaus, S. Optimal design of superelastic-friction base isolators for seismic protection of highway bridges against near-field earthquakes. *Earthq. Eng. Struct. Dyn.* **2011**, *40*, 273–291. [CrossRef]

12. Alhan, C.; Gavin, H. A parametric study of linear and non-linear passively damped seismic isolation systems for buildings. *Eng. Struct.* **2004**, *26*, 485–497. [CrossRef]

13. Nagarajaiah, S.; Saharabudhe, S. Seismic response control of smart sliding isolated buildings using variable stiffness systems: Experimental and numerical study. *Earthq. Eng. Struct. Dyn.* **2006**, *35*, 177–197. [CrossRef]

14. Shook, D.A.; Roschke, P.N.; Ozbulut, O.E. Superelastic semi-active damping of a base-isolated structure. *Struct. Control Health Monit.* **2008**, *15*, 746–768. [CrossRef]

15. Madhekar, S.N.; Jangid, R.S. Variable dampers for earthquake protection of benchmark highway bridges. *Smart Mater. Struct.* **2009**, *18*, 1–18. [CrossRef]

16. Bitaraf, M.; Hurlebaus, S. Semi-active adaptive control of seismic excited 20-story nonlinear building. *Eng. Struct.* **2013**, *56*, 2107–2118. [CrossRef]

17. Das, D.; Datta, T.K.; Madan, A. Semiactive fuzzy control of the seismic response of building frames with MR dampers. *Earthq. Eng. Struct. Dyn.* **2012**, *41*, 99–118. [CrossRef]

18. Kim, H.S.; Roschke, P.N. GA-fuzzy control of smart base isolated benchmark building using supervisory control technique. *Adv. Eng. Softw.* **2007**, *38*, 453–465. [CrossRef]

19. Ozbulut, O.E.; Stefan, H. Fuzzy control of piezoelectric friction dampers for seismic protection of smart base isolated buildings. *Bull. Earthq. Eng.* **2010**, *8*, 1435–1455. [CrossRef]

20. Ozbulut, O.E.; Bitaraf, M.; Hurlebaus, S. Adaptive control of base-isolated structures against near-field earthquakes using variable friction dampers. *Eng. Struct.* **2011**, *33*, 3143–3154. [CrossRef]

21. Zhao, D.; Li, Y. Fuzzy control for seismic protection of semiactive base-isolated structures subjected to near-fault earthquakes. *Math. Probl. Eng.* **2015**, *2015*, 675698. [CrossRef]

22. Hu, J.; Xie, L. Review of rupture directivity related concepts in seismology. *J. Earthq. Eng. Eng. Vib.* **2011**, *31*, 1–8.

23. Sehhati, R.; Rodriguez-Marek, A.; ElGawady, M.; Cofer, W.F. Effects of near-fault ground motions and equivalent pulses on multi-story structures. *Eng. Struct.* **2011**, *33*, 767–779. [CrossRef]

24. Ng, C.L.; Xu, Y.L. Semi-active control of a building complex with variable friction dampers. *Eng. Struct.* **2007**, *29*, 1209–1225. [CrossRef]

25. Ribakov, Y. Base-isolated structures with selective controlled semi-active friction dampers. *Struct. Des. Tall Spec. build.* **2011**, *20*, 757–766. [CrossRef]

26. Zhao, D.; Li, H. Shaking table tests and analyses of semi-active fuzzy control for structural seismic reduction with a piezoelectric variable-friction damper. *Smart Mater. Struct.* **2010**, *19*, 1–9. [CrossRef]

27. Narasimhan, S.; Nagarajaiah, S.; Johnson, E.A.; Gavin, H.P. Smart base-isolated benchmark building. Part I: Problem definition. *Struct. Control Health Monit.* **2006**, *13*, 573–588. [CrossRef]

applied sciences

[MDPI]

Article

The Bivariate Empirical Mode Decomposition and Its Contribution to Grinding Chatter Detection

Huanguo Chen [1,*], Jianyang Shen [1], Wenhua Chen [1], Chuanyu Wu [1], Chunshao Huang [2], Yongyu Yi [1] and Jiacheng Qian [1]

[1] Zhejiang Province's Key Laboratory of Reliability Technology for Mechanical and Electrical Product, Hangzhou 310018, China; slylgdx1992@126.com (J.S.); chenwh@zstu.edu.cn (W.C.); cywu@zstu.edu.cn (C.W.); yiyongyu123@163.com (Y.Y.); 15958315572@163.com (J.Q.)
[2] Hangzhou Hangji Machine Tool Co., Ltd., Hangzhou 311305, China; -h-cs@163.com
*   Correspondence: hgchen@zstu.edu.cn; Tel.: +86-571-8684-3369

Academic Editor: Gangbing Song
Received: 2 November 2016; Accepted: 25 January 2017; Published: 8 February 2017

Abstract: Grinding chatter reduces the long-term reliability of grinding machines. Detecting the negative effects of chatter requires improved chatter detection techniques. The vibration signals collected from grinders are mainly nonstationary, nonlinear and multidimensional. Hence, bivariate empirical mode decomposition (BEMD) has been investigated as a multiple signal processing method. In this paper, a feature vector extraction method based on BEMD and Hilbert transform was applied to the problem of grinding chatter. The effectiveness of this method was tested and validated with a simulated chatter signal produced by a vibration signal generator. The extraction criterion of true intrinsic mode functions (IMFs) was also investigated, as well as a method for selecting the most ideal number of projection directions using the BEMD algorithm. Moreover, real-time variance and instantaneous energy were employed as chatter feature vectors for improving the prediction of chatter. Furthermore, the combination of BEMD and Hilbert transform was validated by experimental data collected from a computer numerical control (CNC) guideway grinder. The results reveal the good behavior of BEMD in terms of processing nonstationary and nonlinear signals, and indicating the synchronous characteristics of multiple signals. Extracted chatter feature vectors were demonstrated to be reliable predictors of early grinding chatter.

Keywords: bivariate empirical mode decomposition (BEMD); Hilbert transform; multiple signals; synchronous characteristic; real-time variance; instantaneous energy

1. Introduction

Grinders, which are often high-precision, large-tonnage machines, are essential in the equipment manufacturing industry. Monitoring their condition and diagnosing their faults ensures their safe operation, which has great practical and economic value [1,2]. Most notably, grinders can enter a "chatter" state during machining operations, which causes a series of negative effects [3] such as machined surface undulations, reduced tool life, poor surface finish, noise, and reduced productivity.

Friction, regeneration and mode coupling are three well-known mechanisms that cause grinding chatter. Typically, regeneration chatter is the most common type of chatter encountered in machining processes. It occurs when oscillations in the grinder cause undulations in the machined surface that are regenerated with subsequent passes of the grinder. As numerous experiments have shown, the amplitude of vibration signals increases substantially when a stable grinding state turns into a chatter state [4]. Additionally, there is a transitional phase in this process, which contains considerable information about grinding status [5]. Thus, we can discover the mechanism of grinding chatter and predict it by analyzing transition-phase signals. It is worth noting that the transitional phase occurs within a very short time and is easily affected by many random factors. In order to alleviate the problem of chatter, reliable chatter detection and identification methods are essential so that effective chatter suppression can be applied in a timely manner [6].

Vibration detection in grinding is usually performed by using accelerometers, acoustic emission (AE) sensors or dynamometers. A considerable amount of research work describes these different methods. For example, Tansel et al. used acceleration signals from a lathe tailstock to detect chatter during turning operations, and used an s-transformation to extract a damping index as a descriptive feature of chatter [7]. Yao et al. proposed two-dimensional feature vectors for chatter detection based on the standard deviation of wavelet transform of drilling machining, which allowed rapid chatter identification [8]. Furthermore, Devillez et al. equipped a lathe with eddy current sensors to monitor vibration. These were processed by a fuzzy method to classify the tests [9]. Moreover, many researchers have predicted cutting chatter by using dynamic forces. Gradisek et al. employed a coarse-grained entropy rate as a chatter index in a cutting process, which exhibited a drastic drop at the onset of chatter [10]. Nari et al. presented the usage of permutation entropy, a computer-based, rapid measurement of chatter onset using sound signals recorded by a unidirectional microphone [11]. Yang et al. proposed a method using feature vectors incorporating signal variance and a one-step self-correlation function as a chatter indicator for detecting cutting state [12]. Acoustic emission sensors were used by Li et al. to detect chatter; the specific peculiarities of acoustic emission sensors can effectively reject external disturbances [13].

In summary, wavelet transform [14,15], s-transformation [16], singular value decomposition (SVD) [17], and artificial intelligence [18], are the most popular methods used to process vibration signals. These analysis methods are mostly based on the theory of Fourier transform. However, grinder vibration signals are mostly nonstationary and nonlinear, such that these conventional methods cannot effectively extract signal features [19]. Subsequently, empirical mode decomposition (EMD), proposed by Huang et al., has been demonstrated as an appropriate approach for overcoming the shortcomings of conventional techniques [20]. The EMD approach can decompose signals into several intrinsic mode functions (IMFs) which distribute from high frequency to low frequency. All IMFs derived from grinding signals are characterized as zero-mean oscillations. However, in practice, grinder vibration signals are usually multidimensional signals, yet current methods are only useful with one-dimensional, real-value time series [21]. They are unfavorable for executing information fusion functions and accurately reflecting real-world situations when processing multiple signals.

In order to overcome the aforementioned problems, Rilling et al. proposed an extension to bivariate time series, which generalizes the rationale underlying EMD to the bivariate framework, namely, bivariate empirical mode decomposition (BEMD) [22]. Since being proposed, it has been applied to image processing such as image segmentation [23], image fusion [24], image compression [25] and image watermarking [26]. At the third session of the HHT International Conference, BEMD was reported to have been successfully applied to wind turbine condition monitoring [27]. It has been shown that the BEMD technique has inherited all the merits of the one-dimensional EMD method. However, in contrast to EMD, BEMD aims at decomposing a complex-valued signal into a collection of zero-mean rotating components, which represent fast to slow rotations in the signal. Another greater advantage is that BEMD is more powerful in extracting weak features from nonstable and nonlinear signals [27]. The authors of this paper have made a comparison between the use of EMD and BEMD for extracting features from grinding chatter signals, and demonstrated the superior performance of BEMD [28]. These findings can be summarized as follows: (1) EMD is initially applied to a one-dimensional signal and extracts zero-mean oscillating components, whereas BEMD is applied to a bivariate signal and extracts zero-mean rotating components; (2) BEMD has computational efficiency due to the simultaneous processing of complex-value signals, and it only computes the upper envelope using the maximum points. Meanwhile, EMD can only decompose signals one by one and obtains both upper and lower envelopes by connecting the extreme points; (3) the number of IMFs derived from signals by EMD are unequal. They cannot reveal any synchronous characteristics and phase shifting, nor can EMD extract an information fusion function. In contrast, the number of IMFs derived by BEMD is normally the same, and BEMD can extract an information fusion function and preserve phase differences; (4) BEMD facilitates the establishment of purified shaft vibration orbits and fully guarantees the correctness of results, which EMD cannot.

The objective of this paper is to assess BEMD's performance in processing grinding chatter signals. The remainder of the paper is organized as follows. In Section 2, the algorithms of BEMD and Hilbert transform are reviewed, as well are the extraction criteria of true IMFs. Moreover, real-time variance and instantaneous energy are presented as feature vectors for grinding chatter. In Section 3, a simulated chatter signal is generated from a chatter signal generator (Simulink, MATLAB, MathWorks Company, Natick, MA, USA), then processed by BEMD and Hilbert transform. Moreover, a selection method for the most ideal number of projection directions is developed. In Section 4, the benefits of combining BEMD and Hilbert transform are experimentally validated by processing chatter signals collected from a real grinder (KD4020X16). Section 5 concludes the paper.

2. Feature Vector Extraction Method Based on Bivariate Empirical Mode Decomposition (BEMD) and Hilbert Transform

2.1. Review and Further Discussion of BEMD

2.1.1. The BEMD Algorithm

The main concept of BEMD is to extract a series of IMFs from multiple signals. This method assumes that multiple signals can be described as the sum of a fast rotation component superimposed on a slower rotation component. Each of the IMFs must satisfy the following conditions:

1. The number of extrema and the number of zero crossings must be equal, or differ at most by one.
2. The mean value of the envelope at any point defined by the local maximum points and the envelope defined by the local minima must be zero.

Moreover, the BEMD process requires projecting a multiple signal on a set of directions and then applying the sifting process of standard EMD to the projected components. This gives a bivariate signal, $s(t) = x(t) + iy(t)$, and a set of projection directions, $\varphi_m = 2m\pi/N, 1 \leq m \leq N$. The fundamental process of BEMD can be depicted as follows:

1. Project the signal $s(t)$ on directions φ_m:

$$p_{\varphi_m}(t) = \text{Re}[e^{-i\varphi_m} s(t)], \tag{1}$$

2. Extract all partial maximum points of $p_{\varphi_m}(t):\{(t_j^m, p_j^m)\}$, where j indicates the number of individual partial maxima.
3. Interpolate the set of points $\{(t_j^m, e^{i\varphi_m} p_j^m)\}$ by cubic spline interpolation to obtain the partial envelope curve in direction φ_m named $e_{\varphi_m}(t)$.
4. Repeat steps 1–3 until the envelope curves in all N projections are obtained.
5. Compute the mean of all envelope curves:

$$\overline{m}(t) = \frac{1}{N} \sum_{r=1}^{N} e_r(t), \tag{2}$$

6. Subtract the mean $\overline{m}(t)$ from $s(t)$ to obtain $g(t)$:

$$g(t) = s(t) - \overline{m}(t), \tag{3}$$

7. Test if $g(t)$ is an IMF, if not, replace $s(t)$ with $g(t)$ and repeat the procedure from Step 1 until $g(t)$ is an IMF. If yes, record the obtained IMF and remove it from $g(t)$, i.e., $c_1(t) = g(t)$, $r_1(t) = s(t) - c_1(t)$.

8. Consider $r_1(t)$ as the original signal and repeat the above procedure until the second IMF, $c_2(t)$, is obtained, and the residual component $r_2(t) = r_1(t) - c_2(t)$.

9. Iterate the previous process until all IMFs are obtained. Referring to the sifting process of BEMD, the term $s(t)$ can be expressed by the procedure:

$$s(t) = \sum_{k=1}^{n} h_k(t) + r_n(t), \qquad (4)$$

where $h_k(t)$ denotes the kth complex-valued IMF and $r_n(t)$ denotes a non-zero mean low-degree polynomial residue. Generally, the residue is mostly recognized as the vibration trend of the signal.

2.1.2. Extraction Criterion of True Intrinsic Mode Functions (IMFs)

The majority of vibration signals will be decomposed into excessive IMFs, which are described as spurious vibratory components that directly affect researchers' capacity to analyze signal features. These spurious IMFs also result in enormous interferences, reflecting the peculiarities of the chatter and allowing diagnosis of the mechanical faults of the grinder. One of the major deficiencies is that the estimation of the authenticity of IMF heavily depends on the user's experience. Thus, a reliable method is required to identify and remove spurious components, which is of great importance in extracting the actual vibration mode and the corresponding features of the time-frequency domain.

Spurious IMFs may be caused by several factors: (1) IMFs defined by BEMD algorithms are based on numerical analysis only and do not consider the vibration characteristics of system in the physical sense; (2) the stopping criterion of sifting results in a phenomenon of excessive decomposition and end effects are not well processed; (3) vibration signals mixed with noise or pulse interference will produce a series of spurious components with high frequency in the decomposition process. Since IMFs are recognized as the orthogonal expression of the signals, true IMFs have a higher correlation with original signals than spurious IMFs. Mathematically, a correlation coefficient is used as a statistical indicator to reflect the degree of correlation between variables. It is based on the deviation of two variables and their respective average mean, computed by the method of product comments, as shown in Equation (5).

$$\rho = \frac{Cov(X, Y)}{\sqrt{D(X)}\sqrt{D(Y)}} = \frac{E[(X - E(X))(Y - E(Y))]}{\sqrt{D(X)}\sqrt{D(Y)}} \qquad (5)$$

where $Cov(X, Y)$ denotes the covariance between the original signal and the IMF, $E(X)$ and $D(X)$ denote the expectation and variance of the original signal, respectively, and $E(Y)$ and $D(Y)$ denote the expectation and variance of the kth IMF, respectively. Therefore, it is reasonable to employ the correlation coefficient as an index to exactly distinguish the spurious components from true IMFs and then classify them as a part of residual [16,29]. In view of this, if the correlation coefficient of each IMF, ρ_k ($k = 1, 2, \ldots, n$), is computed, the true IMFs can be extracted by the following method (Table 1).

Table 1. The extraction criterion of true intrinsic mode functions (IMFs) based on the correlation coefficient.

If $\lvert \rho_k \rvert \geq \lambda$,
Reserve the kth IMF c_k,
Else,
Estimate kth IMF, and $r_n = r_n + c_k$.

The variable λ in Table 1 is a fixed threshold that is generally adopted as a ratio of the maximum correlation coefficient, wherein η is a ratio coefficient larger than 1.

$$\lambda = \max(\lvert \rho_k \rvert / \eta), \ k = 1, 2 \ldots n, \qquad (6)$$

2.1.3. Investigation of Projection Direction Number

Remember that the number of projection directions must be selected in advance because the BEMD algorithm is sensitive to it. Moreover, a large number of directions may be interesting insofar that it reduces the dependence of the final decomposition with respect to rotations of the spatial coordinates. However, excessive directions will reduce the computational efficiency of BEMD. Thus, projection direction number is further investigated by assessing the indicators of computational time and stability degree (SD) which change with the direction number.

The stability degree can be defined as,

$$SD = \frac{\sum\limits_{j=1}^{N} (X_j - \mu)^4}{(\sum\limits_{j=1}^{N} (X_j - \mu)^2)^2} \tag{7}$$

where $X = \{X_1, X_2, X_3, ..., X_N\}^T$ ($X_j \in R$, $j = 1, 2, ..., N$, T is a transpose operator), N represents the sampling points, and μ represents the mean of the signal.

Moreover, the SD is an ideal standard for expressing the fluctuation of signal amplitude. When there is a transient mutation in a signal, the stability degree will significantly improve. The purpose of this paper is to find out the most ideal number of projection directions, so that the precise signal features can be obtained as soon as possible.

2.2. Hilbert Transform

Through the BEMD method, the chatter signal can be decomposed into a number of complex-valued IMF components. Thereby, the Hilbert transform can be used to obtain the Hilbert and marginal spectrum distributions [30]. The Hilbert transform is defined as,

$$x_k(t) = \frac{1}{\pi} P \int\limits_{-\infty}^{\infty} \frac{\text{Re}(c_k(\tau))}{t - \tau} d\tau \text{ and } y_k(t) = \frac{1}{\pi} P \int\limits_{-\infty}^{\infty} \frac{\text{Im}(c_k(\tau))}{t - \tau} d\tau \tag{8}$$

where P is the Cauchy principal value.

Define $X(t)$ and $Y(t)$ from $c(t)$ as follows:

$$X_k(t) = \text{Re}(c_k(t)) + ix_k(t) \text{ and } Y_k(t) = \text{Im}(c_k(t)) + iy_k(t) \tag{9}$$

so that the amplitude and phase functions can be defined:

$$Z_k(t) = a_k(t) + ib_k(t) = \sqrt{\text{Re}^2(c_k(t)) + X_k^2(t)} + i\sqrt{\text{Im}^2(c_k(t)) + Y_k^2(t)}, \tag{10}$$

$$\psi_k(t) = \theta_k(t) + i\varphi_k(t) = \tan^{-1} \frac{X_k(t)}{\text{Re}(c_k(t))} + i\tan^{-1} \frac{Y_k(t)}{\text{Im}(c_k(t))} \tag{11}$$

where $\theta_k(t)$ represents the phase of the real-part IMF, and $\varphi_k(t)$ represents the phase of the imaginary-part IMF.

Then, the instantaneous frequency can be expressed as follows:

$$f_k(t) = \omega_k(t) + i\delta_k(t) = \frac{d\theta_k(t)}{dt} + i\frac{d\varphi_k(t)}{dt} \tag{12}$$

Thus, the Hilbert spectrum can be obtained:

$$H(\omega, \delta, t) = \text{Re}(\sum_{k=1}^{n} a_k(t)e^{i\int \omega_k(t)dt}) + i\text{Re}(\sum_{k=1}^{n} b_k(t)e^{i\int \delta_k(t)dt}) \tag{13}$$

The Hilbert spectrum represents the signals as a function that combines time and frequency. If $H(\omega,\delta,t)$ is integrated over time, the corresponding marginal spectrum is obtained:

$$h(\omega,\delta) = \int_{-\infty}^{\infty} H(\omega,\delta,t)dt \tag{14}$$

The marginal spectrum expresses the amplitude of each frequency in space, and represents the accumulated amplitude in a statistical sense.

Although the Hilbert transform can analyze the instantaneous frequency, its applicability is restricted by the form of the original signal. For poorly formed signals, such as very irregular signals, the results produced by Hilbert transform differ significantly from the actual physical status. However, after BEMD decomposition, the IMFs have superior characteristics with a small bandwidth, reasonable symmetry, and smooth envelope changes, which are suitable for the Hilbert transform [31]. Thus, the BEMD method and Hilbert transform are highly connected.

2.3. Feature Vectors for Grinding Chatter—Real-Time Variance and Instantaneous Energy

Another major objective of this paper is to accurately detect and identify the onset of grinding chatter before chatter makes damage the workpiece, so that suitable techniques for chatter suppression can be employed as fast as possible. According to the experimental results, the amplitude of chatter signals exhibits a substantial expansion when a stable grinding state changes into a chatter state. Consequently, early grinding chatter can be detected by comparing changes in the time-domain statistical parameters of the signal mathematically. Therefore, appropriate feature vectors should be defined where real-time variance and instantaneous energy are recognized ideal feature vectors.

The real-time variance (R_v) indicates the degree of deviation from the mean chatter signal, which in a sense reflects the trend of signal oscillation. The real-time variance can be described as:

$$R_v = \hat{R} - \overline{X} = \frac{1}{N}\sum_{j=1}^{N} X_j^2 - \frac{1}{N}\sum_{j=1}^{N} X_j, \tag{15}$$

where $X = \{X_1, X_2, X_3, ..., X_N\}^T$ ($X_j \in R$, $j = 1, 2, ..., N$, T is a transpose operator), N represents the sampling points, and $\overline{X}$ represents the mean of the signal.

In practice, any internal fluctuation may result in a vibration of instantaneous energy, which means that the change in instantaneous energy has a direct relation to abnormal system operation. Thus, according to Equations (10) and (11), instantaneous energy can be defined as follows:

$$E_k(t) = \frac{1}{2}\alpha_k^2(t), \tag{16}$$

where $\alpha_k(t)$ represents the instantaneous amplitude of the kth IMF.

By these definitions, the R_v and E of each IMF can be calculated out in real time.

3. Application of a Feature Vector Extraction Method for Simulating Chatter Signals

In order to evaluate whether BEMD and Hilbert transform can be a reliable solution for detecting chatter, an experimental program based on BEMD and Hilbert transform was designed (Figure 1). The process consists of the following steps: (1) signal acquisition and pre-processing, such as signal denoising and reconstruction; (2) applying the BEMD and extraction criterion of true IMFs to obtain all true IMFs; (3) applying the Hilbert transform to true IMFs to obtain Hilbert and marginal spectrum distributions; (4) calculating the real-time variance and instantaneous energy of IMFs, and then superposing and normalizing the feature vectors, respectively; and (5) predicting grinding chatter by contrasting the changes of the feature vectors.

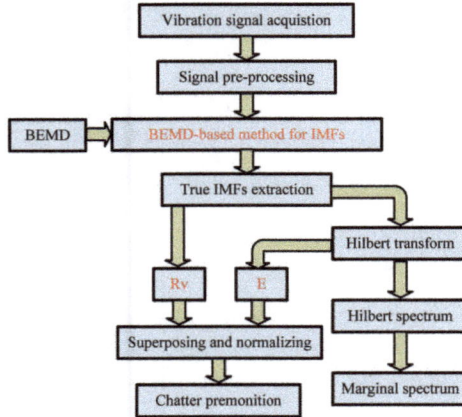

Figure 1. The process of extracting feature vectors from grinding chatter signals.

3.1. The Chatter Vibration Signal Generator

For the sake of demonstration, a BEMD is adopted to process a simulated chatter signal. According to the mechanism of the chatter and the characteristics of frequency and time domains in the grinding process, a chatter vibration signal generator was established through the Simulink module of MATLAB (2010a, MathWorks Company, Natick, MA, USA). The sampling frequency was set as 800 Hz, as shown in Figure 2.

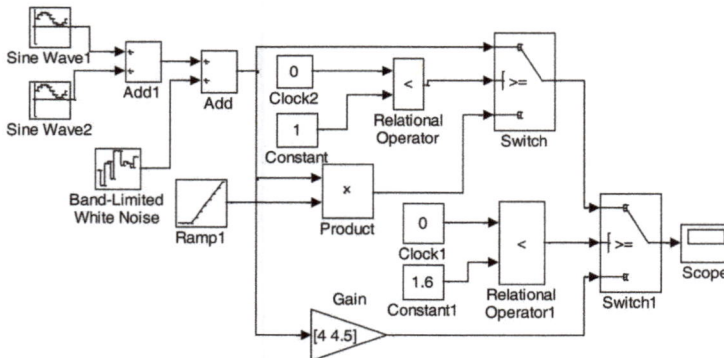

Figure 2. Grinding chatter signal generator.

The principle of this chatter generation system can be described as follows. A vibration signal is generated by two sine wave generators and one white noise generator. The frequency of the first sine signal consists of 50 rad/s and 120 rad/s, and the other one consists of 40 rad/s and 80 rad/s. The phase of the second sine signal is shifted by 4 rad. Thus, the two sine signals and white noise can be expressed as:

$$x(t) = 4\sin(120t) + 4\sin(50t) + 0.3 \text{ and } y(t) = 2\sin(80t + 4) + 2\sin(40t + 4) - 0.2, \quad (17)$$

$$c_1(t) = rand\,(2001, 1) - 0.5 \text{ and } c_2(t) = 0.4\,rand\,(2001, 1) - 0.2, \quad (18)$$

This system relies on the clock module to control current simulation time. When $t \leq 1$ s, the output is a harmonic vibration signal, in order to simulate stable grinding. When $1 < t \leq 1.6$ s, the output is a harmonic vibration signal multiplied with a slant signal, so as to simulate grinding chatter. Furthermore, when $1.6 < t \leq 2.5$ s, the output is the original harmonic vibration signal multiplied by a gain coefficient, in order to simulate stable chattering. For the sake of simplicity, this chatter generation system can be expressed in a mathematical form:

$$s(t) = \begin{cases} x(t) + c_1(t) + i(y(t) + c_2(t)) & 0 \leq t \leq 1 \\ (x(t) + c_1(t))(1 + 4.5(t-1)) + i((y(t) + c_2(t))(1 + 6(t-1)) & \\ 1 < t \leq 1.6 & \\ 4(x(t) + c_1(t)) + i4.5(y(t) + c_2(t)) & 1.6 < t \leq 2.5 \end{cases} \tag{19}$$

The chatter signal generated by this system is shown in Figure 3. It is clearly seen that signal amplitude is small during stable grinding, but significantly increases after 1 s. Then after 1.6 s, the amplitude becomes steady when the grinder reaches stable chatter. Hence the trend and distribution are similar to experimental chatter images in [7,32], meaning that this system simulates chatter signals well.

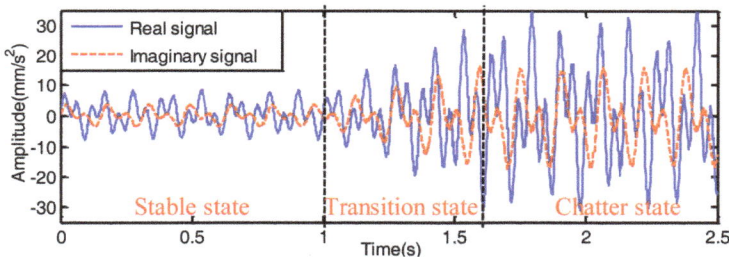

Figure 3. Complex-valued chatter signal generated by chatter generation system.

3.2. The Application of BEMD to Simulated Chatter Signal

Decomposing this chatter signal by BEMD sets 64 projection directions and 10 iterations. The extraction criterion of true IMFs introduced in Section 2.1.2 can be applied to the decomposed IMFs, and the correlation coefficients of each IMF are shown in Table 2.

Table 2. Correlation coefficients of IMFs of the simulated chatter signal.

IMF No.	IMF1	IMF2	IMF3	IMF4	IMF5
Correlation Coefficient	0.7180	0.7175	0.1222	0.0366	0.0363

Compared with the data in Table 2, only the first two IMFs, which have relatively high correlation with the original signals, should be reserved. The other three IMFs have to be removed and classified as a part of the residual. The true IMFs derived from the simulated chatter signal are shown in Figure 4.

In Figure 4, it is clearly seen that the phase shifting and synchronization information between the real part and imaginary components are well preserved and detected by BEMD. The true IMFs (i.e., IMF1 and IMF2) also clearly reveal the amplitude and frequency components of the signals which were initially set in the chatter signal generator.

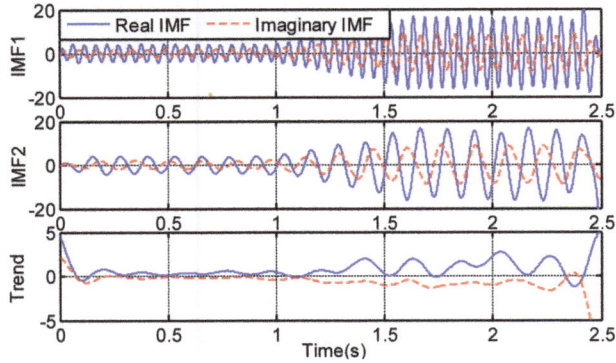

Figure 4. True IMFs of the simulated chatter signal.

3.3. *The Application of Hilbert Transform to Simulated Chatter Signals*

By employing the Hilbert transform to true IMFs, the Hilbert and marginal spectra in Figures 5 and 6 are obtained.

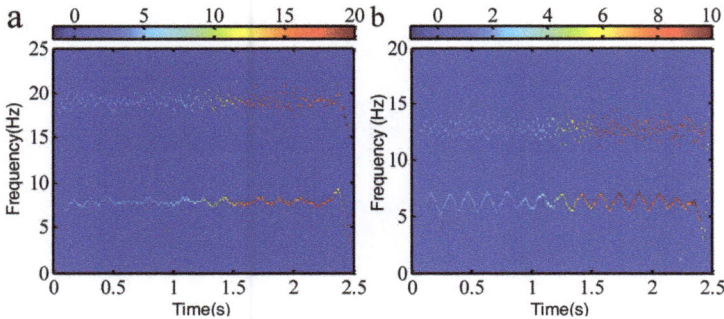

Figure 5. (**a**) Hilbert spectrum of real component; (**b**) Hilbert spectrum of imaginary component. Colors indicate the amplitude magnitude.

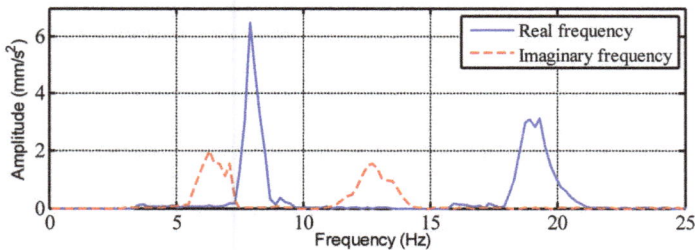

Figure 6. Marginal spectrum of simulated chatter signal.

In Figure 5, it can be clearly seen that the main amplitudes of the real IMFs are distributed around 8 and 19 Hz, and the imaginary IMFs are around 6.4 and 12.8 Hz. However, note that after 1 s, the amplitude increases and reaches a maximum after 1.6 s due to the onset of chatter. That is to say, chatter results in increased amplitude. In Figure 6, the frequency components are accurately revealed;

the peak amplitude values correspond with the frequency components (40, 50, 80 and 120 rad/s, i.e., 6.4, 8, 12.8 and 19 Hz) as set in the chatter generation system. It is clear that the amplitude at low frequency is larger than at high frequency, because the main frequency of vibration becomes low at the onset of grinding chatter. Consequently, the BEMD and Hilbert transform can be successfully combined to extract the actual features from the chatter signals.

3.4. Selection of the Ideal Number of Projection Directions

According to the simulated signal, variation in the correlation coefficient, frequency and calculation time, along with the change in direction number, can be calculated as shown in Figure 7. The trends of correlation coefficient and frequency variation tend to be stable when the maximum projection number is set to 32. Higher numbers are not discussed in this paper.

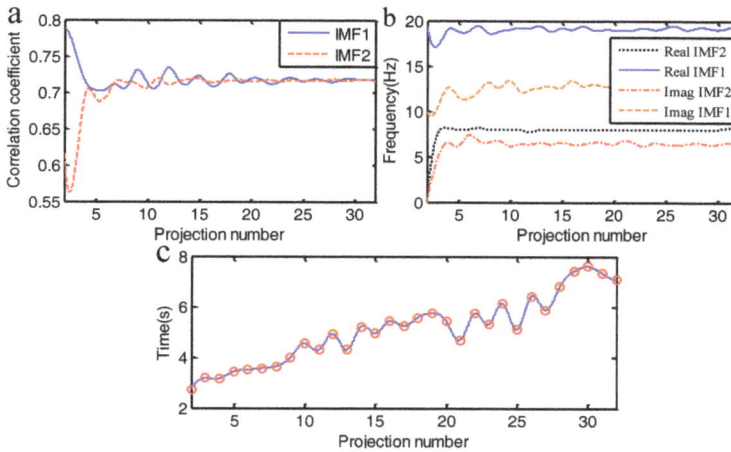

Figure 7. (a) The correlation coefficient of each true IMF; (b) the frequency of each true IMF; (c) calculation time for the BEMD procedure.

The degree of stability of the correlation coefficient and frequency for the true IMFs is shown in Figure 8.

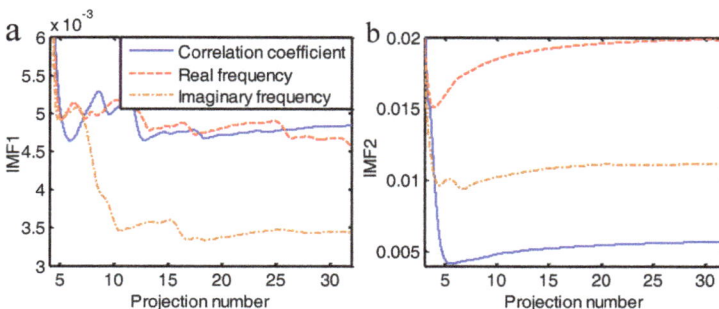

Figure 8. (a) The stability degree of correlation coefficient and frequency from IMF1; (b) the stability degree of correlation coefficient and frequency from IMF2. Blue solid lines represent the correlation coefficient, red dashed lines represent the frequency components of the real part of IMF1, yellow dotted lines represent the frequency components of the imaginary part of IMF2.

In light of the indicators from the above figures, it is clearly seen that the correlation coefficient and frequency components of the signals are almost stable when the direction number is greater than 16. On the other hand, the computational efficiency of the BEMD algorithm gradually reduces with increasing projection number. In general, the ideal number of projection directions is 16, which provides a basis for subsequent BEMD operations.

3.5. Chatter Feature Vector Extraction Based on BEMD and Hilbert Transform

By BEMD decomposition and extraction of true IMFs, the chatter signal is decomposed into a series of true IMFs and a residue. According to Equations (15) and (16), the real-time variance and instantaneous energy of each true IMF can be calculated and then superposed, respectively, to construct synthetic chatter feature vectors for chatter detection and identification. For the sake of comparison and analysis, the chatter feature vectors should be normalized to keep their value between 0 and 1 [33]. The superposed and normalized real-time variance and instantaneous energy of IMFs are shown in Figure 9.

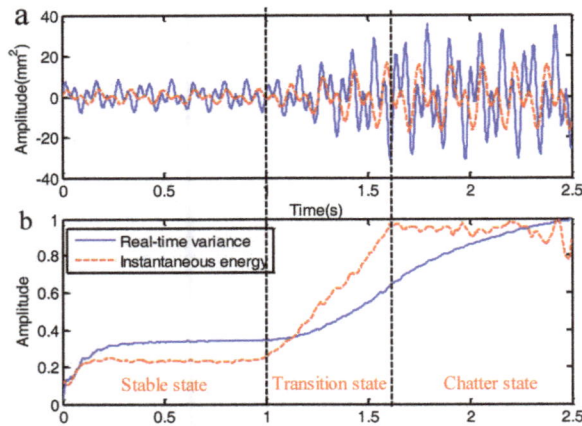

Figure 9. (a) The simulated chatter signal; (b) superposed and normalized real-time variance and instantaneous energy.

In order to achieve good recognition in the above figure, the maximum and mean values of real-time variance and instantaneous energy in various grinding states can be determined by removing the interference of endpoints (Figure 10).

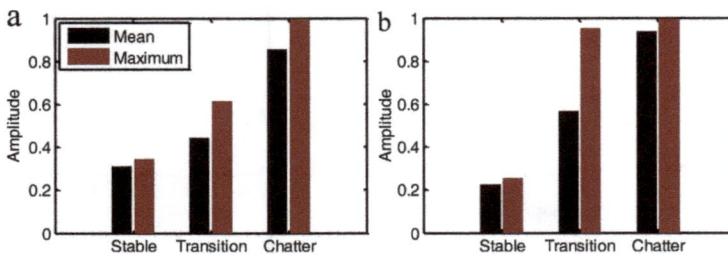

Figure 10. (a) Maximum and mean values of real-time variance; (b) maximum and mean values of instantaneous energy.

From Figures 9 and 10, it can be clearly seen that real-time variance and instantaneous energy in various grinding states exhibit different behaviors. The amplitude of these feature vectors is almost steady in the stable grinding stage, and the maximum and mean values are both very close. While the amplitude drastically increases once the signal enters the transition stage, the mean value of two feature vectors increases 1.4 and 2.5 times respectively, and the maximum increases 1.8 and 3.7 times. Moreover, when the signal is in a chatter stage, the instantaneous energy fluctuates within a certain range that the mean value and maximum are close. However, as real-time variance increases continuously with time, the mean and maximum increase 1.9 and 1.63 times when compared to the transition stage. Though it is difficult to distinguish the transition and chatter states exactly, it is feasible to clearly determine the onset of grinding chatter. Therefore, these two feature vectors can be applied as early chatter detectors.

4. Grinding Experiments Validation

4.1. Grinding Experiments

In order to further validate the feasibility of BEMD for grinding chatter detection, an experimental method was designed to acquire various vibration signals from different grinding parameters for a CNC guideway grinder (KD4020X16, Hangzhou Hangji Machine Tool Co., Ltd., Hangzhou, China). An integral electronic piezoelectric (IEPE) accelerometer sensor with a supporting dynamic signal test and analysis system (TST5912) were used to collect grinding vibration signals (Figure 11).

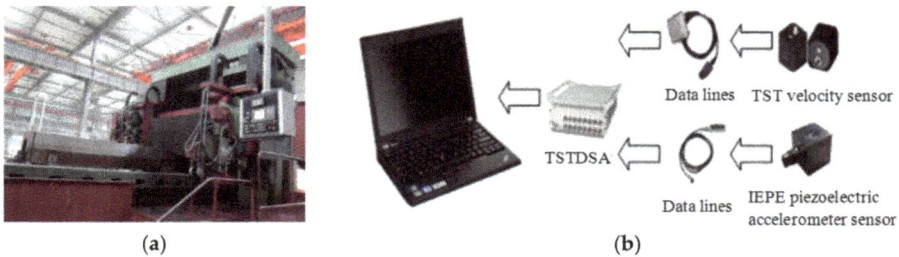

(a)   **(b)**

Figure 11. (a) CNC guideway grinder KD4020X16; (b) TST5912 analysis system.

In practice, the grinder is more sensitive to rotational speed, feeding speed and grinding depth of the grinding wheel, which actually contribute to imbalances in grinding vibration. Thus, this experiment was carried out with the following steps:

- Keeping the workpiece feeding speed and grinding wheel depth constant while gradually increasing the rotational speed;
- Keeping the feeding speed and rotational speed constant while gradually increasing the grinding depth;
- Keeping the rotational speed and depth of grinding constant while gradually increasing the feeding speed.

Thus, the parameters are listed in Table 3.

Table 3. Grinding parameters.

Grinding Wheel Material	Size of Wheel (Diameter × Width, mm)	Work-Piece Material	Size of Workpiece (L × W × D, mm)	Rotational Speed (r·min^{-1})	Feeding Speed (m·s^{-1})	Grinding Depth (μm)
Green silicon carbide	600 × 150	Gray cast iron 250	3050 × 500 × 500	800~1000	0.381 0.254 0.210	5 10 15

According to the experimental conditions shown in Table 3, eight piezoelectric acceleration sensors were used to test vibration acceleration signals from the grinding wheel spindle, motor and machine column in various directions [34]. The position of the sensors and their corresponding sensitivities are given in Table 4.

Table 4. The position of sensors and the corresponding sensitivities.

Sensor Number	1	2	3	4	5	6	7	8
Sensitivity (mv/g)	9.9	10.6	10.4	10.1	10.4	10.5	10.2	10.1
Position	Column Z	Column X	Spindle Z	Spindle Y	Motor Z	Motor X	Motor Y	Column Y

As per the above steps, 80 groups of grinding signals were obtained, of which 45 were in a stable grinding state, while 35 were in a chatter state. Based upon practical experience and considerable experimental data, it has been demonstrated that the chatter of wheel spindles occurs more in the x- and z-directions than in the y-direction. Thus, for convenience of this presentation, this paper selects parts of x- and z-direction chatter data as an original complex-valued signal. However, in the signal acquisition stage, the disturbing noise will inevitably be mixed in grinding vibration signal duo to the influence of many random factors, directly affecting detection accuracy of grinding chatter. It is reasonable to carry out pre-processing to eliminate noise for signals. Moreover, the wavelet de-noising technique, which has the characteristics of time domain localization, multi-resolution and flexibility of selecting wavelet basis, has been widely employed in signal processing [35,36]. Thus, in this paper, the noise is removed from the original complex-valued signal based on wavelet de-noising. Meanwhile, we resample the complex-valued signal based on Nyquist sampling theory [37]. The new, reconstructed complex-valued signal is shown in Figure 12.

From Figure 12, it is found that grinding chatter emerges at about 5–14 s and the transitional phase lasts almost 9 s. It is clearly seen that the amplitude of the vibration signal rapidly expands when the grinder starts chattering, then the amplitude becomes steady when the grinder enters stable chatter. However, the signal vibrates more markedly compared with the stable grinding state. Moreover, note that the abnormal signals derived from the joints of the workpiece will not greatly affect the overall signal analysis.

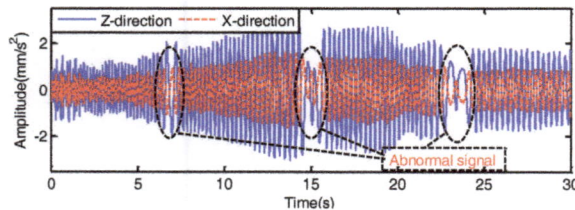

Figure 12. Newly reconstructed complex-valued signal from experimental data.

4.2. The Application of BEMD and Hilbert Transform to Experimental Chatter Signals

As previously mentioned, BEMD demonstrates its powerful capability in terms of processing the two-dimensional simulation chatter signal and extracting true IMFs from it. Moreover, the simulation results well illustrate the phase information and synchronization between real and imaginary parts of the IMFs, which is of significant benefit to applying the BEMD method to real-world chatter. Thus, the BEMD is applicable for decomposing the experimental signal above, which sets 16 projection directions (according to the selection of the ideal projection direction number, Section 3.4) and sets 10 iterations. Then, the extraction criterion based on the correlation coefficient can be applied to the extracted IMFs. The correlation coefficients of each IMF are shown in Table 5.

Table 5. Correlation coefficients of each experimental IMF.

IMF No.	IMF1	IMF2	IMF3	IMF4	IMF5	IMF6	IMF7
Correlation coefficient	0.9694	0.2355	0.0835	0.0242	0.0351	0.0144	0.0264

From Table 5, it is clearly seen that the first three IMFs, which have relatively high correlation with the original signals, should be reserved. The other four IMFs have to be removed and classified as a part of the residual. So, the first three IMFs can be recognized as the true IMFs, which contain the main frequency components of the signal. The eventual decomposition result is shown in Figure 13.

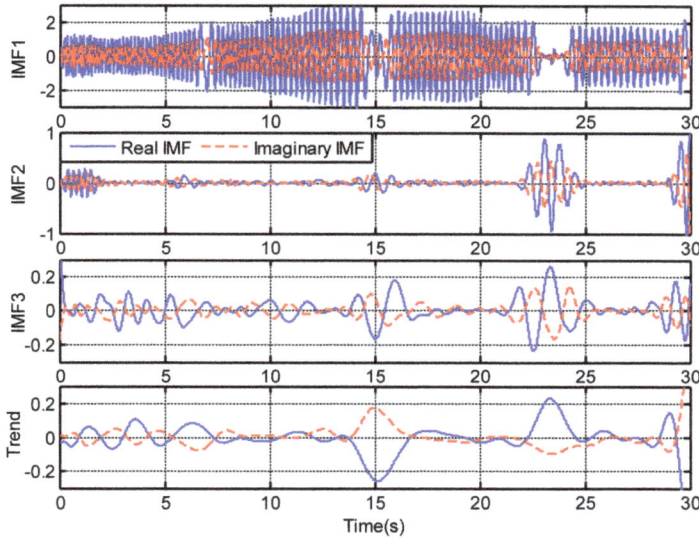

Figure 13. The true IMFs of experimental chatter signal.

Then, the Hilbert transform is performed on each true IMF to obtain the marginal spectrum, as shown in Figure 14.

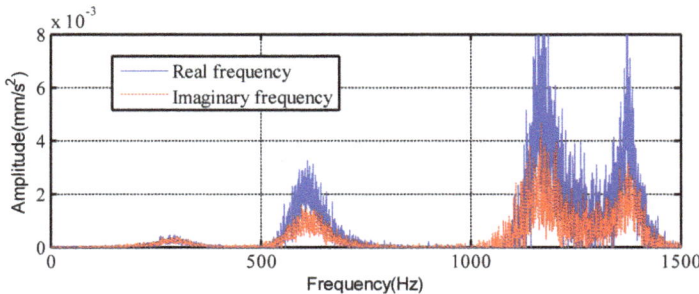

Figure 14. Marginal spectrum of x-direction IMFs.

In Figure 13, the marginal spectrum of both x- and z-directions shows the same frequency components, of about 300, 580, and 1200 Hz. Yet the z-direction IMFs have larger amplitudes relative to the x-direction, as signals tend to vibrate more in the z-direction according to previous studies.

4.3. Chatter Feature Vector Extraction from Experimental Chatter Signals

Based on the same principle and method in Section 3.5, the feature vectors can also be extracted from the experimental true IMFs. The superimposed and normalized real-time variance and instantaneous energy vectors with time are shown in Figure 15.

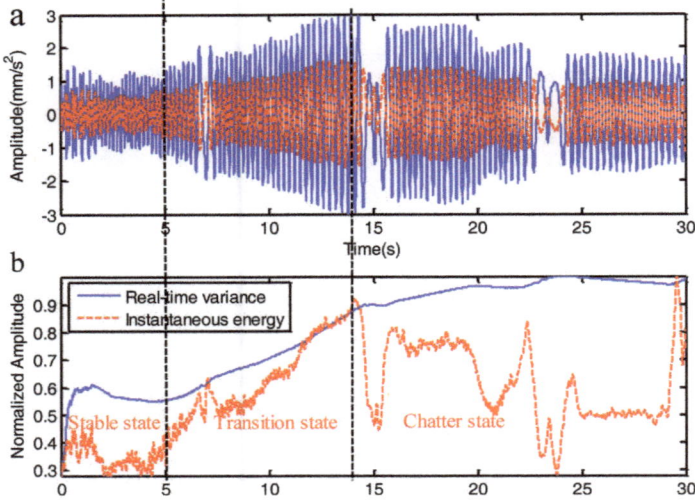

Figure 15. (a) Experimental chatter signal; (b) Superposed and normalized real-time variance and instantaneous energy.

We can see that the behavior of these two feature vectors exhibits the same change regulation as the simulation chatter signal described previously in Section 3.5. However, in order to detect and identify the onset of grinding chatter in time, it is crucial to extract feature vectors during the early stage of chatter development. It shows that the onset time of chatter is nearly 5 s, and that appropriate suppression methods should be carried out to overcome the grinding chatter.

5. Conclusions

In this paper, bivariate empirical mode decomposition (BEMD) and Hilbert transform were demonstrated as reliable methods for processing simulated and experimental chatter signals. Moreover, the extraction criterion of intrinsic mode functions (IMFs) based on correlation coefficients successfully distinguished true IMFs from spurious components. The method of selecting the ideal number of projection directions was investigated. Additionally, real-time variance and instantaneous energy were recognized as effective predictors of chatter. Accordingly, the following conclusions can be made:

- The BEMD and Hilbert transform methods were combined to decompose a simulated chatter signal derived from a grinding vibration signal generator. This was validated with experimental data collected from a CNC guideway grinder (KD4020X16). The results illustrate the good behavior of BEMD in terms of processing nonstationary and nonlinear signals, and providing good indicators of signal phase shifting and synchronization information. Meanwhile, Hilbert and marginal spectra accurately revealed the actual characteristics of signals.
- The extraction criterion of true IMFs based on correlation coefficients is a reliable technique that successfully identifies and estimates the spurious components and reserves the main frequency bands of great import for extracting the actual vibration mode and corresponding features of the time and frequency domains.

- The most ideal number of projection directions was investigated by assessing the indicators of computational time and stability degree (SD) which change along with direction number. According to the results, the ideal number of projection directions is 16.

- Real-time variance and instantaneous energy were proposed as appropriate feature vectors for chatter detection and identification. These can be applied as predictors of chatter in order for chatter suppression to be applied in a timely manner.

Acknowledgments: The work described in this paper was supported by the National Natural Science Foundation of China (Grant No. 51475432), the Zhejiang Provincial National Natural Science Foundation of China (Grant No. LZ13E050003), and the State Key Program of National Natural Science of China (Grant No. U1234207).

Author Contributions: Huanguo Chen and Chunshao Huang conceived and designed the experiments; Yongyu Yi and Jiacheng Qian performed the experiments; Wenhua Chen analyzed the data; Chuanyu Wu contributed reagents/materials/analysis tools; Jianyang Shen wrote the paper.

Conflicts of Interest: The authors declare no conflict of interest.

References

1. Zhao, R.; Zhang, H.; Fan, L.; Li, C.; Cai, Y. The analysis of the remote fault diagnosis expert system and research of key techniques for the NC grinding machine. *Res. CNC Grinder Syst. Online Detect. Mechatron.* **2005**, *1*, 73–74.

2. Huang, C.; Chen, J. Development of Dual-Axis MEMS Accelerometers for Machine Tools Vibration Monitoring. *Multidiscip. Digit. Publ. Inst.* **2011**, *9*, 1293–1306. [CrossRef]

3. Han, Z.; Zhang, Y. Studies and Developments about Grinding Chatter of Machine Tools. *Precis. Manuf. Autom.* **2001**, *4*, 16–18.

4. Ya, W. *Chatter and Control of Machine Tool Cutting System*; Science Publishing Company: New York, NY, USA, 1993; pp. 154–196.

5. Wang, L.; Liu, X.; Jia, Q. Studies and Developments about Cutting Chatter of Machine Tools. *Mach. Tools Hydraul.* **2004**, *11*, 1–5.

6. Meyer, J.; Harrington, B.; Agrawal, B.; Song, G. Vibration Suppression of a Spacecraft Flexible Appendage Using Smart Material. *Smart Mater. Struct.* **1998**, *2*, 95–104. [CrossRef]

7. Tansel, I.N.; Wang, X.; Chen, P.; Yenilmez, A.; Ozcelik, B. Transformations in machining, Part 2. Evaluation of machining quality and detection of chatter in turning by using s-transformation. *Int. J. Mach. Tools Manuf.* **2006**, *46*, 43–50. [CrossRef]

8. Yao, Z.; Mei, D.; Chen, Z. On-line chatter detection and detection based on wavelet and support vector machine. *J. Mater. Process. Technol.* **2010**, *210*, 713–719. [CrossRef]

9. Devillez, A.; Dudzinski, D. Tool vibration detection with eddy current sensors in machining process and computation of stability lobes using fuzzy classifiers. *Mech. Syst. Signal Process.* **2007**, *21*, 441–456. [CrossRef]

10. Gradisek, J.; Govekar, E.; Grabec, I. Using coarse-grained entropy rate to detect chatter in turning. *J. Sound Vib.* **1998**, *214*, 830–841. [CrossRef]

11. Nair, U.; Krishna, B.M.; Namboothiri, V.N.N.; Nampoori, V.P.N. Permutation entropy based real-time chatter detection using audio signal in turning process. *Int. J. Adv. Manuf. Technol.* **2010**, *46*, 61–68. [CrossRef]

12. Mei, Z.; Shi, H.; Liu, J. An Effective and Comprehensive Criterion for Early Stage Diagnosis of Chatter in Metal Cutting Process. *J. Cent. China Inst. Technol.* **1985**, *5*, 87–94.

13. Liu, X.; Jiang, X.; Xue, B. Synthetic Neural Network Based Method for Diagnosing Multi-Complex Abnormalities in Machining Operation. *Chin. J. Mech. Eng.* **1996**, *3*, 70–75.

14. González-Brambila, O.; Rubio, E.; Jáuregui, J.C.; Herrera-Ruiz, G. Chattering detection in cylindrical grinding processes using the wavelet transform. *Int. J. Mach. Tools Manuf.* **2006**, *46*, 1934–1938. [CrossRef]

15. Zhang, L.; Wang, C.; Song, G. Health Status Monitoring of Cuplock Scaffold Joint Connection Based on Wavelet Packet Analysis. *Shock Vib.* **2015**, *2015*. [CrossRef]

16. Zhang, X.; Qi, Y.; Zhu, M. Characteristic Analysis of White Gaussian Noise in S-Transformation Domain. *J. Comput. Commun.* **2014**, *2*, 20–24. [CrossRef]

17. Berger, B.S.; Minis, I.; Rokni, M.; Papadopoulos, M.; Deng, K.; Chavalli, A. Cutting state identification. *J. Sound Vib.* **1997**, *200*, 15–29. [CrossRef]

18. Ibrahim, N. Tansel. Modelling 3-D cutting dynamics with neural networks. *Int. J. Mach. Tools Manuf.* **1992**, *32*, 829–853.

19. Li, L.; Song, G.; Ou, J. Hybrid active mass damper (AMD) vibration suppression of nonlinear high-rise structure using fuzzy logic control algorithm under earthquake excitations. *Struct. Control Health Monit.* **2011**, *6*, 698–709. [CrossRef]

20. Huang, N.E.; Shen, Z.; Long, S.R.; Wu, M.C.; Shih, H.H.; Zheng, Q.; Yen, N.C.; Tung, C.C.; Liu, H.H. The empirical mode decomposition and the Hilbert spectrum for nonlinear and non-stationary time series analysis. *Proc. R. Soc. A Math. Phys. Eng. Sci.* **1998**, *454*, 903–995. [CrossRef]

21. Ma, H.; Fan, H.; Mao, Q.; Zhang, X.; Xing, W. Noise Reduction of Steel Cord Conveyor Belt Defect Electromagnetic Signal by Combined Use of Improved Wavelet and EMD. *Multidiscip. Digit. Publ. Inst.* **2016**, *9*, 2–11. [CrossRef]

22. Rilling, G.; Flandrin, P. Bivariate empirical mode decomposition. *IEEE Signal Process. Lett.* **2007**, *14*, 936–939. [CrossRef]

23. Jun, Z.; Ming, L.; Yuming, L.; Xiaoqin, Y. Fatigue Fracture Image Segmentation Based on BEMD. *Fail. Anal. Prev.* **2011**, *6*, 70–73.

24. Xu, X.; Li, H.; Wang, A. The application of BEMD to multi-spectral image fusion. In Proceedings of the 2007 International Conference on Wavelet Analysis and Pattern Recognition, Beijing, China, 2–4 November 2007; Volume 1, pp. 448–452.

25. Feng, L.; Qiang, X.; Hui, L. Image compression research based on BEMD. *Comput. Eng. Appl.* **2010**, *46*, 183–185.

26. Huang, W.; Sun, Y. A New Image Watermarking Algorithm Using BEMD Method. In Proceedings of the 2007 International Conference on Communications, Circuits and Systems, Kokura, Japan, 11–13 July 2007; pp. 558–592.

27. Yang, W.; Court, R.; Tavner, P.J. Bivariate empirical mode decomposition and its contribution to wind turbine condition monitoring. *J. Sound Vib.* **2011**, *330*, 3766–3782. [CrossRef]

28. Shen, J.; Chen, H.; Yi, Y.; Wu, J.; Li, Y.; Huang, C. Comparison between EMD and BEMD in Extracting Features for Grinding Chatter Signals. In Proceedings of the 2016 11th International Conference on Reliability, Maintainability and Safety, Hangzhou, China, 26–28 October 2016.

29. Chen, H.G.; Yan, Y.J.; Chen, W.H.; Jiang, J.S.; Yu, L.; Wu, Z.Y. Early damage detection in composite wingbox structures using Hilbert-Huang Transform and Genetic Algorithm. *Int. J. Sturct. Health Monit.* **2007**, *6*, 281–297. [CrossRef]

30. Li, X.; Li, Z.; Wang, E.; Feng, J.; Kong, X.; Chen, L.; Li, B.; Li, N. Analysis of natural mineral earthquake and blast based on Hilbert–Huang transform (HHT). *J. Appl. Geophys.* **2016**, *128*, 79–86. [CrossRef]

31. Yu, X.; Ding, E.; Chen, C.; Liu, X.; Li, L. A Novel Characteristic Frequency Bands Extraction Method for Automatic Bearing Fault Diagnosis Based on Hilbert Huang Transform. *Multidiscip. Digit. Publ. Inst.* **2015**, *15*, 27869–27893. [CrossRef] [PubMed]

32. Ming, W.; Meng, F.; Ziliang, Y.; Tao, J. Prediction of Grinding Chatter Based on the ARIMA. *J. Beijing Univ. Technol.* **2016**, *42*, 609–613.

33. Han, H.; Cai, C. Comparison study of normalization of feature vector. *Eng. Appl.* **2009**, *45*, 117–119.

34. Yi, T.H.; Li, H.N.; Song, G.; Zhang, X.D. Optimal sensor placement for health monitoring of high rise structure using adaptive monkey algorithm. *Struct. Control Health Monit.* **2015**, *22*, 667–681. [CrossRef]

35. Cui, H.; Qiao, Y.; Yin, Y.; Hong, M. An investigation of rolling bearing early diagnosis based on high-frequency characteristics and self-adaptive wavelet de-noising. *Neurocomputing* **2016**, *216*, 649–656. [CrossRef]

36. Guo, X.; Li, Y.; Suo, T.; Liang, J. De-noising of digital image correlation based on stationary wavelet transform. *Opt. Lasers Eng.* **2016**, *90*, 161–172. [CrossRef]

37. Li, H. The selection of Non-Nyquist Samping Frequency. *J. Sens. Technol.* **1998**, *4*, 15–19.

![applied sciences logo] *applied sciences*

MDPI

Article

Optimization Design of Coupling Beam Metal Damper in Shear Wall Structures

Zhe Zhang [1], Jinping Ou [2], Dongsheng Li [2,*] and Shuaifang Zhang [3]

[1] Faculty of Vehicle Engineering and Mechanics, Dalian University of Technology, State Key Laboratory of Structural Analysis for Industrial Equipment, Dalian 116024, China; zhangzhe@mail.dlut.edu

[2] Faculty of Infrastructure Engineering, Dalian University of Technology, Dalian 116024, China; oujinping@dlut.edu.cn

[3] Department of Mechanical Engineering, Penn State University, State College, PA 16802, USA; zhangsf1988@gmail.com

*   Correspondence: lidongsheng@dlut.edu.cn; Tel.: +86-411-8470-6416

Academic Editor: Gangbing Song
Received: 10 November 2016; Accepted: 22 January 2017; Published: 3 February 2017

Abstract: The coupling beam damper is a fundamental energy dissipation component in coupling shear wall structures that directly influences the performance of the shear wall. Here, we proposed a two-fold design method that can give better energy dissipation performance and hysteretic behavior to coupling beam dampers. First, we devised four in-plane yielding coupling beam dampers that have different opening types but the same amount of total materials. Then the geometry parameters of each opening type were optimized to yield the maximum hysteretic energy. The search for the optimal parameter set was realized by implementing the Kriging surrogate model which iterates randomly selected input shape parameters and the corresponding hysteretic energy calculated by the infinite element method. By comparing the maximum hysteretic energy in all four opening types, one type that had the highest hysteresis energy was selected as the optimized design. This optimized damper has the advantages of having a simple geometry and a high dissipation energy performance. The proposed method also provided a new framework for the design of in-plane coupling beam dampers.

Keywords: coupling beam dampers; hysteretic behavior; carrying capacity; Kriging surrogate model

1. Introduction

Based on the assumptions of ductility design, coupled shear wall structures should satisfy the features of a "strong wall with a weak coupling beam". Such a setup will guarantee that a structure has enough stiffness to allow the shear wall to not fail first during rare earthquake events. In these scenarios, a plastic hinge would dissipate the energy and ensure that the shear wall will not yield too early. The ductility of coupling beams is one of the most important factors that influence the ultimate bearing capacity, and both shear force and deformation requirements should be met by a coupling beam under reciprocating load which occurs in earthquakes. A great deal of research has been done around the globe to explore approaches to satisfy the ductility requirements of shear wall beam structures.

Paulay et al. [1] improved the behavior of commonly-used reinforcement concrete using diagonal crossing reinforcement, which can remarkably improve anti-seismic features. Coull [2] proposed placing stiff coupling beams on the top of a shear wall structure in order to strengthen the structure's integrity and to improve lateral stiffness. Kelly and Skinner et al. [3,4] raised the idea of metallic dampers, which, made out of metal, could have the advantage of the material's plastic yield feature to dissipate earthquake energy inputs. Fortney et al. [5–7] developed the concept of a changeable steel coupling beam, which features a weakened middle part (namely the safety wire) that can dissipate energy via shear yielding and is easy to switch after damage. Chung et al. [8] used bending as a damper energy dissipative form and set up a frictional damping device in the middle of a coupling beam that can decrease the response of the shear wall structure during earthquakes. Kim et al. [9,10] developed a compound energy-dissipation damper, in which high-damping rubber material and two U-shaped steel plates were combined to acquire high energy dissipation. Lyons [11] developed a new type of visco-elasticity damper which could potentially replace the coupling beam damper. Mao et al. [12] proposed a new shape for memory alloy dampers that restore their original shape after damage, and do not require replacement as coupling beam dampers do. Although the above-mentioned designs of dampers have led to progress at some points, they also made the structure coupling beams complicated enough to hardly justify the improvements in the energy dissipation capacity. In addition, parameter optimization which can potentially improve damper performance is often overlooked in the design of coupling beam dampers.

To this end, we proposed a two-fold method for the design of a simple and easy-to-make metallic damper that features a high energy dissipation capability. Based on the existing dampers, four types of openings in the damper are developed and studied. A constitutive law is provided to establish the relationship between design parameters, including the opening size and thickness, and the energy dissipation capacity. A Kriging surrogate model is then introduced with a shuffled complex evolution (SCE) intelligent optimization algorithm to search the optimized design parameters that can lead to the maximum hysteresis energy. The proposed work can also be used as a platform for opening parameter design with high surrogate accuracy and global optimization efficiency.

2. Coupling Beam Metal Dampers

2.1. Typical Pores in Coupling Beam Metal Dampers

Following the principle of equivalent strength, large pores should open in the middle part of the slab with no weakening at the two ends. We chose the following four types of dampers with a length of 280 mm and a width of 160 mm, which can be seen in Figure 1. Based on the premise of equal materials of the same size, the limit boundary of the pore geometrical parameters can be determined using pore porosity limitations of 10%–25%. The four kinds of models are described as follows:

Plate with split elliptic pores (SP1). The independent variables are elliptical semi-major axis a and semi-minor axis b, and the dependent variable is the thickness of plate t; a is in the interval of 53.42 to 84.46 and b is in the interval of 26.71 to 42.23.

Plate with one central elliptic pore (SP2). The independent variables are elliptical semi-major axis a and semi-minor axis b, which have the same upper and lower bounds as SP1. The dependent variable is the thickness of plate t.

Plate with single column of uniform pores (SP3). The independent variables are the length of row pore a and the distance between two rows b, respectively. The dependent variables are the width of row pores c and the thickness of plate t; a is in the interval [74, 67, 118, 06] and b is in the interval [19, 87, 27, 26].

Plate with two columns of uniform pores (SP4). The distance between the two columns is fixed at 60 mm. The independent variables and the dependent variables are similar to those of SP3, where the range of a is reset to [30, 68, 48, 50].

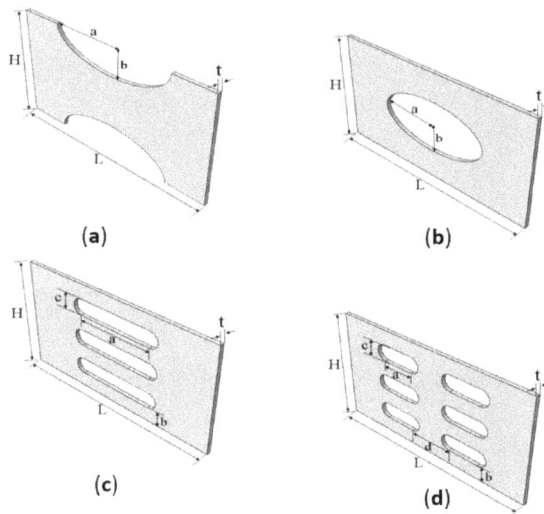

Figure 1. Section dimensions of the components: (**a**) Plate with split elliptic pores (SP1); (**b**) Plate with one central elliptic pore (SP2); (**c**) Plate with single column of uniform pores (SP3); (**d**) Plate with two columns of uniform pores (SP4).

2.2. Hysteresis Energy Calculation

The bilinear kinematic hardening model of the material was used as shown in Figure 2. The elastic modulus was 2.07×10^5 MPa, Poisson's ratio was 0.3, and the yield strength was 235 MPa. The elastic modulus after yielding was 0.004 times that of the initial modulus. The component was fixed at the left end, and was freely constrained in the X (horizontal) and Z (out-of-plane) direction at the right end. Vertical displacement was applied in the Y direction, and the right end of the component was applied to the vertical reciprocating displacement. The incremental displacement was 5 mm to control the loading rate. Then cycles of loading were done in total and the maximum displacement was 50 mm, and the displacement load is shown in Figure 3.

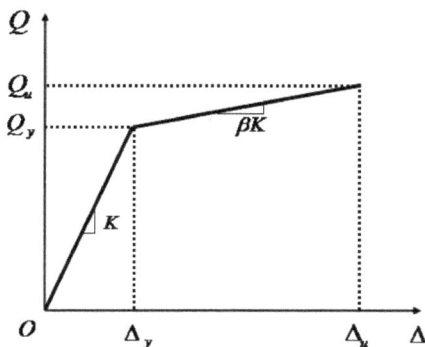

Figure 2. Curve of constitutive relation.

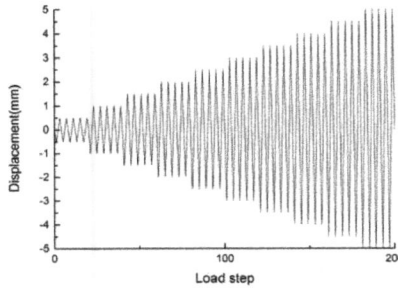

Figure 3. Load–displacement curve.

3. The Relationship between Pore Geometrical Parameters and Hysteresis Energy

3.1. Sampling Selection

Fifty samples of the each kind of metal damper were selected using the Latin hypercube sampling method. Then, the dependent variables, c, in SP3 and SP4SP4 could be deduced under the precondition of equal materials being used which are of the same size:

$$c = (160 - 4b)/3 \tag{1}$$

For the plate with elliptic pores, the thickness could be calculated using the following equation:

$$t = \frac{A}{280 \times 160 - \pi ab} \tag{2}$$

in which A represents the total material, and a and b are the semi-major axis and semi-minor axis, respectively. For SP3 and SP4SP4, the thickness of the plate could be written as the following

$$t = \frac{A}{280 \times 160 - 3(ac + \pi(\frac{c}{2})^2)} \tag{3}$$

$$t = \frac{A}{280 \times 160 - 6(ac + \pi(\frac{c}{2})^2)} \tag{4}$$

The sampling results for each kind of metal damper are shown in Figure 4.

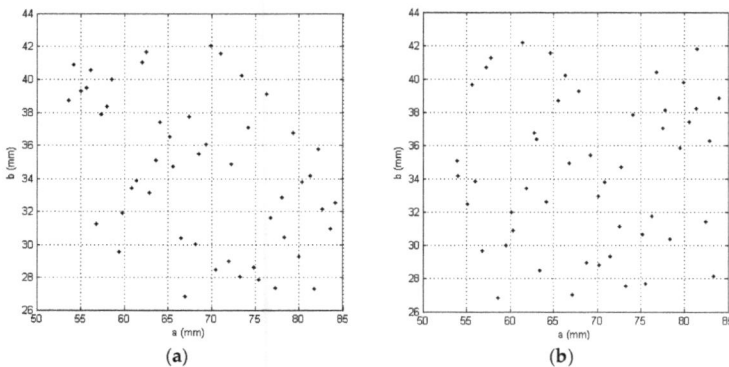

(a)

(b)

Figure 4. *Cont.*

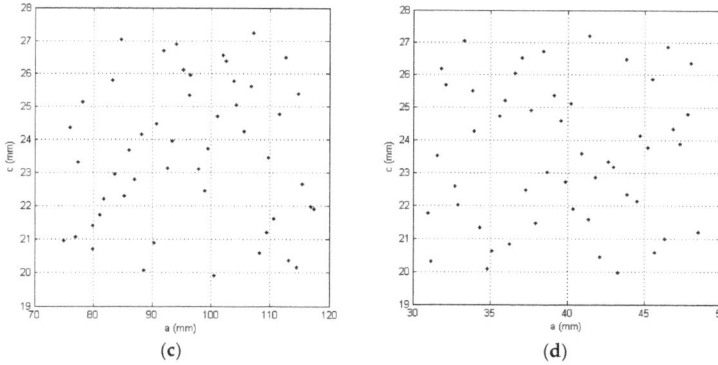

(c) (d)

Figure 4. Parameter sampling for coupling beam metal dampers: (**a**) SP1; (**b**) SP2; (**c**) SP3; and (**d**) SP4.

3.2. Construction of the Kriging Surrogate Model

Kriging surrogate model–based optimization has been applied in many fields over the past few decades. Simpson [13] applied the method to the design of the space shuttle, and compared it with the calculation accuracy and efficiency of the response surface. Lee [14] used the Kriging surrogate model to optimize the design of the cylindrical member crashing problem. Gao [15] used the Kriging model in order to reduce the warping of injection molding process components in order to optimize the design.

Kriging interpolation, as a semi-parameterized technique, including a regression portion and a non-parametric part, consists of two parts: polynomials and random distributions.

$$y(x) = F(\beta, x) + z(x) = f^T(x)\beta + z(x) \tag{5}$$

where beta is the regression coefficient, $f(x)$ is a polynomial in X, an in the design space, the simulated global approximation can be zero-order, first-order, or second-order polynomials; $z(x)$ is a random distribution error, providing the approximation of the analog local deviations, with statistical characteristics as follows:

$$E[z(x)] = 0 \tag{6}$$

$$Var[z(x)] = \sigma_z^2 \tag{7}$$

$$cov[Z(x_i), Z(x_j)] = \sigma_z^2 [R_{ij}(\theta, x_i, x_j)] \tag{8}$$

in which x_i, x_j are any two points in the training samples, $R_{ij}(\theta, x_i, x_j)$ is the correlation function with parameter θ to characterize the spatial correlation between the training sample points. Thus, the Kriging agent model treated any corresponding value as a random variable following a normal distribution, and the model is not limited to a particular form and is strongly flexible.

The response surface of the four kinds of metal dampers could be obtained, as shown in Figure 5.

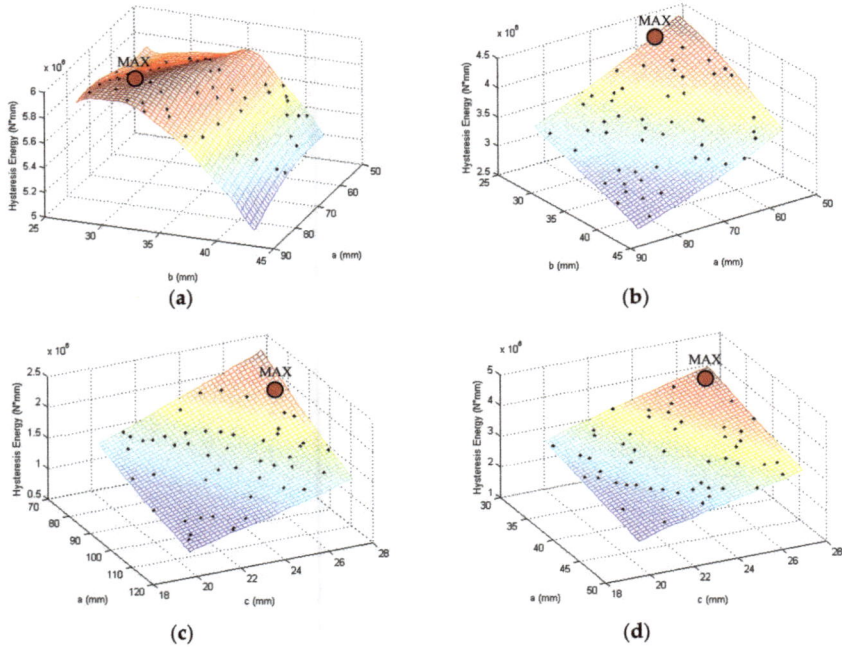

Figure 5. The Kriging model for four kinds of coupling beam metal dampers: (**a**) SP1; (**b**) SP2; (**c**) SP3; and (**d**) SP4.

3.3. Design Parameters and Hysteretic Energy Dissipation Relationship

The thickness of all kinds of metal dampers changes when independent variables change on the response surface. It can be seen that hysteresis energy decreases along with the bounds of the independent variables in the SP2, SP3, and SP4 cases, but the response surface for SP1 seems to be parabolic instead of a monotonic decay trend. The existing research results [16] also show that when the thicknesses of the four kinds of metal dampers are equal, the same pattern (above) can be seen.

4. Optimization of Pore Parameters Based on the Kriging Surrogate Model

4.1. Optimization Problem

The problem of choosing pore parameters of the metal dampers, in order to maximize hysteresis energy, arises:

$$\text{Find } x_1 \text{ and } x_2$$

$$Max(Y(x_1, x_2))$$

Subjected to Equations (1) and (2), also

$$lb_1 \leq x_1 \leq ub_1$$

$$lb_2 \leq x_2 \leq ub_2$$

in which Y is the hysteresis energy function of independent variables x_1 and x_2, and lb and ub are the lower and upper bounds.

The current optimal design problems need to be solved using an optimization algorithm. Classical optimization algorithms, such as the simplex method, the steepest descent method, and the sequential quadratic programming method, are all determined by local information (e.g., derivatives), so they are effective in a depth search but cannot effectively conduct a breadth search. These methods cannot jump out of the local optimum for classical optimization algorithms.

The SCE algorithm was proposed by Duan [17] to solve for the continuity of watershed hydrological model parameter selection problems. It combines certain complex search methods and the biological evolution principle in nature, in which every generation can be divided into several complexes that evolve independently. After some evolution, a new compound will be created by means of random restructuring in order to ensure the quality of the whole group's overall improvement.

4.2. Process of Pore Parameter Optimization Based on the Kriging Model

The Kriging model can not only predict responses using the new input parameters, but can also estimate optimal parameters satisfied with equality or inequality response constraints in both linear and nonlinear systems. When the original surrogate model is constructed, the quality of the model can be assessed according to the accuracy of the predictions:

$$SC = 1 - \frac{\sum_{j=1}^{q} [\hat{y}_j - y_j]^2}{\sum_{j=1}^{q} [y_j - \tilde{y}]^2} \tag{9}$$

where $\hat{y}_j$ and y_j are the jth component of the response vector of the surrogate model and the true value calculated via FE analysis, respectively; $\tilde{y}$ is the mean of all true values. Some scholars proposed to maximize the expected increasing (EI) add criteria [15], multiple-spot add criteria [18], experience semivariogram add criteria [19], etc. Here the EI sample adding criteria is used.

4.3. Maximizing the Expected Increasing (EI) Add Criteria

Maximizing the expected increase (EI) is considered as a forecast variance weighted method. In the design, the point x, prior to its response value calculated in response to the value of $y(x)$, is also unknown, but the Kriging agent model is able to predict its mean value $\tilde{y}$ and the variance sigma $\sigma^2(x)$. If the response to this optimal design value is y_k^*, it would improve the point target response value to $I(x) = y_k^* - y(x)$ for the minimization problem, which follows a normal distribution, so the probability density function is:

$$\frac{1}{\sqrt{2\pi}\sigma(x)} exp \left[-\frac{(y_k^* - I(x) - \tilde{y}(x))^2}{2\sigma^2(x)} \right] \tag{10}$$

Additionally, the response to the value of the target's increased expectations is:

$$E[I(x)] = \int_{I=0}^{I=\infty} I \left\{ \frac{1}{\sqrt{2\pi}\sigma(x)} exp \left[-\frac{(y_k^* - I(x) - \tilde{y}(x))^2}{2\sigma^2(x)} \right] \right\} dI \tag{11}$$

After integration:

$$E[I(x)] = \sigma(x)[u\Phi(u) + \phi(u)] \tag{12}$$

$$u = \frac{y_k^* - \tilde{y}(x)}{\sigma(x)} \tag{13}$$

in which Φ and ϕ are the positive probability distribution function and the probability density function, respectively.

Equation (10) is the sum of the two terms. The first is the predictive value $\tilde{y}$ of point x, multiplied by the probability of improving the current optimal response, which is the difference. A larger x will cause the predicted variance to be small, which means that it will find little predictive value and a more accurate prediction point. Forecasting the variance of the probability density function of the product, it is clear that when the predicted value is relatively large, the value is large, but the probability density function of a limited predictive value is not far from the current optimal response. In Equation (10) the probability distribution function and the probability density function play the role of a "penalty" when a point is smaller than the current optimum of the forecast variance, which is expected to be very small and close to 0 or to have a negative value.

In summary, maximizing the expected increase aims to find a point to make the predictive value smaller than the current optimal response.

4.4. Optimization Process

The process of optimization based on the Kriging model can be described as follows (Figure 6):

1. Based on the initial sample points X and the response values Y, build agents for the Kriging model;
2. Use the SCE optimization algorithm to obtain the EI maximum;
3. Calculate the current optimal solution and its corresponding hysteresis energy, and set $k = 1$;
4. Run the convergence criterion test; if the convergence criteria meet the requirements, then the object problem optimal solution is acquired. If it is not satisfied, the value will be added to the sample of the current optimal design for the next optimization modeling until the convergence criteria is met.

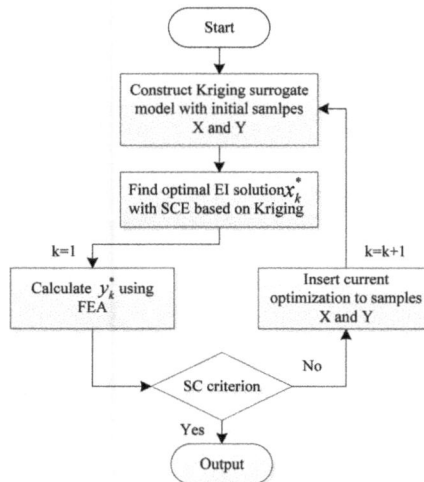

Figure 6. Optimization process.

4.5. Optimal Design Results and Discussion

The optimization results of the dampers, based on four different opening shapes, are shown in Table 1. As shown in the table, maximum hysteresis energy exists in the feasible region for SP1, while the other dampers are shown in the feasible region boundary under the conditions of having the same amount of material and the same length and width in the plane. The result shows that thickness has very little influence on the hysteresis energy for these three types of plates, besides SP1. This table also shows that the larger the hysteresis energy with a smaller opening area is, there is a restriction relationship between the parameters of SP1 and the thickness of the steel plate. The hysteresis energy for the four kinds of coupling beam metal dampers, after optimization, is compared in Figure 7, which shows that SP1 has the maximum hysteresis energy while SP3 has the minimum capacity of hysteresis energy.

Table 1. Optimization results for sample optimal solution (SOS) and optimization model solution (OMS).

Opening Shapes	Design Variables	x_1 (mm)	x_2 (mm)	Hysteresis Energy (N*mm)	Hysteresis Energy Improvement
SP1	SOS	76, 72	31, 64	5,940,777	0.064%
	OMS	74, 74	31, 78	5,944,571	
SP2	SOS	58, 60	26, 83	4,252,911	5.3%
	OMS	53, 42	26, 71	4,477,092	
SP3	SOS	84, 66	27, 03	2,141,277	13.7%
	OMS	74, 67	27, 27	2,436,392	
SP4	SOS	32, 04	26, 9	3,950,700	6%
	OMS	30, 68	27, 27	4,084,100	

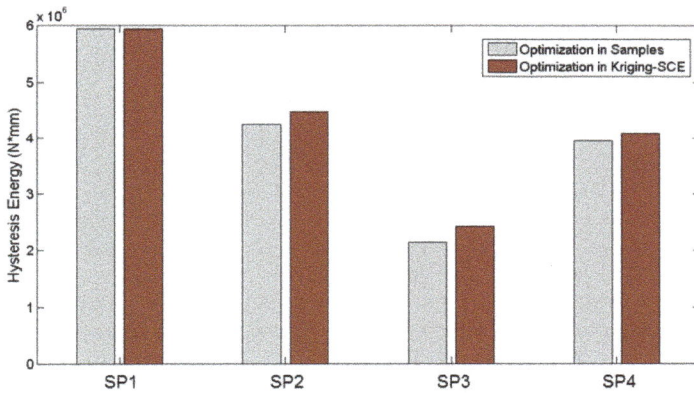

Figure 7. Hysteresis energy of four kinds of optimized coupling beam metal dampers.

SP1 is selected to analyze the influence of the numbers of the initial samples on the optimization design efficiency. Four different initial samples (25, 50, 75, 100) were chosen to build the initial Kriging surrogate model, as shown in Figure 8. The shapes and types of all four different initial surrogate models were almost the same except for some minor local differences.

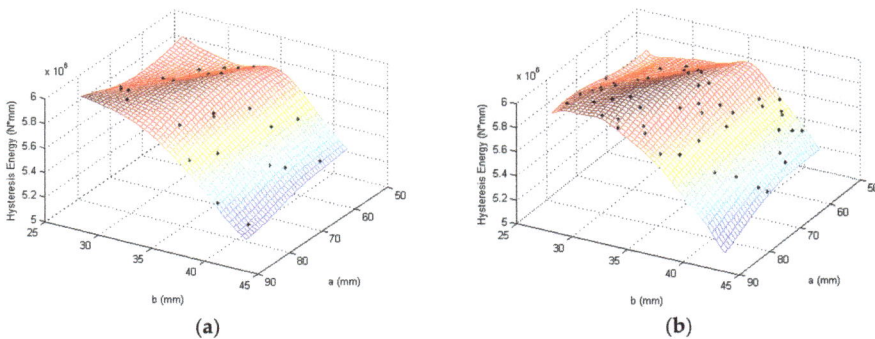

(a)

(b)

Figure 8. *Cont.*

179

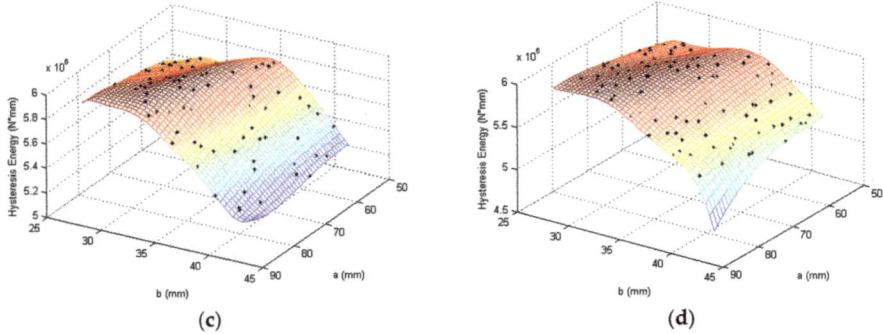

Figure 8. Kriging models with different initial samples for SP1: (**a**) 25 samples; (**b**) 50 samples; (**c**) 75 samples; and (**d**) 100 samples.

In order to illustrate the optimization design efficiency of the Kriging surrogate model, SP1 was optimized with the SCE optimization design. The calculation times for the different initial samples were compared using the same computer (Intel Core(™)i7-2600@3.40 GHz, 16 G RAM, Santa Clara, CA, USA), as shown in Figure 9. The combination of the Kriging surrogate model with the SCE optimization algorithm could highly improve the parameter optimization design efficiency. Meanwhile, the initial sample had an influence on the design optimization efficiency. In general, as the initial samples increased, the surrogate model became more accurate and needed fewer iteration steps. However, the iteration steps of the SCE optimization algorithm may increase due to randomness during the period of searching for the optimum, as well as in the selection of a threshold at the end of the iteration. The dimensions of the optimal problem and the complexity of the engineering conditions, as well as the method of selecting initial samples, could influence the iteration numbers, as well as the calculation time. It should be noted that it also took time to calculate the hysteresis energy and it took longer to calculate the energy when the initial sample was increased. Thus, a larger initial sample is not better for the optimal design case.

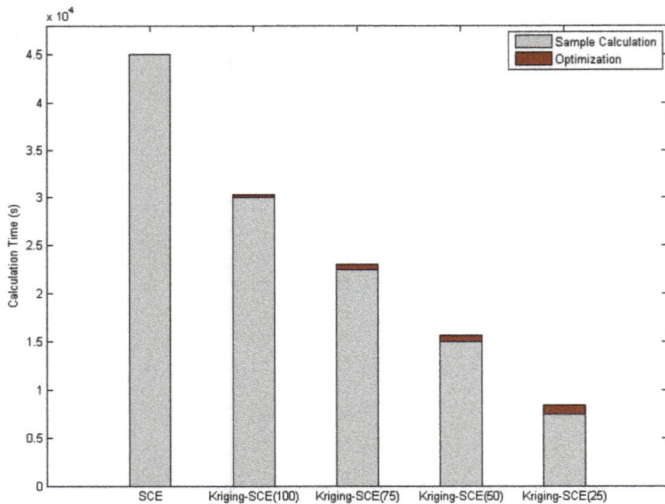

Figure 9. Calculation times.

5. Conclusions

In this paper, we studied four different coupling shear wall dampers with varying opening shapes and compared them for energy dissipation performance. By establishing the relationships between the opening size parameters and hysteresis energy based on the Kriging surrogate model, we optimized the parameters to allow the beam damper to achieve the maximum hysteresis energy. Two appealing conclusions could be reached as follows:

(1) The maximum hysteresis energy exists in the feasible region for the plate with split elliptic pores, SP1, while the hysteresis energy is a monotone function in the feasible region. This suggests that the maximum hysteresis energy lies at the boundary of the feasible region for the plate with one central elliptic pore, and the ones with uniform row pores and uniform column-row pores.

(2) The parameter of the optimum design method, based on the combination of the Kriging surrogate model with an intelligence algorithm, could significantly reduce the calculation time. Additionally, the initial sample conditions do influence the speed of the iterations.

The designed coupling shear beam dampers with split elliptic pores, SP1, have many advantages, which include a high hysteresis energy dissipation performance, facile fabrication, and low costs. We believe the optimal SP1 opening shape we found in this work will have a promising future in manufacturing and applications.

Acknowledgments: This work is supported by the National Natural Science Foundation of China (NSFC91315301-12).

Author Contributions: Jinping Ou and Dongsheng Li conceived the idea and provided the support of the experimental setup, Zhe Zhang designed the experiments and analyzed the data, Zhe Zhang and Shuaifang Zhang performed the experiments, Dongsheng Li and Zhe Zhang wrote the paper, Shuaifang Zhang helped to modify the grammar in this paper.

Conflicts of Interest: The authors declare no conflicts of interests.

References

1. Paulay, T. Simulated Seismic Loading of Spandrel Beams. *J. Struct. Div. ASCE* **1971**, *97*, 2047–2419.
2. Choo, B.S.; Coull, A. Stiffening of Laterally Loaded Coupled shear Walls on Elastic Foundations. *Build. Environ.* **1984**, *4*, 251–256. [CrossRef]
3. Kelly, J.M.; Skinner, R.I.; Heine, A.J. Mechanisms of Energy Absorption in Special Devices for Use in Earthquake Resistant Structures. *Bull. Environ. Contam. Toxicol.* **1972**, *5*, 63–73.
4. Skinner, R.I.; Tyler, R.G.; Heine, A.J. Hysteretic Dampers for Earthquake-Resistant Structures. *Earthq. Eng. Struct. Dyn.* **1975**, *3*, 287–296. [CrossRef]
5. Fortney, P.J.; Shahrooz, B.M.; Rassati, G.A. Large-Scale Testing of a Replaceable "Fuse" Steel Coupling Beam. *J. Struct. Eng.* **2007**, *133*, 1801–1807. [CrossRef]
6. Rassati, G.A.; Fortney, P.J.; Shahrooz, B.M.; Johnson, P.W., III. Performance Evaluation of Innovative Hybrid Coupled Core Wall Systems. In *Composite Construction in Steel and Concrete VI*; American Society of Civil Engineers: Reston, VA, USA, 2011; pp. 479–492.
7. Fortney, P.J.; Shahrooz, B.M.; Rassati, G.A. The Next Generation of Coupling Beams. In Proceedings of the Fifth International Conference on Composite Construction in Steel and Concrete, Mpumalanga, South Africa, 18–23 July 2004; pp. 619–630.
8. Chung, H.S.; Moon, B.W.; Lee, S.K.; Park, J.H.; Min, K.W. Seismic performance of friction dampers using flexure of RC shear wall system. *Struct. Des. Tall Spec. Build.* **2009**, *18*, 807–822. [CrossRef]
9. Kim, H.J.; Choi, K.S.; Oh, S.H.; Kang, C.H. Comparative study on seismic performance of conventional RC coupling beams and hybrid energy dissipative coupling beams used for RC shear walls. In Proceedings of the 15th World Conference on Earthquake Engineering, Lisbon, Portugal, 24–28 September 2012.
10. Choi, K.Y.; Kim, H.J.; Kang, C.H. Experimental validation on dynamic response of RC shear wall systems coupled with hybrid energy dissipative devices. In Proceedings of the 15th World Conference on Earthquake Engineering, Lisbon, Portugal, 24–28 September 2012.

11. Lyons, R.M.; Montgomery, M.S. Enhancing the seismic performance of RC coupled wall high-rise buildings with visco elastic coupling dampers. In Proceedings of the 15th World Conference on Earthquake Engineering, Lisbon, Portugal, 24–28 September 2012.

12. Mao, C.; Dong, J.; Li, H.; Ou, J. Seismic performance of RC shear wall structure with novel shape memory alloy dampers in coupling beams. In Proceedings of the Society of Photo-Optical Instrumentation Engineers (SPIE) Conference Series, San Diego, CA, USA, 11 March 2012; pp. 304–320.

13. Simpson, T. W.; Lin, D.K.; Wei, C. Sampling Strategies for Computer Experiments: Design and Analysis. *Int. J. Reliab. Saf.* **2001**, *2*, 209–240.

14. Lee, K.H.; Park, G.J. A global robust optimization using the Kriging based Approximation model. *JSME Int. J.* **2006**, *49*, 779–788. [CrossRef]

15. Gao, Y.; Wang, X. Surrogate-based process optimization for reducing warpage in injection molding. *J. Mater. Process. Technol.* **2009**, *209*, 1302–1309. [CrossRef]

16. Zhang, Z.; Ou, J.; He, Z. Optimization Design for Coupling Beam Dampers of Shear Walls. *Appl. Mech. Mater.* **2013**, *444*, 115–121. [CrossRef]

17. Duan, Q.; Sorooshian, S.; Gupta, V. Effective and efficient global optimization for conceptual rainfall-runoff models. *Water Resour. Res.* **1992**, *28*, 1015–1031. [CrossRef]

18. Sakata, S.; Ashida, F.; Zako, M. An efficient algorithm for Kriging approximation and optimization with large-scale sampling data. *Comput. Methods Appl. Mech. Eng.* **2004**, *193*, 385–404. [CrossRef]

19. Song, X.M.; Xia, J. Integration of a statistical emulator approach with the SCE-UA method for parameter optimization of a hydrological model. *Chin. Sci. Bull.* **2012**, *57*, 3397–3403. [CrossRef]

![applied sciences logo] *applied sciences*

MDPI

Article

Wind Turbine Gearbox Fault Diagnosis Based on Improved EEMD and Hilbert Square Demodulation

Huanguo Chen, Pei Chen, Wenhua Chen *, Chuanyu Wu, Jianmin Li and Jianwei Wu

Faculty of Mechanical Engineering & Automation, Zhejiang Sci-Tech University, Hangzhou 310018, China; hgchen@zstu.edu.cn (H.C.); 15700197537@163.com (P.C.); cywu@zstu.edu.cn (C.W.); ljmzrz@163.com (J.L.); wujianwei0410@126.com (J.W.)
* Correspondence: chenwh@zstu.edu.cn; Tel.: +86-186-5883-1322

Academic Editor: Gangbing Song
Received: 2 November 2016; Accepted: 6 January 2017; Published: 26 January 2017

Abstract: The rapid expansion of wind farms has accelerated research into improving the reliability of wind turbines to reduce operational and maintenance costs. A critical component in wind turbine drive-trains is the gearbox, which is prone to different types of failures due to long-term operation under tough environments, variable speeds and alternating loads. To detect gearbox fault early, a method is proposed for an effective fault diagnosis by using improved ensemble empirical mode decomposition (EEMD) and Hilbert square demodulation (HSD). The method was verified numerically by implementing the scheme on the vibration signals measured from bearing and gear test rigs. In the implementation process, the following steps were identified as being important: (1) in order to increase the accuracy of EEMD, a criterion of selecting the proper resampling frequency for raw vibration signals was developed; (2) to select the fault related intrinsic mode function (IMF) that had the biggest kurtosis index value, the resampled signal was decomposed into a series of IMFs; (3) the selected IMF was demodulated by means of HSD, and fault feature information could finally be obtained. The experimental results demonstrate the merit of the proposed method in gearbox fault diagnosis.

Keywords: wind turbine gearbox; fault diagnosis; EEMD; Hilbert square demodulation

1. Introduction

In order to harvest wind energy more efficiently, wind turbines are becoming larger and more complex. As a result, the fault rates of wind turbines are increasing, which impacts enormously on the cost of wind energy [1]. The gearbox is the major component of a wind turbine drive train and is costly and vulnerable to failure, inevitably causing the unit to stop working [2]. Accordingly, investigation into fault diagnosis for the wind turbine gearbox is becoming a popular field of research. Typically, gear tooth damages and bearing faults are the most frequent faults. Since vibration signals carry much information related to the system's dynamical characteristics, vibration analysis is a common approach for wind turbine gearbox fault diagnosis, especially with respect to the rotation parts [3,4]. However, the wind turbine working environment is strict and poor, which makes the vibration signals non-linear and non-stationary. Although traditional time domain and frequency analysis techniques, such as energy analysis, kurtosis, crest factor and spectrum analysis, have been widely used in fault diagnosis, these methods have only been effective in a stationary signal process. When it comes to non-stationary signal analyzing, the diagnostic performance has usually been unsatisfactory.

For this reason, many time-frequency analysis techniques, such as wavelet packet transform (WPT) [5,6], empirical model decomposition (EMD) [7–9] and independent component analysis (ICA) [10], have been applied to deal with the non-linear and non-stationary characteristics exhibited in the vibration signals. Wang et al. integrated EEMD and ICA for gearbox bearing fault diagnosis [11].

Appl. Sci. **2017**, *7*, 128

Law et al. (2000) proposed a method based on wavelet packet decomposition and Hilbert–Huang transform (WPD-HHT) and successfully applied it to spindle bearings condition monitoring [12]. He et al. (2007) used ICA to detect signal transients caused by localized gear damage [13]. Amirat et al. (2013) developed an EEMD-based wind turbine bearing failure detecting method using the generator stator current homopolar component [14]. Although those techniques made some progress into fault diagnosis, they had their own disadvantages. WPT could not process signals self-adaptively and needed a massive amount of data to ensure accurate results. EMD, a self-adaptive signal processing method, could decompose the non-stationary signal into several intrinsic mode functions (IMFs), which were almost orthogonal. However, mode mixing was a huge shortcoming, which restricted the application of EMD in many engineering situations. ICA extracted the transient feature without the need for prior information; however, this algorithm required redundant information measured from multiple sensors. Advanced signal processing techniques needs to be developed for this challenging task.

EEMD, an improvement of EMD, was presented by Wu and Huang (2009) to overcome mode mixing [15]. Not only did EEMD have the virtue of self-adaptability, but it also eliminated mode mixing by adding noise to the original signal. It had an absolute advantage in dealing with non-stationary and nonlinear signals. However, the performance of EEMD was also affected by parameters, such as the amplitude of added noise and the number of ensemble trials. Several efforts have been made to explore choosing proper values of these parameters to evaluate the performance of EEMD. Zhang et al. (2010) investigated the effect of the above-mentioned parameters pertinent to EEMD and improved it by replacing white noise with a finite bandwidth noise [16]. Both Stevenson et al. (2005) and Ring investigated the effect of sampling frequency on EMD and proposed a sampling limit [17,18]. However, the factor of added noise made the effect of sampling frequency on EEMD different from EMD. In this paper, the effect of sampling frequency on EEMD is quantitatively analyzed, and a criterion is proposed on how to select a resampling frequency according to the analysis result. By selecting a proper resampling frequency, the accuracy of EEMD has been further increased. The vibration data were decomposed into several IMFs by means of improved EEMD. Then, the fault-related signal was extracted by selecting the IMF with the largest kurtosis [11,19]. After fault-related signal extraction, fault information should normally be identified to provide guidance for maintenance. However, in the diagnosis of a gearbox, the amplitude modulation occurs in a measured signal, which means the fault information cannot be obtained by spectrum analysis directly. The phenomenon of amplitude modulation occurs because a high-frequency carrier signal is varied by a low-frequency modulating signal. The modulating signal results from the impacts caused by defects of a bearing or gear impulses appearing every time the tooth or rolling element crosses the defected area, which leads to amplitude modulation [20,21]. To deal with this phenomenon, Hilbert square demodulation (HSD) techniques are introduced. HSD can derive the low-frequency modulating signal from the modulated signal. Lastly, spectrum analysis is applied on the demodulated signal, and fault-related information is obtained.

In this paper, a novel hybrid method based on EEMD and HSD is presented for wind turbine gearbox fault diagnosis. The paper is organized as follows. The review of EEMD and Hilbert square demodulation is presented in Section 2. The proposed method for gearbox fault diagnosis is discussed in Section 3. The experimental and practical validations are presented in Section 4. A discussion and a conclusion are given in Section 5.

2. Review of Ensemble Empirical Mode Decomposition and Hilbert Square Demodulation

2.1. Ensemble Empirical Mode Decomposition

EEMD is an adaptive data-driven signal processing method, which substantially improves on EMD in overcoming the problem of mode mixing. The procedures of EEMD are as follows:

1.   Given that $x(t)$ is an original signal, add random white noise signal $randn_j(t)$ to $x(t)$:

$$x_j(t) = x(t) + randn_j(t) \quad j = 1, 2, \ldots, M \tag{1}$$

where $x_j(t)$ is the noise-added signal and M is the number of trials.

2.   Decompose $x_j(t)$ into a series of IMFs $c_{ij}(t)$ utilizing EMD as follows:

$$x_j(t) = \sum_{i=1}^{N} c_{ij}(t) + r_{N_j}(t) \tag{2}$$

where $c_{ij}(t)$ stands for the IMF obtained each trial, $r_{N_j}(t)$ denotes the residue of the j-th trial and N_j is the IMFs number of the j-th trial.

3.   If $j < M$, repeat Steps 1 and 2, and add different random white noise signals each time.

4.   Since the noise series in each decomposition step is statistically different and low correlation exists among the various series, the noise will be canceled out in the ensemble means, provided that the sufficient number of steps has been taken. The ensemble means of the corresponding IMFs can be obtained as the final IMFs:

$$c_i(t) = \left(\sum_{j=1}^{M} c_{ij}(t) \right) / M \tag{3}$$

where $i = 1, 2, 3, \ldots, I$ and $I = \min(N_1, N_2, \ldots, N_M)$.

5. $c_i(t)$ is the ensemble mean of corresponding IMF of the decomposition.

2.2. Hilbert Square Demodulation

Once a localized defect emerges in a bearing or gear, impulses occur every time the tooth or rolling element crosses the defective area; thus, the amplitude modulation occurs in measured signals. HSD, a type of non-stationary signal processing technique based on Hilbert transform (HT), has been investigated for feature extraction of the amplitude modulation signal [22]. The following is the basic principle of HSD.

The reason for the phenomenon of amplitude modulation is that a high-frequency carrier signal is varied by a low-frequency modulating signal. Thus, the modulated signal could be the product of the modulating signal with the carrier signal. The modulating signal results from the impacts caused by defects of a bearing or gear, whereas the carrier signal is a combination of the resonance frequencies of the bearing. Therefore, the mathematical model of modulated signal can be described as:

$$x(t) = s(t)f(t) \tag{4}$$

where $f(t)$ is the high-frequency carrier signal and $s(t)$ is the low-frequency modulating signal.

Given:

$$Z(t) = x^2(t) + H^2[x(t)] \tag{5}$$

where $H[x(t)]$ is the Hilbert transform of $x(t)$.

According to the Bedrosian product theorem [23], $H[x(t)]$ can be written:

$$H[x(t)] = H[s(t)f(t)] = s(t)H[f(t)] \tag{6}$$

Thus, $Z(t)$ can be expressed as:

$$Z(t) = s^2(t)\left\{ f^2(t) + H^2[f(t)] \right\} \tag{7}$$

Generally, the high-frequency carrier signal $f(t)$ is a harmonic signal:

$$f(t) = a_1 \cos(\omega_1 t) \tag{8}$$

so the Hilbert transform of $f(t)$ can be written as follows:

$$H[f(t)] = a_1 \sin(\omega_1 t) \tag{9}$$

The second term of Equation (7) on the right-hand side can be transformed as follows:

$$f^2(t) + H^2[f(t)] = a_1{}^2 \cos^2(\omega_1 t) + a_1{}^2 \sin^2(\omega_1 t) = a_1{}^2 \tag{10}$$

so $Z(t)$ finally becomes:

$$Z(t) = a_1{}^2 s^2(t) \tag{11}$$

It is clear that only the low-frequency modulating signal is left in Equation (11), so we can get the fault characteristic frequency by means of spectrum analysis.

3. The Proposed Method

3.1. Criterion of Resampling Frequency Selection

To examine the effect of sampling frequency, a simulated signal with transient impulse was constructed as shown in Equation (12). The reason for using this kind of simulated signal was that impulses would appear in the acceleration signals when gear tooth damage or a bearing fault occurred in the gearbox of a wind turbine. The simulated signal, its spectrum and components, are shown in Figure 1.

$$
\begin{aligned}
x_1(t) &= e^{-400t_1} \sin(2\pi 800 t), \ t_1 = mod(t, 1/33) \\
x_2(t) &= \sin(2\pi 180 t) \\
x_3(t) &= \sin(2\pi 50 t) \\
x_4(t) &= 0.16 random(n, 1), n = length(t) \\
x(t) &= x_1(t) + x_2(t) + x_3(t) + x_4(t)
\end{aligned}
\tag{12}
$$

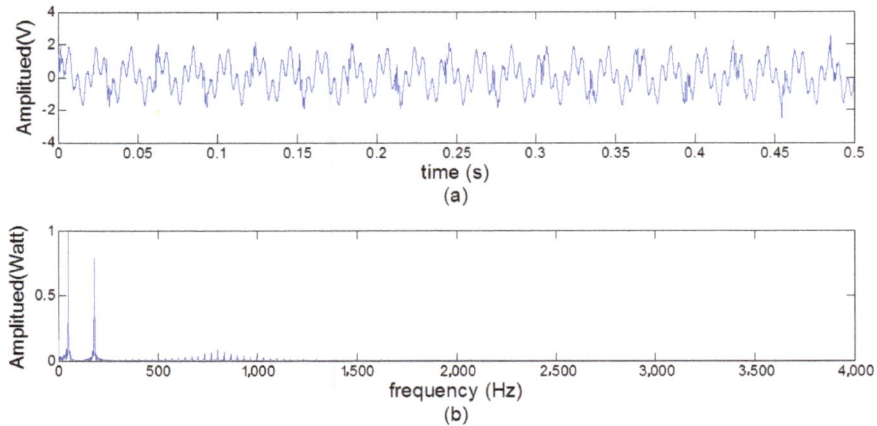

Figure 1. *Cont.*

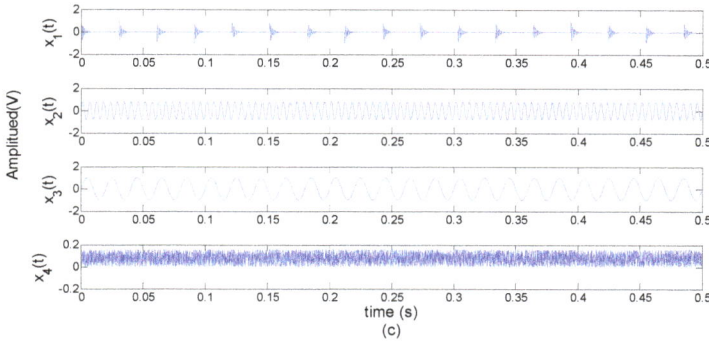

Figure 1. Simulated signal and its components. (**a**) Waveform of the simulated signal; (**b**) its spectrum; (**c**) source components of the simulated signal.

To facilitate quantitative evaluation, three quantities were utilized to analyze the performance of the EEMD. These were the successive IMF orthogonality (SIO), the IMF coherence (IC) and the residual energy (RE). The last two measures were only utilized when the number of components in the decomposition was known. The SIO was a measure equivalent to that stated in [17], but was only implemented on consecutive IMFs. This measure was defined as:

$$\text{SIO} = \sum_{i=1}^{I-1} \frac{1}{N} \sum \text{IMF}_i(n)\text{IMF}_{i+1}(n) \tag{13}$$

where I was the number of IMFs, N the length of signal and n the discrete time.

The IC was the measure of correlation between the valid or expected number of components (ENC) of the decomposition. This measured the ability of the EMD to decompose the signal into physically meaningful components and was defined as:

$$\text{IC} = \frac{1}{\text{ENC}} \sum_{i}^{\text{ENC}} cor(\text{IMF}_i(n), s_i(n)) \tag{14}$$

where $s_i(n)$ was the source component of the raw signal corresponding to $\text{IMF}_i(n)$.

The RE was the energy in the IMFs that were outside the expected number of components. This was another measure of the interpretability of EEMD and was defined as:

$$\text{RE} = \frac{1}{N} \left(s(n)^2 - \sum_{i}^{\text{ENC}} \sum_{n=1}^{N} \text{IMF}_i(n)^2 \right) \tag{15}$$

where $s(n)$ was the signal under analysis.

According to the Nyquist criteria, the sampling frequency must be more than twice the maximum frequency component of the signal; therefore, the sampling frequency of the simulated signal varied from 1650 Hz to 14,100 Hz by steps of 50 Hz. As shown in Figure 2, these performance measure results present a periodic variation trend with the increase of sampling frequency. As the value of sampling frequency was about 2400 Hz, 4800 Hz and 9200 Hz, which corresponded to 3-times, 6-times and 12-times the maximum frequency of the simulated signal respectively, SIO and RE became the local minimum value, while IC became the local maximum value. When SIO and RE take the minimum value, decomposition results had a good orthogonality and maintained the integrity of the information of the raw signal. When IC take the maximum value, the decomposition result was physically meaningful, which meant the EEMD decomposition results were relatively satisfactory. From the view of the information integrity, accuracy and efficiency of the decomposition results, the resampling frequency should be selected as about 12-times in the maximum frequency of a vibration signal.

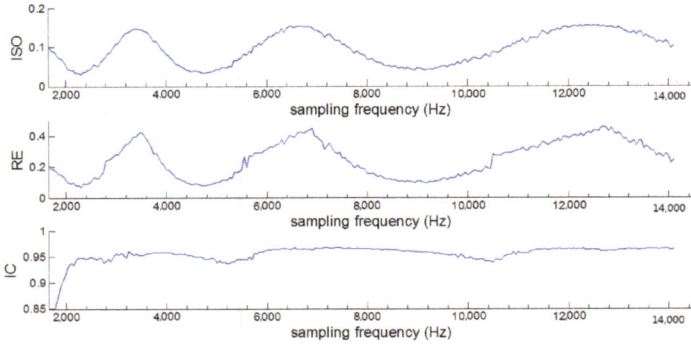

Figure 2. The effect of sampling on ensemble empirical mode decomposition (EEMD) performance. ISO, successive IMF orthogonality; RE, residual energy; IC, IMF coherence.

The decomposition results of the simulated signal with different sampling frequencies of 9450 Hz and 12,100 Hz are illustrated in Figure 3.

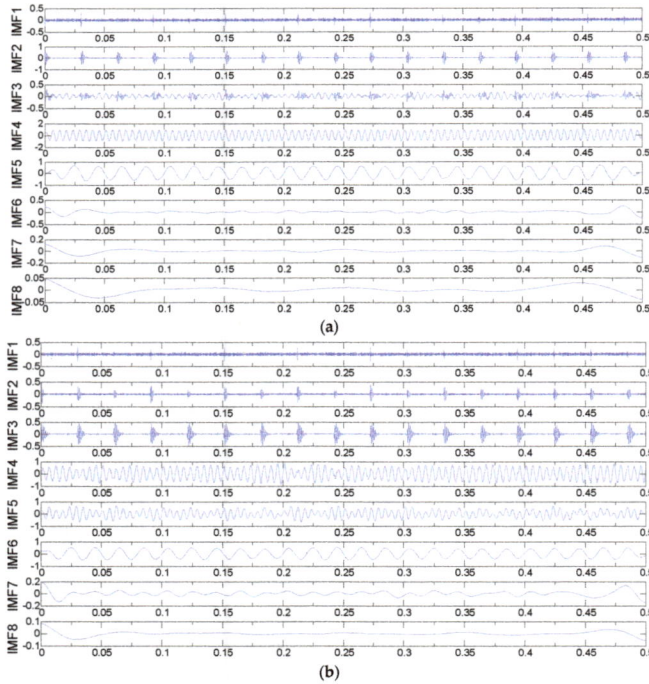

Figure 3. EEMD decomposition result with different sampling frequencies. (**a**) Decomposition result of EEMD with a sampling frequency of 9450 Hz; (**b**) decomposition result of EEMD with a sampling frequency of 14,100 Hz.

As shown in Figure 3a, IMF2, IMF4 and IMF5 corresponded to $x_1(t)$, $x_2(t)$ and $x_3(t)$, respectively. However, there were great deviations between IMF4 and $x_2(t)$ in Figure 3b. By comparing the two decomposition results, we saw that a higher sampling frequency did not necessarily provide a better decomposition result, affirming the value of the criterion of resampling frequency selection.

3.2. Improved EEMD

As mentioned above, the decomposition result of EEMD was directly affected by the sampling frequency. Too high of a sampling frequency did little to help increase the accuracy. On the contrary, it decreased the computational efficiency. Therefore, a proper resampling frequency should be selected according to the frequency characteristic of the vibration signal. In this paper, we show how EEMD was improved by replacing the raw vibration signal with a resample signal. The procedures of the improved EEMD are as follows:

1. In order to ascertain the frequency components of the vibration signal, Fourier transform was first performed to obtain its frequency spectrum.
2. Twelve-times the maximum frequency was subtracted from the sampling frequency:

$$f_e = f_s - f_r = f_s - 12 \times f_{max} \tag{16}$$

where f_s was the sampling frequency of vibration, f_r was the resampling frequency calculated by the resampling frequency criterion, f_{max} was the maximum frequency of the vibration signal and f_e was the frequency to determine whether or not to execute the resample algorithm.

If $f_e \geq f_{max}$, we resampled the raw signal with the resampling frequency f_r and output the resample signal; otherwise, we output the raw signal directly.

3. Perform EEMD on the signal obtained in Step 2.

3.3. The Proposed Method

Typical wind turbine systems have a complex structure and combine many different components. Therefore, the vibration signal measured from wind turbine gearboxes is characteristically nonlinear and non-stationary. The useful fault feature in the vibration is very weak and dominated by strong gear meshing frequencies. The low signal to noise ratio (SNR) and transient nature pose significant difficulties and challenges to fault diagnosis of wind turbine gearboxes. Moreover, gear tooth damage leads to a reduction in gear tooth stiffness intermittently throughout the rotation of the gear. Changes due to the gear damage appear in the vibration spectrum as amplitude modulation. To overcome the above limitation, a fault diagnosis method based on improved EEMD and HSD has been proposed. Firstly, the frequency components of the raw vibration signal were calculated to determine whether a resample was needed. Secondly, the raw vibration signal or resampled signal were decomposed into a series of IMFs using EEMD. Then, the fault-related signal, which had the highest kurtosis value, was selected and processed by HSD. Finally, spectrum analysis was applied on the demodulation signal and a satisfactory extraction result was obtained. The complete process of the proposed approach is shown in Figure 4, which includes the following steps:

1. Spectrum analysis to extract the distributions of the frequencies of the vibration signal, using the resampling frequency criterion to calculate the resampling frequency value f_r.
2. Raw vibration signal resample. Subtract the resampling frequency from the sampling frequency:

$$f_e = f_s - f_r \tag{17}$$

If $f_e \geq f_{max}$, resample the raw vibration signal with resampling frequency f_r, and output the resampled signal; otherwise, output the raw vibration signal directly.

3   The resampled signal decomposes by means of EEMD. Use EEMD to decompose the vibration signal obtained in Step 2 into a set of IMFs.

4   Sensitive IMF selection with the biggest kurtosis index value.

5   Feature information extraction from sensitive IMF by HSD. Use HSD to demodulate sensitive IMF to extract feature information from the demodulated signal.

6   Fault type identified by means of spectrum analysis.

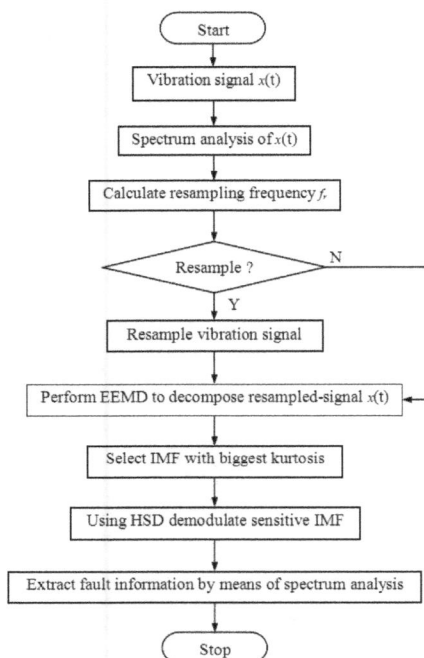

Figure 4. The flow chart of the proposed method. HSD, Hilbert square demodulation.

4. The Proposed Method for Gearbox Fault Diagnosis of a Wind Turbine

Two different experimental signals were used to verify the performance of the proposed method. The results demonstrate its effectiveness and robustness for wind turbine gear box fault diagnosis.

4.1. Bearing Experimental Evaluation

An experimental study on the fault diagnosis of a rolling bearing was firstly employed to show how the proposed method worked and to validate its effectiveness and suitability. The experimental dataset by the Case Western Reserve University (CWRU) Bearing Data Center [24] has become a standard reference in the bearing diagnostics field. The proposed method was applied to the CWRU dataset. The basic layout of the test rig is shown in Figure 5. It consisted of a 2-hp motor (left), a torque transducer (center), a dynamometer (right) and control electronics (not shown). Two bearings were installed in the motor-driven mechanical system, one at the drive end of the motor and the other at the fan end. In both bearings, three types of faults (outer race, inner race and ball faults) and various levels of fault severity (7–28 mils) were introduced using electro-discharge machining. Each bearing was tested under four different loads, 0–3 hp. The motor rotational speed was varied between 1730 and 1797 RPM depending on the load. Data was collected at 12,000 samples per second and at 48,000 samples per second for drive end bearing experiments. All fan end bearing data were collected at 12,000 samples per second. A more detailed description of the experimental setup and the apparatus involved can be found at the Case Western Reserve University's website [24].

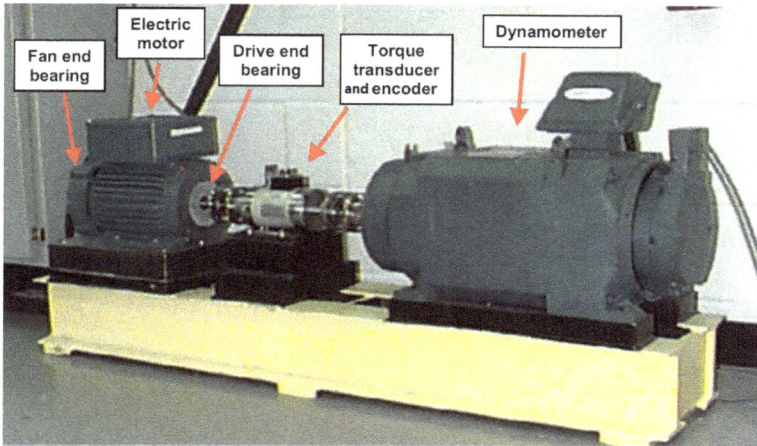

Figure 5. Bearing test rig.

In this study, we employed the drive end bearing data collected at 1719 RPM ($f_r = 28.5$ Hz) using an accelerometer. The sampling frequency f_s was 48 kHz. The fault was seeded at the outer race with a 21 mils fault diameter. Fault characteristic frequency was the multiple of the rotating frequency, the coefficient of the driven end bearing outer ring and the coefficient of the drive end bearing and is listed in Table 1 [24]. Therefore, the fault characteristic frequency could be obtained by Equation (18):

$$f_{ep} = \frac{1719}{60} \times 3.5848 = 102.7 \, \text{Hz} \tag{18}$$

Table 1. Coefficient of drive end bearing: 6205-2RS JEM SKF, deep groove ball bearing.

Inner Ring	Outer Ring	Cage Train	Rolling Element
5.4152	3.5848	0.39828	4.7135

The time domain waveform of the experimental vibration signal and its FFT spectrum are shown in Figure 6, from which the fault characteristic frequency cannot be identified. The proposed method was applied to this vibration signal.

As show in Figure 6a, the vibration signal presented a periodic change with the increasing of experiment times. Spectrum analysis has been made, and the result is shown in Figure 6b. However, it is difficult to find out any obvious fault information, for the feature with gear fault information is drawn in the background vibration signals.

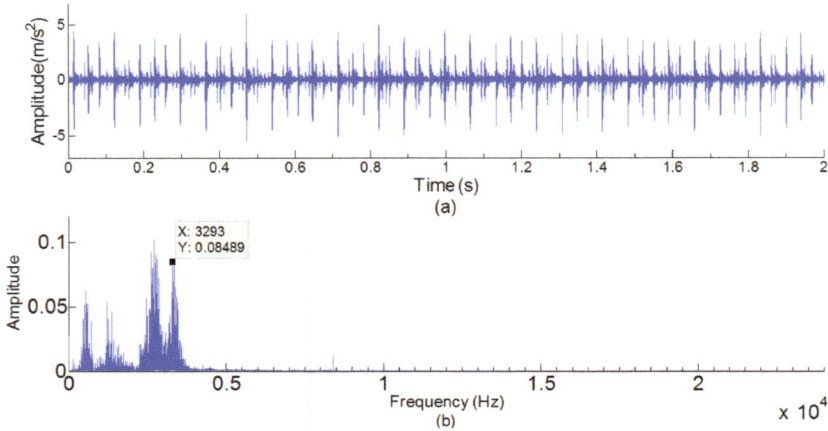

Figure 6. Experimental bearing vibration signal and its spectrum. (**a**) Experimental bearing vibration signal; (**b**) spectrum of the bearing vibration signal.

The proposed method in this article is then introduced to analyze the vibration signal. According to the resample criterion presented in Section 3.2, if we subtract 12-times the maximum frequency from sampling frequency, the value of the result was bigger than the maximum frequency. Therefore, the resample should be taken to improve the accuracy and computational efficiency of EEMD.

$$f_e = f_s - f_r = 48000 - 12 \times 3293 = 8484 \text{ Hz} \geq 3293 \text{ Hz} \tag{19}$$

The resample was first applied on the vibration signal with a frequency of 39,516 Hz. The first eight IMFs of the resample signal are shown in Figure 7a. The IMF1 was selected for further analysis because it had the highest kurtosis value compared with the other IMFs. The demodulation spectrum obtained from applying the HSD on IMF1 is illustrated in Figure 7b, which shows that the identified frequency matched the calculated fault frequency $f_{ep} = 102.7$ Hz. The side-bands' frequencies were also clearly identified.

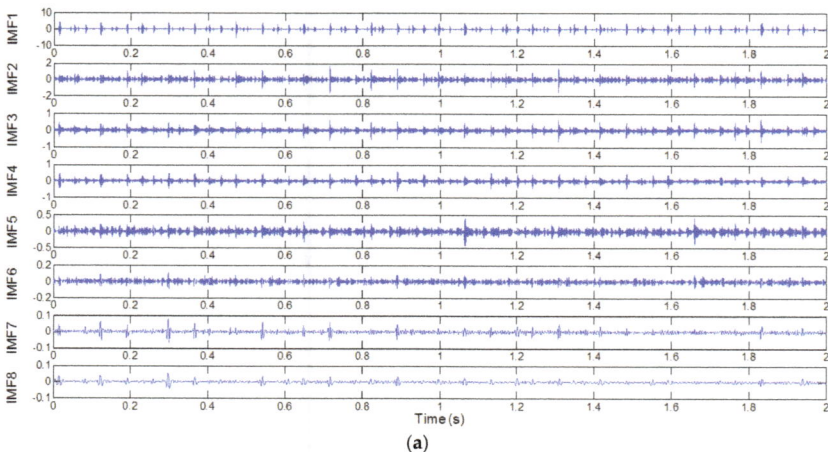

Figure 7. *Cont.*

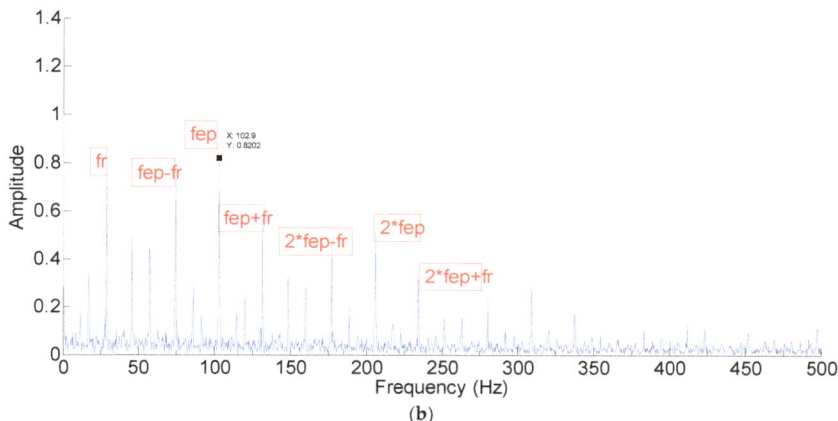

(b)

Figure 7. The decomposition results of vibration with improved EEMD and the spectrum of the demodulated signal. (**a**) The decomposition results of vibration with improved EEMD; (**b**) the spectrum of the demodulated signal.

4.2. Gear Experimental Evaluation

A set of gear fault vibration signals was kindly provided by the Reliability Research Lab in the Department of Mechanical Engineering at the University of Alberta [25]. The diagram of the experimental system is displayed in Figure 8 [26,27]. The system included a gearbox, a 3-hp AC motor for driving the gearbox and a magnetic brake for loading. The motor rotating speed was controlled by a speed controller, which allowed the tested gear to operate under various speeds. The load was provided by the magnetic brake connected to the output shaft, and the torque could be adjusted by a brake controller. As shown in Figure 8b, the gearbox was driven by the motor through a timing belt, and there were three shafts inside the gearbox, which were mounted to the gearbox housing by rolling element bearings. Gear 1 on Shaft 1 has 48 teeth and meshes with Gear 2 with 16 teeth. Gear 3 on Shaft 2 has 24 teeth and meshes with Gear 4, which was on the output shaft (Shaft 3) and has 40 teeth. Gear 3 was the tested gear. Gears with different levels of crack faults were simulated in the experimental system. As shown in Figure 9a, α was the crack angle, a one half of the chordal tooth thickness and b the face width. The crack thickness was 0.4 mm in the experiment based on the measurement of the thinnest knife of the machine tools in the lab.

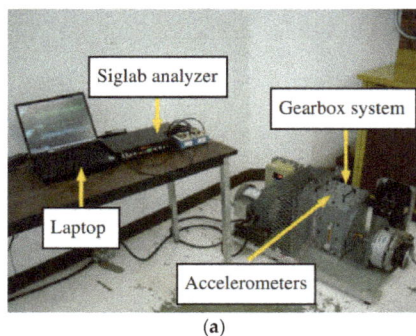

(a)

Figure 8. *Cont.*

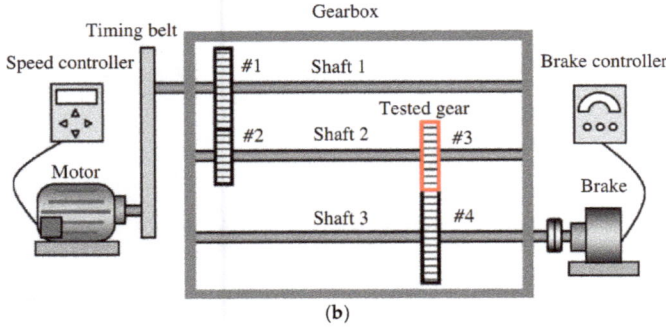

(b)

Figure 8. Gear test rig. (**a**) Experimental system; (**b**) diagram of the system.

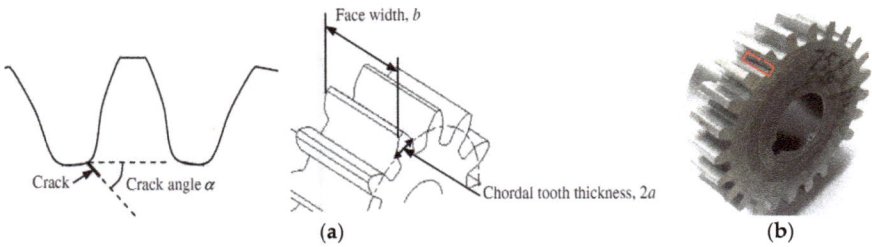

Figure 9. Gear crack angle, face width and chordal tooth thickness and experiment fault gear. (**a**) Crack angle, face width and chordal tooth thickness of a gear; (**b**) 75% level crack in the gear.

In this study, we employed the gear with 75% crack level as shown in Figure 9b. The vibration was measured at 2200 RPM with the sampling frequency $f_s = 12,800$ Hz. We used two acceleration sensors produced by PCB Electronics with Model Number 352C67. The meshing frequencies are summarized in Table 2 [26]. From the configuration of the gearbox system, the fault characteristic frequency was equal to the rotating frequency of Shaft 2.

$$f_{eq} = f_2 = 26.1 \text{ Hz} \tag{20}$$

Table 2. Motor speed and characteristic frequencies of the gearbox.

Motor Speed (RPM)	f_1 (Hz)	f_{12} (Hz)	f_2 (Hz)	f_{34} (Hz)	f_3 (Hz)
2200	8.73	418.89	26.19	628.56	15.72

Note: f_1, f_2 and f_3 are the rotating frequencies of Shaft 1, Shaft 2 and Shaft 3, respectively. f_{12} and f_{34} are the meshing frequencies of Gears 1 and 2 and Gears 3 and 4, respectively.

The time domain waveform of the experimental vibration signal and its spectrum are illustrated in Figure 10, from which the fault characteristic frequency cannot be identified. The vibration signal was analyzed with the proposed method.

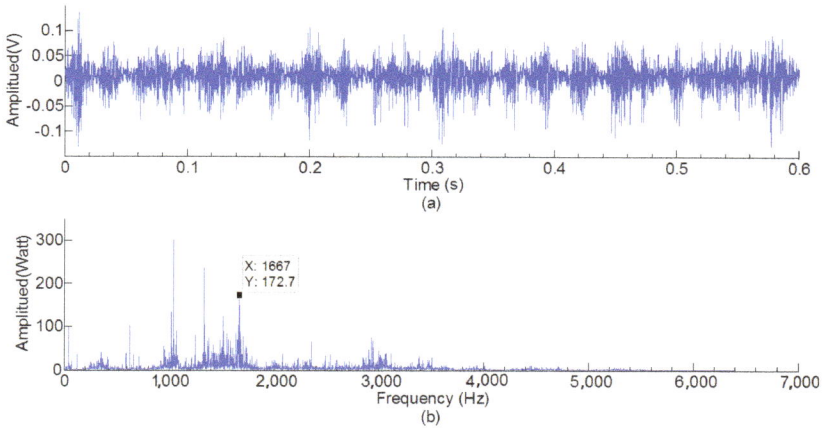

Figure 10. Gear experimental vibration signal and its spectrum. (**a**) Gear experimental vibration signal; (**b**) spectrum of gear vibration signal.

According to the resample criterion presented in Section 3.2, subtract 12-times the maximum frequency with the sampling frequency, and the value of the result is less than the maximum frequency. Therefore, the resample algorithm is unable to be carried out.

$$f_e = f_s - f_r = 12800 - 12 \times 1667 = -7204 \text{ Hz} \leq 1667 \text{ Hz} \tag{21}$$

The results of the decomposition of the vibration signal directly with EEMD are shown in Figure 11a. The IMF2 was selected for further analysis because it had the biggest kurtosis value compared with the other IMFs. HSD was applied on IMF2, and the obtained demodulation spectrum is shown in Figure 11b. It was found that the identified frequency component matched the gear fault characteristic frequency $f_{eq} = 26.1$ Hz accurately.

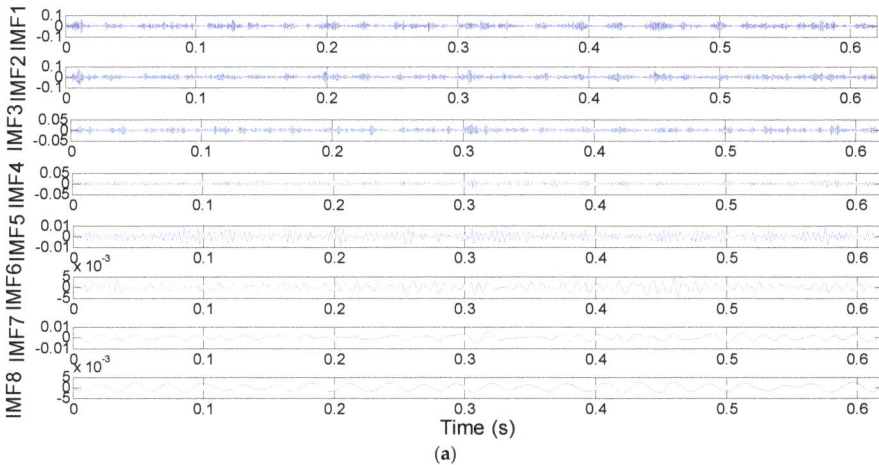

Figure 11. *Cont.*

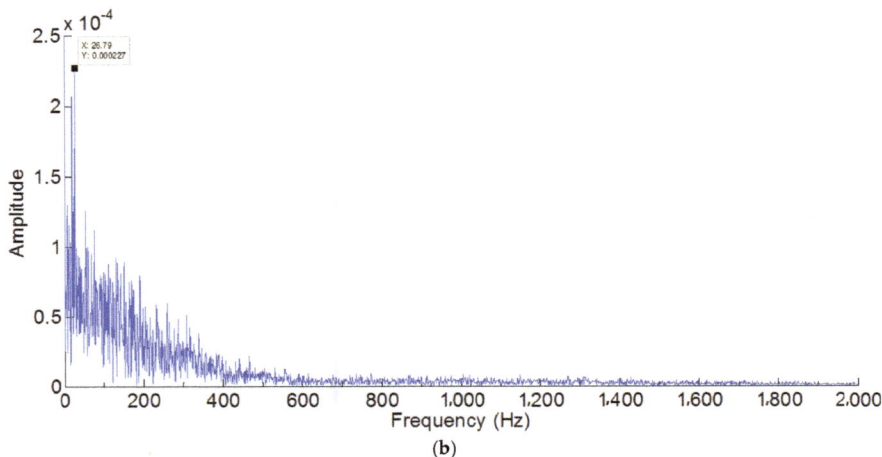

Figure 11. The decomposition results of vibration with improved EEMD and spectrum of the demodulated signal. (**a**) The decomposition results of vibration with improved EEMD; (**b**) the spectrum of the demodulated signal.

5. Discussion and Conclusions

In order to detect gearbox faults of wind turbines as early as possible, we proposed a novel fault diagnosis method based on improved EEMD and HSD. Firstly, the frequency components of the raw vibration signal were calculated to determine whether a resample was needed. In order to increase the accuracy of EEMD, a resampling criterion for the raw signal was developed. Secondly, a raw signal or resample signal was decomposed into several IMFs using improved EEMD. Then, using the IMF with the highest kurtosis value, the fault feature was selected for demodulation by the Hilbert square demodulation technique. Finally spectrum analysis was applied and fault information obtained. Experimental studies have demonstrated the effectiveness of the proposed method in fault diagnosis for the gear and bearing of the wind turbine gearbox.

Acknowledgments: This work is supported by the National Natural Science Foundation of China (Grant No. 51475432), the Zhejiang Provincial National Natural Science Foundation of China (Grant No. 2013C24005) and the Key Program for International S&T Cooperation Projects of China (Grant No. 2015DFA71400).

Author Contributions: H.C. and W.C. designed experiment verification and validation schemes. P.C. and J.W. gathered experimental data and analyzed experimental results. P.C., H.C., C.W. and J.L. Studied the signal processing method. H.C. and P.C. wrote the manuscript.

Conflicts of Interest: The authors declare no conflict of interest.

References

1. Song, G.; Li, H.; Gajic, B.; Zhou, W.; Chen, P.; Gu, H. Wind turbine blade health monitoring with piezoceramic-based wireless sensor network. *Int. J. Smart Nano Mater.* **2013**, *4*, 150–166. [CrossRef]
2. Antoniadou, I.; Manson, G.; Staszewski, W.J.; Barszcz, T.; Worden, K. A time-frequency analysis approach for condition monitoring of a wind turbine gearbox under varying load conditions. *Mech. Syst. Signal Process.* **2015**, *64*, 188–216. [CrossRef]
3. Zheng, J.; Cheng, J.; Yang, Y. Generalized empirical mode decomposition and its applications to rolling element bearing fault diagnosis. *Mech. Syst. Signal Process.* **2013**, *40*, 136–153. [CrossRef]
4. Lei, Y.G.; He, Z.J.; Yanyang, Z.; Chen, X. New clustering algorithm-based fault diagnosis using compensation distance evaluation technique. *Mech. Syst. Signal Process.* **2008**, *22*, 419–435. [CrossRef]

5. Nikolaou, N.; Antoniadis, I. Rolling element bearing fault diagnosis using wavelet packets. *NDT & E Int.* **2002**, *35*, 197–205.

6. Zhang, L.; Wang, C.; Song, G. Health Status Monitoring of Cuplock Scaffold Joint Connection Based on Wavelet Packet Analysis. *Shock Vib.* **2015**, *2015*, 1–7. [CrossRef]

7. Wang, Z.; Lu, C.; Wang, Z.; Ma, J. Health assessment of rotary machinery based on integrated feature selection and Gaussian mixed model. *J. Vibroeng.* **2014**, *16*, 1753–1762.

8. Lu, C.; Hu, J.; Liu, H. Application of EMD-AR and MTS for hydraulic pump fault diagnosis. *J. Vibroeng.* **2013**, *15*, 761–772.

9. Lu, C.; Yuan, H.; Tang, Y. Bearing performance degradation assessment and prediction based on EMD and PCA-SOM. *J. Vibroeng.* **2014**, *16*, 1387–1396.

10. He, Q.; Feng, Z.; Kong, F. Detection of signal transients using independent component analysis and its application in gearbox condition monitoring. *Mech. Syst. Signal Process.* **2007**, *21*, 2056–2071. [CrossRef]

11. Wang, J.; Gao, R.X.; Yan, R.Q.; Wang, L. An integrative computational method for gearbox diagnosis. *Procedia CIRP* **2013**, *12*, 133–138. [CrossRef]

12. Koizumi, T.; Tsujiuchi, N.; Matsumura, Y. Diagnosis with the correlation integral in time domain. *Mech. Syst. Signal Process.* **2000**, *14*, 1003–1010. [CrossRef]

13. Liang, X.H.; Zuo, M.J.; Hoseini, M.R. Vibration signal modeling of a planetary gear set for tooth crack detection. *Eng. Fail. Anal.* **2015**, *48*, 185–200. [CrossRef]

14. Amirat, Y.; Choqueuse, V.; Benbouzid, M. EEMD-based wind turbine bearing failure detection using the generator stator current homopolar component. *Mech. Syst. Signal Process.* **2013**, *41*, 667–678. [CrossRef]

15. Wu, Z.; Huang, N.E. Ensemble empirical mode decomposition: A noise assisted data analysis method. *Adv. Adapt. Data Anal.* **2009**, *1*, 1–14. [CrossRef]

16. Zhang, J.; Yan, R.; Gao, R.; Feng, Z. Performance enhancement of Ensemble Empirical Mode Decomposition. *Mech. Syst. Signal Process.* **2010**, *24*, 2104–2123. [CrossRef]

17. Huang, N.E.; Shen, Z.; Long, S.R.; Wu, M.C.; Shih, H.H.; Zheng, Q.; Yen, N.-C.; Tung, C.C.; Liu, H.H. The Empirical Mode Decomposition and the Hilbert Spectrum for Nonlinear and Non-stationary Time Series Analysis. *R. Soc. Lond. Proc.* **1998**, *454*, 903–995. [CrossRef]

18. Stevenson, N.; Mesbah, M.; Boashash, B. A sampling limit for the empirical mode decomposition. In Proceedings of the 8th International Symposium on Signal Processing and its Applications, Sydney, Australia, 28–31 August 2005; pp. 647–650.

19. Wang, H.; Chen, J.; Dong, G. Feature extraction of rolling bearing's early weak fault based on EEMD and tunable Q-factor wavelet transform. *Mech. Syst. Signal Process.* **2014**, *48*, 103–119. [CrossRef]

20. Stack, J.R.; Harley, R.G.; Habetler, T.G. An amplitude modulation detector for fault diagnosis in rolling element bearings. *IEEE Trans. Ind. Electron.* **2004**, *51*, 1097–1102. [CrossRef]

21. Zhang, X.; Liang, Y.; Zhou, J. A novel bearing fault diagnosis model integrated permutation entropy, ensemble empirical mode decomposition and optimized SVM. *Measurement* **2015**, *69*, 164–179. [CrossRef]

22. Yu, X.; Ding, E.; Chen, C.; Liu, X.; Li, L. A Novel Characteristic Frequency Bands Extraction Method for Automatic Bearing Fault Diagnosis Based on Hilbert Huang Transform. *Sensors* **2015**, *15*, 27869–27893. [CrossRef] [PubMed]

23. Bedrosian, E. A product theorem for Hilbert transform. *Proc. IEEE* **1963**, *51*, 868–869. [CrossRef]

24. Case Western Reserve University Bearing Data Center Website. Available online: http://csegroups.case.edu/bearingdatacenter/home (accessed on 25 December 2015).

25. Reliability Research Lab. Available online: http://www.mece.ualberta.ca/groups/reliability/index.html (accessed on 28 December 2015).

26. Lei, Y.; Zuo, M.J. Gear crack level identification based on weighted K nearest neighbor classification algorithm. *Mech. Syst. Signal Process.* **2009**, *23*, 1535–1547. [CrossRef]

27. Lei, Y.; Zuo, M.J.; He, Z.; Zi, Y. A multidimensional hybrid intelligent method for gear fault diagnosis. *Expert Syst. Appl.* **2010**, *37*, 1419–1430. [CrossRef]

applied sciences

MDPI

Article

Investigations on the Effects of Vortex-Induced Vibration with Different Distributions of Lorentz Forces

Hui Zhang *, Meng-ke Liu, Bao-chun Fan, Zhi-hua Chen, Jian Li and Ming-yue Gui

Science and Technology on Transient Physics Laboratory, Nanjing University of Science and Technology, Nanjing 210094, China; lmk369@hotmail.com (M.-k.L.); bcfan@njust.edu.cn (B.-c.F.); chenzh@njust.edu.cn (Z.-h.C.); lijian0628@hotmail.com (J.L.); mygui@njust.edu.cn (M.-y.G.)
* Correspondence: zhanghui1902@hotmail.com; Tel.: +86-25-8430-3929

Academic Editor: Gangbing Song
Received: 31 October 2016; Accepted: 3 January 2017; Published: 7 January 2017

Abstract: The control of vortex-induced vibration (VIV) in shear flow with different distributions of Lorentz force is numerically investigated based on the stream function–vorticity equations in the exponential-polar coordinates exerted on moving cylinder for Re = 150. The cylinder motion equation coupled with the fluid, including the mathematical expressions of the lift force coefficient C_l, is derived. The initial and boundary conditions as well as the hydrodynamic forces on the surface of cylinder are also formulated. The Lorentz force applied to suppress the VIV has no relationship with the flow field, and involves two categories, i.e., the field Lorentz force and the wall Lorentz force. With the application of symmetrical Lorentz forces, the symmetric field Lorentz force can amplify the drag, suppress the flow separation, decrease the lift fluctuation, and then suppress the VIV while the wall Lorentz force decreases the drag only. With the application of asymmetrical Lorentz forces, besides the above-mentioned effects, the field Lorentz force can increase additional lift induced by shear flow, whereas the wall Lorentz force can counteract the additional lift, which is dominated on the total effect.

Keywords: flow control; vortex-induced vibration; electro-magnetic control; hydrodynamic force

1. Introduction

Bluff structures such as offshore spar, marine risers, overhead transmission lines and heat exchangers are subjected to vortex-induced vibration (VIV) when exposed to a flowing fluid, which contributes to the fatigue life reduction of structures and may produce structure damage under certain unfavorable conditions. The interactions between the shedding among various cylinders are much more complicated and they are also relevant to a number of problems. One of the problems where this is critical is concentrated solar power (CSP) collectors, where the shedding from one row affects the next one, leading to mechanical problems due to vibration and fatigue [1,2]. The vibrations arise from the time-periodic fluid force associated with the time-periodic shedding vortex, and subsequently alter the flow field, which will change the flow-induced force in turn. Such fluid–structure interactions increase the complexity of the fluid mechanisms.

Extensive research on VIV has been conducted. The earliest studies have been the subject of interests with lock-in phenomenon of VIV (e.g., [3–10]). The time varying parameters such as cylinder response, flow force, vortex shedding and the influences of cylinder motion on the vortex structure are further studied [11–14], and the vortex-formation modes have been focused on later by [15–18], which paid attention to the interactions with multiple cylinders, and showed that the arrangement or gap had a significant effect on the VIV system responses.

However, a large fluctuation of flow forces, and the increase of drag, acoustic noise and even structure damage are usually generated by the undesirable flow separations and vibrations of the body in the previous cases. Therefore, modern flow control approaches and technologies are considered to suppress the above phenomena. Flow control usually involves two broad categories. For one, called passive control, the flow is modified without external energy input as helical strakes [19], fin [20], rope and bump [21], thick fairings [22] (p. 19377), splitter plates [23], guide foils [24] and so on. For the other, called active control, energy requires to be injected into the flow as momentum injections [25–27], moving-wall [28], synthetic jet actuators [29], suction and blowing [30–34] and so on.

The Lorentz force, one of the active approaches, has been subjected to various studies since the 1960s. In the recent years, the Lorentz force has attracted more attention due to its promising applications in engineering fields. Crawford and Karniadakis [35] has numerically investigated that the Lorentz force can eliminate the flow separation when the flow past a stationary cylinder, and the suppressing effect of the Lorentz force has been confirmed by Weier et al. [36] with both experiments and calculations. Later, in the cases of Kim and Lee [37] and Posdziech and Grundmann [38], it is found that both of the continuous and pulsed Lorentz forces can suppress the force fluctuation and stabilize the flow. Recently, optimal and closed-loop control have been developed by Zhang et al. [39–41] aiming to improve the control efficiency of the cylinder wake, and the VIV of the shear incoming flow was also investigated preliminarily [42]. However, there has not been discussion in the literature on the control of VIV by Lorentz forces in shear flow.

In this paper, control of VIV with Lorentz forces for the shear flow has been numerically investigated. The stream function–vorticity equations, the initial and boundary conditions, distribution of hydrodynamic force and the cylinder motion equation are deduced in the exponential-polar coordinate with the coordinate at the moving cylinder. The Lorentz force can be classified into the field Lorentz force and the wall Lorentz force. On the other hand, the evolution of VIV starting from rest to vibration, and suppression are all presented. The mechanism of fluid–cylinder interactions in the shear flow with different distributions of Lorentz forces is discussed in detail.

2. Governing Equations

A circular cylinder experiences time varying lift and drag, which are related to the vortex shedding, when placed in a flowing fluid. Thus, time varying vibration will occur when the cylinder is constrained to move on flexible supports, which is known as vortex-induced vibration (VIV). Moreover, the fluid around the cylinder is altered by this vibration, which affects the induced hydrodynamic forces in turn and the structure response is then changed. Therefore, this problem is associated with fully coupled fluid–structure interactions.

For control of VIV in an electrically low-conducting fluid, the actuators on the cylinder surface consists of two half cylinders mounted with alternating electrodes and magnets as shown in Figure 1. The momentum of the fluid round the cylinder surface is increased with the application of Lorentz force, which is directed parallel to the cylinder surface.

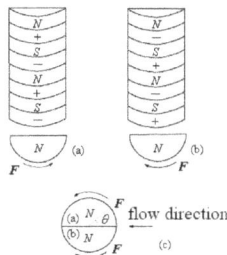

Figure 1. Lorentz force on the cylinder surface.

In this way, the electro-magnetic force **F** can be written in the dimensionless form:

$$\mathbf{F}^* = \mathbf{J}^* \times \mathbf{B}^*,$$

where the superscript "*" refers to the dimensional form, no superscript refers to the dimensionless form. This can be written in dimensionless form [36,38]

$$\mathbf{F}^* = N\mathbf{F}, \tag{1}$$

with

$$F_r = 0,$$

$$F_\theta = e^{-\alpha(r-1)}g(\theta) \text{ with } g(\theta) = \begin{cases} 1 & \text{covered with actuatoron upper surface} \\ -1 & \text{covered with actuator on lower surface} \\ 0 & \text{elsewhere} \end{cases}$$

where r and θ represent polar coordinates, r and θ denote the components in r and θ directions, respectively. α is a constant, denoting the effective depth of Lorentz force in the fluid. The interaction parameter is termed as $N = \frac{j_0 B_0 a}{\rho u_\infty^2}$, with the current density j_0, B_0 the magnetic field, ρ the fluid density, and a the cylinder radius.

The exponential-polar coordinates system (ξ, η) is introduced here defined as $r = e^{2\pi\xi}$, $\theta = 2\pi\eta$. Then, the stream function–vorticity equations in the dimensionless form, which express the flow considering an applied Lorentz force with the coordinate system attached on the moving cylinder, are termed as

$$H\frac{\partial\Omega}{\partial t} + \frac{\partial(U_r\Omega)}{\partial\xi} + \frac{\partial(U_\theta\Omega)}{\partial\eta} = \frac{2}{Re}\left(\frac{\partial^2\Omega}{\partial\xi^2} + \frac{\partial^2\Omega}{\partial\eta^2}\right) + NH^{\frac{1}{2}}\left(\frac{\partial F_\theta}{\partial\xi} + 2\pi F_\theta - \frac{\partial F_r}{\partial\eta}\right), \tag{2}$$

$$\frac{\partial^2\psi}{\partial\xi^2} + \frac{\partial^2\psi}{\partial\eta^2} = -H\Omega. \tag{3}$$

We need to mention that the above equations (Equations (2) and (3)) have the same forms as the equations in the absolute coordinate system. The stream function ψ is defined as $\frac{\partial\psi}{\partial\eta} = U_r = H^{\frac{1}{2}}u_r$, $-\frac{\partial\psi}{\partial\xi} = U_\theta = H^{\frac{1}{2}}u_\theta$, while the vorticity Ω is $\Omega = \frac{1}{H}\left(\frac{\partial U_\theta}{\partial\xi} - \frac{\partial U_r}{\partial\eta}\right)$, with u_r and u_θ indicating the velocity components in r and θ directions, respectively. Moreover, $H = 4\pi^2 e^{4\pi\xi}$, $Re = \frac{2u_\infty a}{\nu}$, u_∞ is the free-stream velocity, ν is the kinematic viscosity, and the dimensionless time is $t = \frac{t*u_\infty}{a}$.

The sketch of shear flow, which shows the change of the velocity in the cross-section plane of the cylinder, with a linear velocity profile $u = u_\infty + Gy$ [43] over a cylinder in two-dimensional domain is shown in Figure 2, where u_∞ is the free-stream velocity at the center-line $\theta = 0$, y is the coordinate in the lateral direction with $y = 0$ at the center of the cylinder, and G is the lateral velocity gradient.

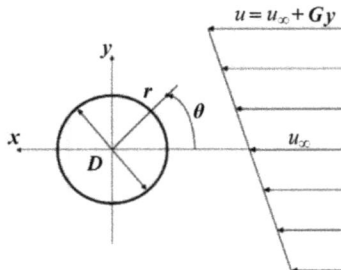

Figure 2. Sketch of shear flow over circular cylinder.

Defining the shear rate K as $K = 2Ga/u_\infty$, a refers to the cylinder radius. Only the circumstance of a positive shear rate, i.e., $K > 0$, needs to be discussed here, indicating that the flow velocity of the upper side prevails over that of the lower side.

Initially, the flow is regarded as inviscid. When the cylinder can only move in a cross flow direction, the initial flow field in the moving frame of reference can be described by

$$\psi = -2sh(2\pi\zeta)\left[\sin(2\pi\eta) + \tfrac{K}{2}(2ch(2\pi\zeta)\cos(4\pi\eta) - e^{2\pi\zeta})\right]$$
$$\text{and } \Omega = K \tag{4}$$

As for $t > 0$, the cylinder is constrained to vibrate along the transverse direction with the effect of the vortex shedding. Based on the Galilean velocity decomposition and the stream function definition, we have

$$\psi = \psi' + \frac{dl(t)}{dt}e^{2\pi\zeta}\cos(2\pi\eta)$$

where the superscript "'" denotes the absolute coordinate (an inertial frame of reference), and no superscript denotes the coordinate (a non-inertial frame of reference) which is attached on the cylinder moving with the velocity $\frac{dl(t)}{dt}$, where l is the cylinder displacement in the transverse direction.

The relative angle of the incoming flow direction is defined as $\theta_0 = \tan^{-1}\left[\frac{dl(t)}{dt}\right]$, then

$$\psi = \psi' + (\tan\theta_0)e^{2\pi\zeta}\cos(2\pi\eta) \tag{5}$$

when $\zeta \to \infty$, $e^{2\pi\zeta} \to 2sh(2\pi\zeta)$, from Equation (5), we have

$$\psi = -2sh(2\pi\zeta)\left[\frac{\sin(2\pi\eta - \theta_0)}{\cos\theta_0} + \frac{K}{2}(2ch(2\pi\zeta)\cos(4\pi\eta) - 2sh(2\pi\zeta))\right]$$

which depends on the shear rate K and the vibration of cylinder, and

$$\Omega = K \tag{6}$$

On $\zeta = 0$, $-\frac{1}{H}\frac{\partial^2\psi}{\partial\zeta^2} = -\frac{1}{H}\frac{\partial^2\psi'}{\partial\zeta^2} - \tan\theta_0\cos(2\pi\eta)$

$$-\frac{1}{H}\frac{\partial^2\psi}{\partial\eta^2} = -\frac{1}{H}\frac{\partial^2\psi'}{\partial\eta^2} + \tan\theta_0\cos(2\pi\eta) \tag{7}$$

and $\Omega' = -\frac{1}{H}\left(\frac{\partial^2\psi'}{\partial\zeta^2} + \frac{\partial^2\psi'}{\partial\eta^2}\right) = -\frac{1}{H}\left(\frac{\partial^2\psi}{\partial\zeta^2} + \frac{\partial^2\psi}{\partial\eta^2}\right) = \Omega$

Since $\psi = 0$, so that

$$\psi' = -(\tan\theta_0)\cos(2\pi\eta) \text{ and } \frac{\partial^2\psi'}{\partial\eta^2} = H\tan\theta_0\cos(2\pi\eta)$$

where $H = 4\pi^2$. From Equation (7),

$$\frac{\partial^2\psi}{\partial\eta^2} = 0$$

Thus, one finally obtains

$$\Omega = -\frac{1}{H}\frac{\partial^2\psi}{\partial\zeta^2} \tag{8}$$

In addition, the applied Lorentz force has no relationship with the flow field, and involves two types which can be defined as the field Lorentz force $F_\theta|_{\zeta>0}$ and the wall Lorentz force $F_\theta|_{\zeta=0}$. The field Lorentz force does not appear in any above equations, but acts as a source term affecting the boundary layer fluid and accordingly leads to the variations of hydrodynamic force. The wall Lorentz force, however, is independent of the flow field due to the non-slip boundary and affects directly on the surface of cylinder.

3. Hydrodynamic Forces

Defining the net hydrodynamic force $\mathbb{F}^\theta$ acting on a cylinder as

$$C_F^\theta = \frac{\mathbb{F}^\theta}{\frac{1}{2}\rho u_\infty^2} = \sqrt{\left(C_\tau^\theta\right)^2 + \left(C_p^\theta\right)^2}$$

If the prime coordinate system indicates the absolute coordinate (the inertial frame of reference) in the previous reference, the shear stress C_τ^θ is written as

$$C_\tau^\theta = \frac{\tau_{r\theta}}{\frac{1}{2}\rho u_\infty^2} = -\frac{4}{ReH}\frac{\partial^2 \psi'}{\partial \xi'^2}$$

As $\frac{\partial^2 \psi'}{\partial \xi'^2} + \frac{\partial^2 \psi'}{\partial \eta'^2} = -H\Omega'$, then

$$C_\tau^\theta = \frac{4}{Re}\left(\Omega' + \frac{1}{H}\frac{\partial^2 \psi'}{\partial \eta'^2}\right)$$

On the cylinder surface,

$$\psi' = -\frac{dl}{dt}\cos(2\pi\eta) \text{ and } \Omega' = \Omega$$

Thus,

$$C_\tau^\theta = C_{\tau F}^\theta + C_{\tau V}^\theta \tag{9}$$

where $C_{\tau F}^\theta = \frac{4}{Re}\Omega$

$$C_{\tau V}^\theta = \frac{4}{Re}\frac{dl}{dt}\cos(2\pi\eta)$$

Therefore, the shear stress involves two components denoted as $C_{\tau F}^\theta$ and $C_{\tau V}^\theta$, where $C_{\tau F}^\theta$ is proportional to vorticity at the wall, whereas $C_{\tau V}^\theta$ induced by the cylinder motion in viscous flow, has no relationship with the vorticity field.

The pressure C_p^θ is termed as based on the definition [41]

$$C_p^\theta = \frac{\mathbb{F}_p^\theta}{\frac{1}{2}\rho u_\infty^2} = \frac{p_\theta - p_\infty}{\frac{1}{2}\rho u_\infty^2} = P_\theta - P_\infty$$

where the pressure is $P = \frac{p}{\rho u_\infty^2/2}$ with p the pressure of the flow field. Then, C_p^θ can be formulated further by following mathematical derivation.

From the momentum equations in the moving coordinate of reference, one obtains

$$P_\theta - P_0 = \frac{4}{Re}\int_0^\eta \frac{\partial \Omega}{\partial \xi}d\eta + 4\pi N\int_0^\eta F_\theta|_{\xi=0}d\eta - 4\frac{d^2l(t)}{dt^2}\sin(2\pi\eta) \tag{10}$$

and

$$P_\infty - P_0 = -4\pi\int_0^\infty \frac{\partial u_r}{\partial t}e^{2\pi\xi}d\xi - 1 - 2\int_0^\infty u_\theta\frac{\partial u_r}{\partial \eta}d\xi + 4\pi\int_0^\infty u_\theta^2 d\xi - \frac{4}{Re}\int_0^\infty \frac{\partial \Omega}{\partial \eta}d\xi \tag{11}$$

Then,

$$C_p^\theta = P_\theta - P_\infty = C_{pF}^\theta + C_{pW}^\theta + C_{pV}^\theta \tag{12}$$

where

$$C_{PF}^\theta = \frac{4}{Re}\int_0^\eta \frac{\partial \Omega}{\partial \xi}d\eta + C_p^0$$

$$\mathcal{C}_p^0 = 1 + 4\pi \int\limits_0^\infty \frac{\partial u_r}{\partial t} e^{2\pi\xi} d\xi + 2 \int\limits_0^\infty u_\theta \frac{\partial u_r}{\partial \eta} d\xi - 4\pi \int\limits_0^\infty u_\theta^2 d\xi + \frac{4}{\mathrm{Re}} \int\limits_0^\infty \frac{\partial \Omega}{\partial \eta} d\xi$$

$$\mathcal{C}_{pW}^\theta = 4\pi N \int\limits_0^\eta F_\theta|_{\xi=0} d\eta$$

$$\mathcal{C}_{pV}^\theta = -4 \frac{d^2 l(t)}{dt^2} \sin(2\pi\eta)$$

Here, pressure $\mathcal{C}_p^\theta$ consists of $\mathcal{C}_{pF}^\theta$ induced by the field Lorentz force, $\mathcal{C}_{pW}^\theta$ induced by the wall Lorentz force and $\mathcal{C}_{pV}^\theta$ induced by the inertial force.

The hydrodynamic force is also regarded as a force consisting of drag force and lift force denoting the force components in the streamwise and the transverse directions, respectively. We have

$$\mathcal{C}_d^\theta = \mathcal{C}_p^\theta \cos(2\pi\eta) + \mathcal{C}_\tau^\theta \sin(2\pi\eta)$$

and

$$\mathcal{C}_l^\theta = \mathcal{C}_p^\theta \sin(2\pi\eta) + \mathcal{C}_\tau^\theta \cos(2\pi\eta)$$

where the subscripts "d" and "l" denote the drag and lift force, respectively.

The total force is attained by integrating the force distribution function along the cylinder surface, by defining the dimensionless form

$$C = \frac{F}{\rho u_\infty^2 a},$$

Then, the drag C_d can be termed as

$$C_d = \int\limits_0^{2\pi} \mathcal{C}_d^\theta d\theta = C_{dF} + C_{dW} \tag{13}$$

where

$$C_{dF} = \frac{2}{\mathrm{Re}} \int\limits_0^1 \left(2\pi\Omega - \frac{\partial \Omega}{\partial \xi} \right) \sin(2\pi\eta) d\eta$$

$$C_{dW} = -2\pi N \int\limits_0^1 F_\theta|_{\xi=0} \sin(2\pi\eta) d\eta$$

The lift C_l is written as

$$C_l = \int\limits_0^{2\pi} \mathcal{C}_l^\theta d\theta = C_{lF} + C_{lW} + C_{lV} \tag{14}$$

where

$$C_{lF} = \frac{2}{\mathrm{Re}} \int\limits_0^1 \left(2\pi\Omega - \frac{\partial \Omega}{\partial \xi} \right) \cos(2\pi\eta) d\eta$$

$$C_{lW} = -2\pi N \int\limits_0^1 F_\theta|_{\xi=0} \cos(2\pi\eta) d\eta$$

$$C_{lV} = -4\pi \frac{d^2 l}{dt^2} - \frac{4\pi}{\mathrm{Re}} \frac{dl}{dt}$$

Hence,

$$C_l = C_{lF} + C_{lW} - 4\pi \frac{d^2l}{dt^2} - \frac{4\pi}{\text{Re}} \frac{dl}{dt} \tag{15}$$

where the first term on the right-hand side C_{lF}, the so-called vortex-induced force, depends on the vorticity and the boundary vorticity flux on the surface of cylinder. The field Lorentz force $N F_\theta|_{\xi>0}$ serves as a source to influences the vorticity field, in turn, to alternate C_{lF}. The second term, C_{lW}, is induced by wall Lorentz force, which has no relationship with flow field. The third term, called inertial force, depends on acceleration of cylinder and the fourth term, called viscous damping force, depends on Reynolds number and cylinder velocity. Therefore, the second to fourth terms have no relationship with the varying flow field.

4. Cylinder Responses

As the cylinder is constrained to experience VIV in the transverse direction, the equation of motion in the absolute coordinate (the inertial frame of reference) may be termed as [42]

$$m \frac{d^2l}{dt^2} + \varsigma \frac{dl}{dt} + m_{vir} \left(\frac{\omega_n}{\omega}\right)^2 \omega^2 l = F_l \tag{16}$$

where m is the dimensionless mass; ς is the dimensionless structure damping; m_{vir} is the virtual mass; and $\omega = 2\pi f$, f is vortex shedding frequency. When the lock-in occurs, the vortex shedding frequency is synchronized with the natural frequency of cylinder and f_n/f remains invariant. Furthermore, F_l is the hydrodynamic force in the transverse direction.

According to Equation (15),

$$F_l = \frac{C_l}{\pi} = \frac{C_{lF}}{\pi} + \frac{C_{lW}}{\pi} - \frac{4}{\text{Re}} \frac{dl}{dt} - 4 \frac{d^2l}{dt^2} \tag{17}$$

5. Numerical Approach and Procedure for Fluid-Structure Coupling

The calculations are performed numerically. The detail process on numerical procedure of fluid–structure interaction is shown in Figure 3. At the beginning, based on the initial flow field described by Equation (4) and the boundary conditions Equations (5)–(8), the lift force exerted on the cylinder can be obtained at $t > t_1$ by Equation (15). Then, the displacement and velocity of cylinder can be obtained from the motion Equation (17). Subsequently, the flow field is advanced to the next time step by the integration of Equations (2) and (3) according to the boundary conditions calculated for update, as does the control process with Lorentz force at $t > t_2$. Therefore, the flow field, the hydrodynamic force, the cylinder motion and so on can be obtained in the whole fluid–structure interactiion process with this method.

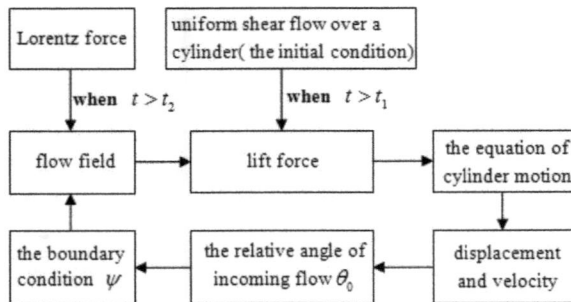

Figure 3. The detailed process of the numerical procedure of fluid–structure interaction.

The exponential mapping in the radial direction allows us to work in a very large physical domain, so that we can avoid the well-known blockage effect (i.e., the effect of the size of apparatus or computational box). On the other hand, the grids near the cylinder are fine enough in a very small physical domain, so that the flow field can be described accurately. In the numerical calculations presented in this paper, the physical circular domain in polar coordinates is transformed into the rectangular (ξ, η) domain through the coordinate transformation $(r, \theta) \to (\xi, \eta)$. The computations presented in this paper are run on a mesh consisting of $N_\xi \times N_\eta = 400 \times 256$ grid points.

The Alternative-Direction Implicit (ADI) algorithm was used to solve the equation of vorticity transport. A Fast Fourier Transform (FFT) algorithm was used to integrate the equation of stream function. Solving Equation (16) using the Runge–Kutta method allows calculating the cylinder motion. For more details on the numerical method and validation of the code, refer to [39–42]. The concrete procedure for simulating the coupling of fluid and structure is the same as [42]. Moreover, the above numerical methods have the accuracy of second order in space and first order in time [42]. The computational step sizes are $\Delta \xi = 0.004$ and $\Delta \eta = 0.004$ for all simulations appearing in the paper. The input parameters are as follows: density of fluid $\rho = 1.0 \times 10^3$ kg/m^3, kinematic viscosity $\nu = 1.0 \times 10^{-6}$ m^2/s, free-stream velocity $u_\infty = 7.5 \times 10^{-3}$ m/s, cylinder radius $a = 1.0 \times 10^{-2}$ m, (so that Re $= \frac{2u_\infty a}{\nu} = 150$, where the flow about the cylinder is fully laminar), density of cylinder $\rho_{cyl} = 2.6\pi \times 10^3$ kg/m^3, and $f = f_n = 0.0675$/s (so that $\rho_{cyl}/\rho = 2.6\pi$). For VIVs in dense fluid, the structural damping is so small that it is negligible [44]. In order to accentuate the effect of viscous damping force ($\frac{4}{\text{Re}}\frac{dl}{dt}$), it may even be assumed zero (i.e., $\varsigma = 0$).

6. Results

From the previous derivations, the Lorentz force is parallel to the cylinder surface along the flow direction, which leads to the acceleration of the boundary layer fluid and improves the capacity of the fluid for overcoming the adverse pressure gradient. Therefore, the flow separation is suppressed, which change the induced fluid forces in turn and subsequently the cylinder response is altered. Moreover, the definition of interaction parameter N is indicated in Equation (1), which represents the strength of Lorentz force.

6.1. Control of VIV with Symmetrical Lorentz Force

In order to describe the differences between the VIV behavior of the cylinder in shear flow and that in uniform flow, the periodical variation of vortex-induced vibration for different shear rate K at the steady VIV with Re = 150 in a shedding cycle are exhibited in Figure 4 by the shaded vorticity contours, where the red refers to the negative vortex, and blue the positive. The cross-hairs mark the equilibrium position of cylinder for shear rate $K = 0$. For convenience, the times at t = 0T/4, 1T/4, 2T/4 and 3T/4 in all the K cases are denoted by A, B, C and D, respectively, where T denotes one period of the cylinder oscillation.

The symmetrical flow field is broken due to the background vorticity, which is generated by shear flow that also causes the increase of upper vortex strength and the decrease of lower vortex strength. With the effect of shear flow, the separation point on the upper side of cylinder moves downstream due to the increase of fluid momentum on the upper side of cylinder while the separation point on the lower side of cylinder moves upstream due to the decrease of fluid momentum on the lower side of cylinder. The vortex street inclines toward the lower side and the inclination of vortex streets increase with the increasing shear rate K, as does the distance of two vortex rows. The equilibrium position of vibration cylinder shifts to the lower side due to the shear flow.

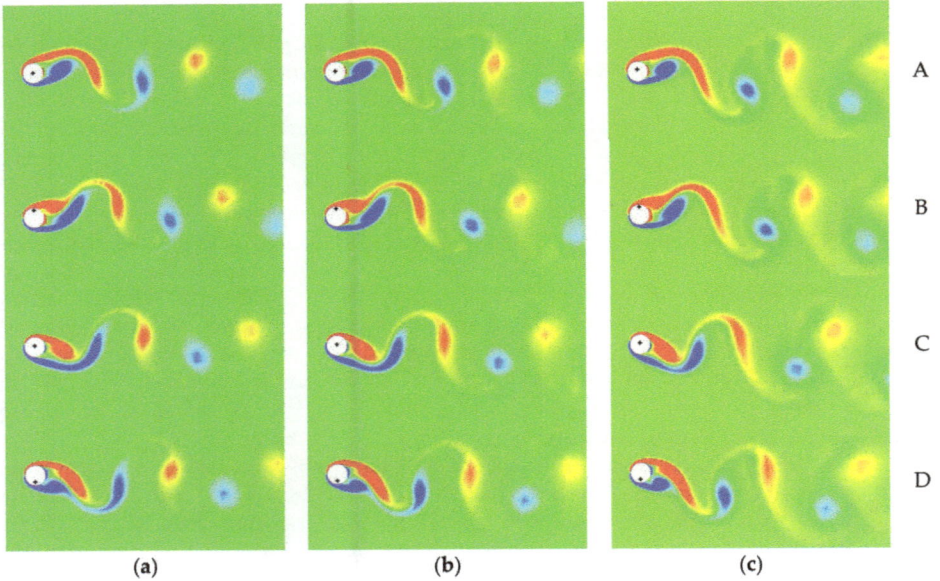

Figure 4. The periodical variation of flow field with different shear rate K where A, B, C and D represent the times at t = 0T/4, 1T/4, 2T/4 and 3T/4, respectively: (**a**) $K = 0$; (**b**) $K = 0.1$; and (**c**) $K = 0.2$.

The variation of $C_{dF} \sim C_{lF}$ phase diagram with shear rate K is shown in Figure 5. The curve moves down due to the shear flow, which means the lift points to the lower side of cylinder. Moreover, the absolute value of lift increases with the increase of shear rate, as do the amplitudes of drag and lift, which lead to the separation of points A and C.

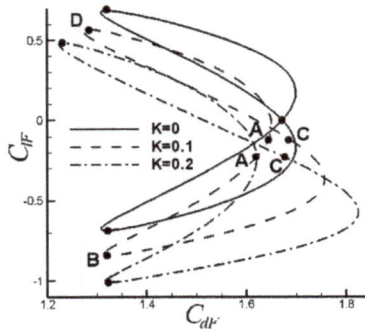

Figure 5. The variation of $C_{dF} \sim C_{lF}$ phase diagram with shear rate K.

The periodic variations of vorticity of VIV with $K = 0.2$ under the Lorentz force control with different values of interaction parameter N are exhibited in Figure 6 by the shaded vorticity contours, where the red represents the negative vortex, and blue the positive. The initial position of the cylinder is marked with cross-hairs. For convenience, the times at t = 0T/4, 1T/4, 2T/4 and 3T/4 in all the N cases are denoted by A, B, C and D, respectively, where T denotes one period of the cylinder oscillation. The interaction parameters are $N = 0$, 0.8 and 1.3 for the three columns in Figure 6. From the figure, the symmetry of flow field is broken as the background vorticity, which is generated by shear flow, increases the strength of upper vortex and decreases the strength of lower vortex. The vortex street slants toward the lower side due to the background vorticity, which is shown in the instance $N = 0$. It is obvious that the vortex shedding and the cylinder vibration are suppressed under the influence of symmetrical Lorentz force and the control effects increase with the increase of interaction parameter N. Despite the control, it is still observed that the front stagnation point moves to the upper side of cylinder and the wake shifts to the lower side, which generally occurs with the effect of shear incoming flow.

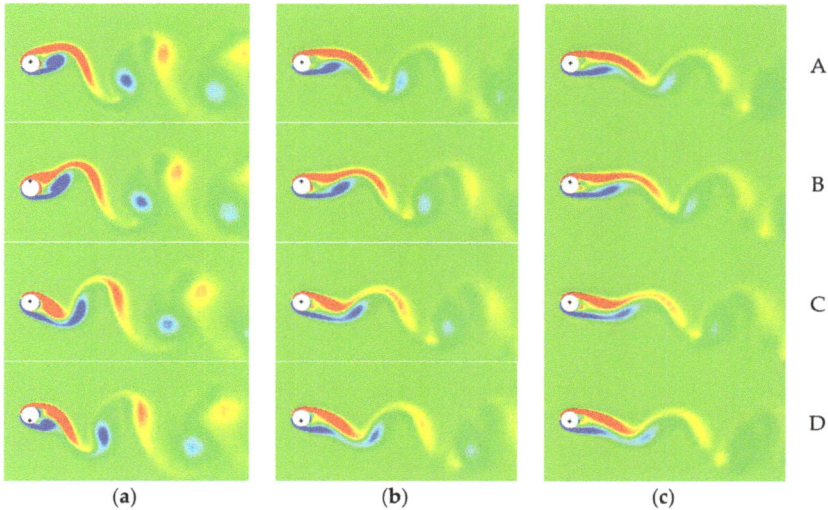

Figure 6. Vorticity evolutions of VIV with different values of interaction parameter N for shear rate $K = 0.2$ where A, B, C and D represent the times at t = 0T/4, 1T/4, 2T/4 and 3T/4, respectively: (a) $N = 0$; (b) $N = 0.8$; and (c) $N = 1.3$.

The distributions of shear stress C_τ^θ with different values of interaction parameter N for the different stages of one cycle are depicted in Figure 7, where the shear rate $K = 0.2$. The symmetrical curve is broken due to the different strength of shear layers on the upper and lower sides, which is generated by shear flow. The boundary layer fluid is hastened by Lorentz force, leading to the increase of shear stress. In addition, the strength of shear stress increases with the increase of interaction parameter N. However, the shift of front stagnation point still exists and the shift amplitude of shear stress C_τ^θ decreases with the increase of Lorentz force (interaction parameter N) on a cycle.

With the Lorentz force applied, the pressure on the surface of vibrating cylinder is composed of three parts

$$C_p^\theta = P_\theta - P_\infty = C_{pW}^\theta + C_{pF}^\theta + C_{pV}^\theta$$

where C_{pW}^θ is the pressure induced by the wall Lorentz force, C_{pF}^θ is the pressure induced by vortex shedding and C_{pV}^θ is the pressure affected by the inertial force.

The effects of interaction parameter N on the distributions of C_{pW}^θ which has no relationship with flow field and cylinder vibration, are discussed firstly based on Figure 8 to examine the VIV control. The curve with positive values is symmetrical about the line $\theta = 180°$. Therefore, the wall Lorentz force only causes the drag decrease, without the effect on the lift.

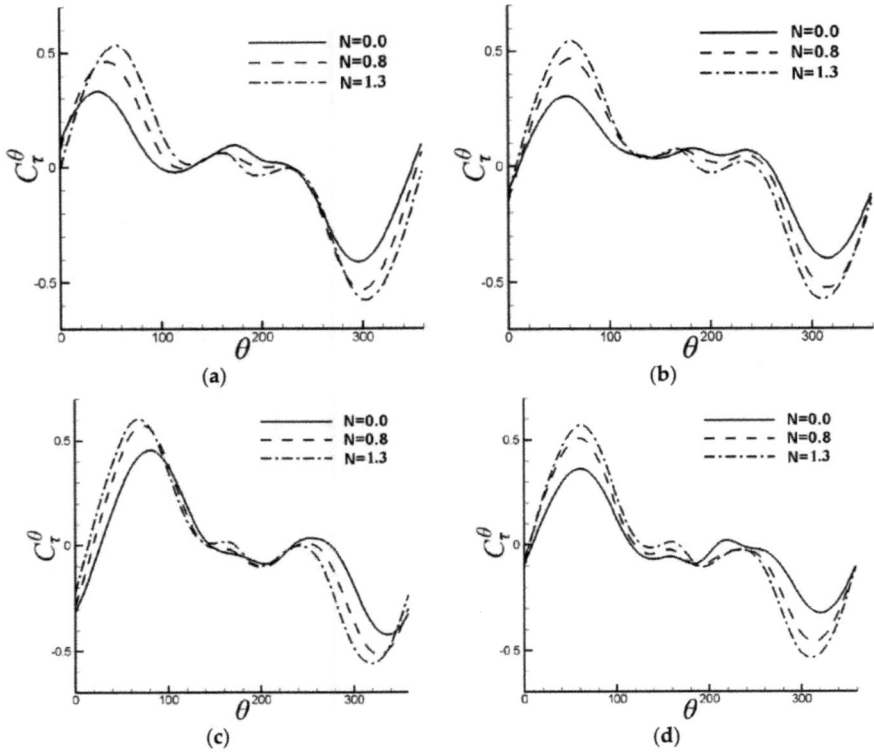

Figure 7. Distributions of shear stress C_τ^θ on the surface of cylinder with different values of interaction parameter N for shear rate $K = 0.2$: (**a**) Time A; (**b**) Time B; (**c**) Time C; and (**d**) Time D.

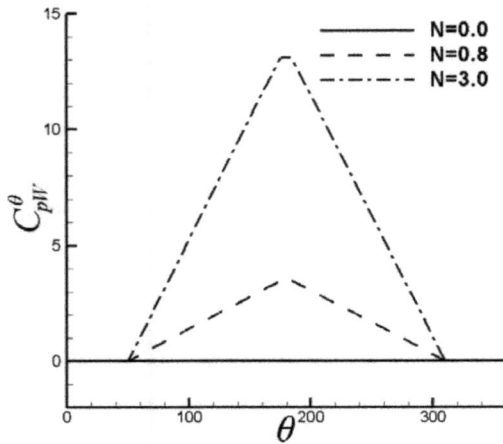

Figure 8. Distributions of pressure coefficient C_{pW}^θ with Lorentz force controls.

The distributions of C_{pF}^{θ} induced by vortex shedding at different stages of one cycle with different values of interaction parameter N are depicted in Figure 9 where the shear rate $K = 0.2$, and A, B, C, and D correspond to those in Figure 6. As the stagnation point shifts to the upper side with the effect of shear flow, the pressure distribution shifts along the clockwise direction as well, which then leads to the increase and decrease of pressure on the upper and lower side, respectively. Obviously, these basic features of VIV in shear flow maintains for the small symmetrical Lorentz force. The pressure on the leeward decreases rapidly with the effect of Lorentz force, leading to the increase of drag and the decrease of the lift oscillating amplitude, in turn, as well as the decline of cylinder vibration. However, the curves are asymmetrical about $\theta = 180°$ due to the shear flow even if Lorentz force is large enough.

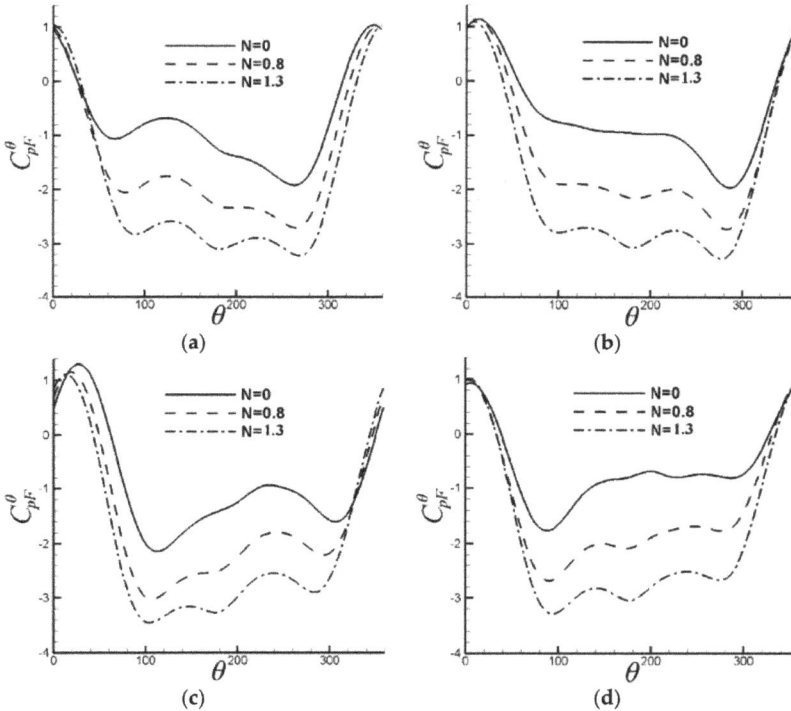

Figure 9. Distribution of C_{pF}^{θ} on the surface of cylinder with different values of interaction parameter N for shear rate $K = 0.2$: (**a**) Time A; (**b**) Time B; (**c**) Time C; and (**d**) Time D.

The inertial force C_{pV}^{θ}, resulting from the acceleration of the cylinder, has influence on the pressure distributions as well. The distributions of C_{pV}^{θ} related with the inertial force for different values of interaction parameter N are depicted in Figure 10. When $N = 0$, the additional lift generated by the background vortex caused by the shear incoming flow leads to the acceleration $\frac{d^2 l(t)}{dt^2}$ at Time B being larger than that at Time D. Namely, the inertial force at Time B is larger than that at Time D. When $N > 0$, the inertial force decreases with the increase of the values of Lorentz force while the inertial force at Time B is still larger than that at Time D.

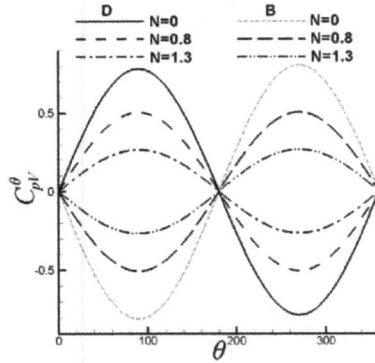

Figure 10. Distributions of $\mathcal{C}_{pV}^{\theta}$ on cylinder surface with $K = 0.2$ for different values of interaction parameter N.

The lift–drag phase diagrams with different values of interaction parameter N for shear rate $K = 0.2$ are depicted in Figure 11. The lift–drag curve shrinks with the increase of the values of Lorentz force, which indicates the decrease of drag and lift vibration. Therefore, the amplitude of vibrating cylinder decreases. Moreover, the lift–drag curve (a Figure 8 shape at $N = 0$) is switched to a teardrop shape by increasing the interaction parameter N. For $N = 0$, the drag at Time C is larger than that at Time A due to the fact that the strength of upper vortex is larger than that of lower vortex with the effect of shear flow (shown in Figure 6) and then the pressure $\mathcal{C}_{pF}^{\theta}$ on the cylinder leeward at Time C is smaller than that at Time A (shown in Figure 9), which leads to Point A and Point C separating in the lift–drag phase diagram. The flow field on the cylinder leeward is gradually dominated by the upper vortex with the increase of Lorentz force. At Time C, the shear layer of upper side is accelerated under the effect of pressure side which leads to the increase of the strength of upper vortex, then the decrease of pressure $\mathcal{C}_{pF}^{\theta}$ on the cylinder leeward and subsequently the increase of drag. The distance between Point A and Point C of lift–drag curve is increased, which leads to the lift–drag phase diagram switching to a teardrop shape gradually. Furthermore, the curve shifts from the left to the right, which implicates the increase of drag. The additional lift still exists under the effect of shear flow.

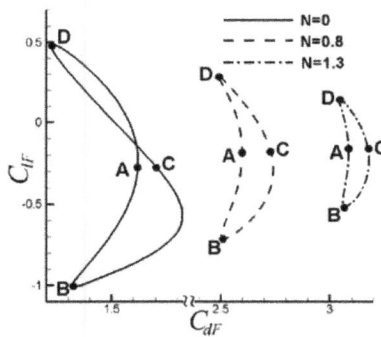

Figure 11. Lift–drag phase diagrams with different values of interaction parameter N for shear rate $K = 0.2$.

The evolution of the cylinder displacement from still to steady vibration, then to a new steady state with symmetric Lorentz force ($N = 0.8$) for $K = 0.2$ is shown in Figure 12. The release of transverse confinement occurs at $t = 446$, which leads to the displacement of cylinder increasing gradually under the effect of lift. Moreover, the equilibrium position which separates from the point $l/a = 0$ and shifts to the lower side. When $t = 640$, the steady vibration is reached. When the symmetric Lorentz force is initiated at time $t_2 = 650$, a well-developed VIV is performed, and the cylinder displacement will decrease over time. However, the equilibrium position shifts because shear flow still exists.

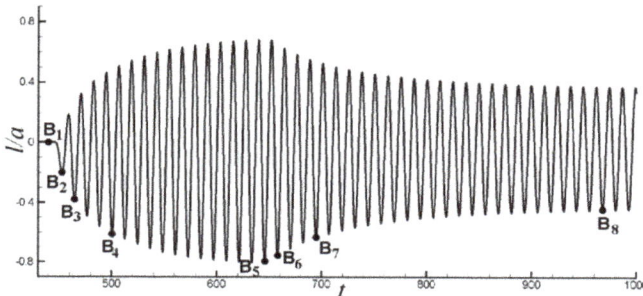

Figure 12. Evolution of the cylinder displacement with the symmetric Lorentz force ($N = 0.8$) for $K = 0.2$.

The variation of vorticity of VIV before and after the application of symmetric Lorentz force ($N = 0.8$) for $K = 0.2$ is shown in Figure 13 where time B_i in this figure corresponds to that in Figure 12 while the cylinder arrives at the lowermost position. The cylinder starts to vibrate with the effect of lift as the confinement of cylinder is released. With the energy transferring from the fluid to the cylinder, there is an increase in the cylinder oscillations. The corresponding vortex patterns are described as $B_1 \sim B_4$. The cylinder vibrates steadily as the total cylinder energy develops into an equilibrium state. The corresponding vortex pattern is described as B_5. When the symmetric Lorentz force is attached on the cylinder at time $t_2 = 650$, the cylinder vibration and the fluid separation are suppressed at some extent ($B_6 \sim B_8$) while the wake vortex is inclined to the lower side.

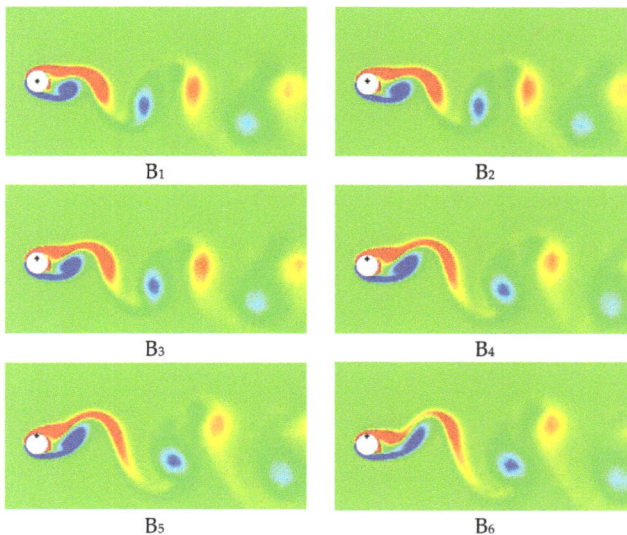

Figure 13. *Cont.*

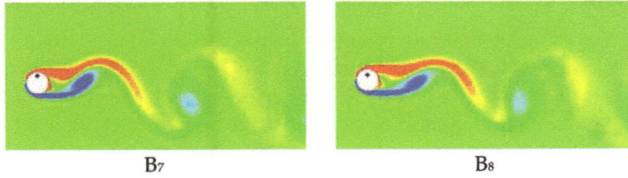

B₇   B₈

Figure 13. (B_1–B_8) Instantaneous vortex patterns of VIV before and after the application of symmetric Lorentz force ($N = 0.8$) for $K = 0.2$.

Time evolutions of lift–drag phase diagram for VIV in a developing and then suppressed process with $N = 0.8$ and $K = 0.2$ are shown in Figure 14. The closed curve $A_1B_1C_1D_1A_1$, which represents the stationary cylinder, is turned right at $180°$ as the cylinder starts to vibrate. Then, Point A and Point C are separated until VIV is well-established. However, Point A and Point C do not coincide any longer, where the phase diagram $C_{dF} \sim C_{lF}$ is depicted by the closed-curve $A_5B_5C_5D_5A_5$. The drag C_{dF} that is generated by the field Lorentz force increases with the application of symmetric Lorentz, though the total drag C_d decreases with the wall Lorentz force effect [41]. The phase diagram therefore shifts to the right significantly. Due to the suppression of flow separation under the effect of symmetric Lorentz force, the vibration of lift/drag decays and the curve shrinks, which then leads to the decrease of the displacement oscillation. Finally, the cylinder vibrates steadily with small amplitude (Figure 14, curve $A_8B_8C_8D_8A_8$).

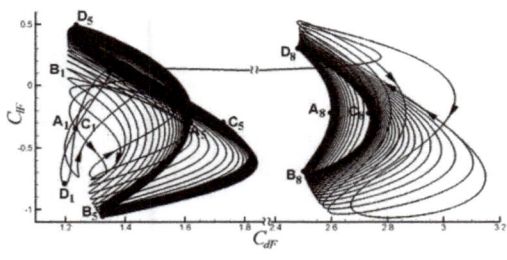

Figure 14. Time evolutions of lift C_{lF} and drag C_{dF} for VIV development and suppression with the symmetric Lorentz force ($N = 0.8$) for $K = 0.2$.

The vibration amplitude varies significantly with the interaction parameter N, as depicted in Figure 15. Notably, the amplitude of cylinder vibration decreases with the increase of N, and the cylinder will finally rest when the interaction parameter N is large enough.

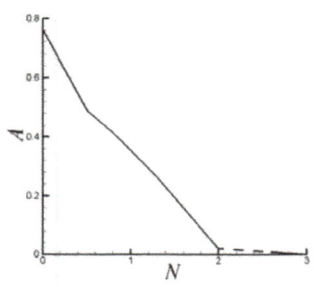

Figure 15. Limiting amplitude of oscillating cylinder displacement after control versus the interaction parameter N for $K = 0.2$.

212

6.2. Control of VIV with Asymmetrical Lorentz Force

For a shear flow past a cylinder, the shift of front stagnation point and additional lift, pointing to lower side for $K > 0$, are induced under the effect of background vorticity, even if symmetrical Lorentz forces are applied. Therefore, the equilibrium position of cylinder controlled by symmetrical Lorentz forces departs from its initial position, as mentioned in Figure 12. However, the lift C_{lL} generated by asymmetrical Lorentz forces can be used to counteract the additional lift, so that the total lift $C_l = 0$ and then the cylinder shifts to the initial position.

The evolution of the cylinder displacement before and after the application of asymmetric Lorentz force (upper surface $N = 3$, lower surface $N = 2$) in a shear flow with $K = 0.2$ is shown in Figure 16. When the asymmetric Lorentz force is initiated at time $t_2 = 650$ when the VIV is well-developed, the lift C_{lL} pointing to the upper side is generated, which can counteract the additional lift to suppress the vibration of cylinder and lead to the equilibrium position of cylinder shifting back to the upper side. Finally, the cylinder is steady on the initial position.

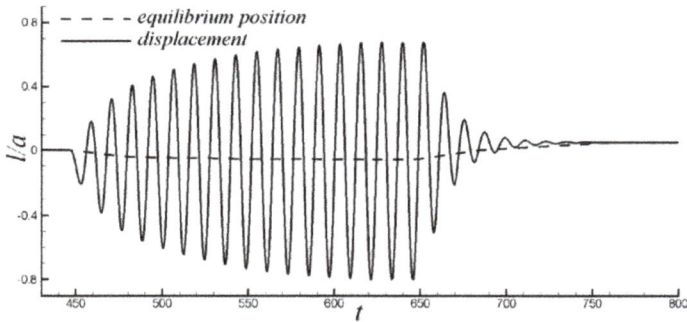

Figure 16. Evolution of the cylinder displacement before and after application of Lorentz force (upper surface $N = 3$, lower surface $N = 2$) in shear flow for $K = 0.2$.

The vorticity of VIV with the asymmetric Lorentz force is shown in Figure 17. The wake suppressed completely by the Lorentz force inclines to the lower side with the influence of shear flow.

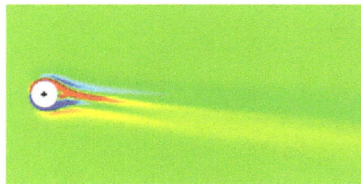

Figure 17. The vorticity of VIV with the Lorentz force (upper surface $N = 3$, lower surface $N = 2$) for $K = 0.2$.

The distribution of shear stress C_τ^θ on the surface of cylinder with the asymmetric Lorentz force is depicted in Figure 18. Due to the higher flow speed on the upper surface, the shear stress on the upper side prevails over that on the lower side obviously.

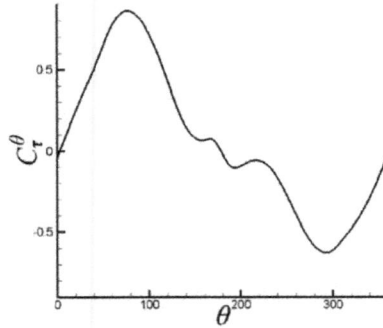

Figure 18. The distribution of shear stress C_τ^θ on the cylinder surface with the asymmetric Lorentz force (upper surface $N = 3$, lower surface $N = 2$) for shear rate $K = 0.2$.

The distribution of pressure coefficient C_{pW}^θ with the asymmetric Lorentz force is shown in Figure 19. The value of C_{pW}^θ is positive and the maximum is achieved at $\theta = 180°$. Meanwhile, the C_{pW}^θ on the lower side is larger than that on the upper side. Thus, the wall Lorentz force results in the amplification of lift with the direction pointing to the upper side and the drag reduction (thrust generated).

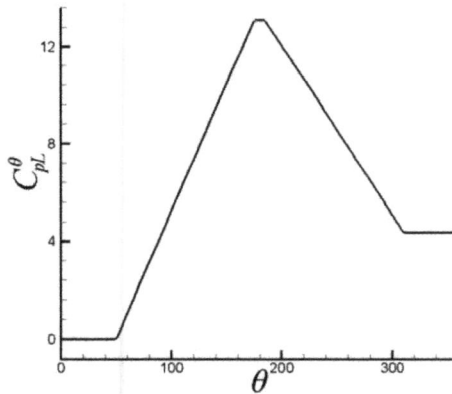

Figure 19. The distributions of pressure coefficient C_{pW}^θ with the asymmetric Lorentz force (upper surface $N = 3$, lower surface $N = 2$) for shear rate $K = 0.2$.

The distribution of C_{pF}^θ on the surface of cylinder with the asymmetric Lorentz force is depicted in Figure 20. With the effect of asymmetric Lorentz force, the curve moves down. The pressure on the upper surface prevails over that on the lower surface, while the decrease of pressure on the leeward side is sharper than that on the windward side. Therefore, the lift decreases and the drag increases with the effect of field Lorentz force, which is opposite with the effect of wall Lorentz force. However, the total effect is dominated by the wall Lorentz force.

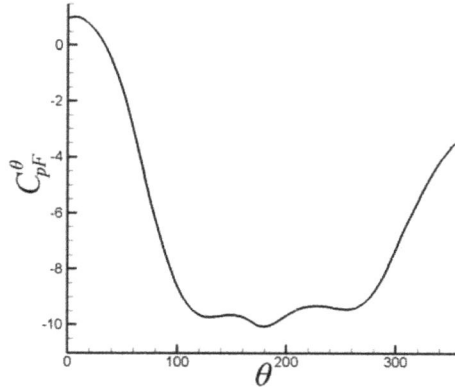

Figure 20. The distribution of pressure induced by vortex C_{pF}^{θ} on the surface of cylinder with the asymmetric Lorentz force (upper surface $N = 3$, lower surface $N = 2$) for shear rate $K = 0.2$.

7. Conclusions

With exponential-polar coordinates attached on a moving cylinder, we have deduced the related equations considering the electro-magnetic force for the shear flow, i.e., the stream function–vorticity equations, the initial and boundary conditions, the hydrodynamic forces on the cylinder surface, and the cylinder motion equation together with the mathematical expressions of the lift force C_l. The applied Lorentz force was not dependent on the flow field and it involves two categories: the field Lorentz force $F_\theta|_{\zeta>0}$ and the wall Lorentz force $F_\theta|_{\zeta=0}$. As a source term, the field Lorentz force affected the flow field in the boundary layer, which leads to the variations of hydrodynamic force. The wall Lorentz force, however, is independent of the flow field due to the non-slip boundary and affects directly on the surface of cylinder.

The momentum of the fluid round the cylinder surface is increased by Lorentz force, which overcomes the effect of the "adverse" pressure gradient. Therefore, the separation points move downstream, and even disappear on the cylinder surface. With the application of symmetrical Lorentz forces, the symmetric field Lorentz force can increase the drag, suppress the flow separation, decay the lift fluctuation, and, in turn, suppress the VIV, whereas the wall Lorentz force decreases the drag only, i.e. no effect on the lift. With the application of asymmetrical Lorentz forces besides above-mentioned effects, the field Lorentz force can strengthen additional lift (negative value) induced by shear flow, whereas the wall Lorentz force can counteract the additional lift, which dominates the total effect. Therefore, it is notable that the asymmetrical Lorentz force, which suppresses the lift vibration and enhances the lift value, can be applied to overcome the lift loss that is caused by the shear flow.

From the above discussion, the aims for the vibration suppression, drag reduction and lift amplification have been obtained with application of Lorentz force. More complicated geometries, such as wing sections, can be controlled with application of this approach. The optimum performance can be obtained with changing the position of actuators in the applications based on the separation points, vortexes, the suction and pressure sides and so on. In addition, the value of Lorentz force varies with time according to the instantaneous flow field, which can be applied to control varying flow conditions, such as dynamically-changing angles of attack.

Acknowledgments: The work was supported by the National Natural Science Foundation of China (Grant Nos. 11672135 and 11202102), the Fundamental Research Funds for the Central Universities (Grant No. 30916011347) and A Foundation for the Author of National Excellent Doctoral Dissertation of PR China (Grant No. 201461).

Author Contributions: Hui Zhang conceived the study, wrote the paper and interpreted the results. Meng-ke Liu participated in data analysis and editing the manuscript. Bao-chun Fan and Zhi-hua Chen reviewed the study plan and corrected the grammar mistakes. Jian Li and Ming-yue Gui helped in the simulation.

Conflicts of Interest: The authors declare no conflict of interest.

References

1. Cachafeiro, H.; de Arevalo, L.F.; Vinuesa, R.; Lopez-Vizcaino, R.; Luna, M. Analysis of vacuum evolution inside Solar Receiver Tubes. *Energy Procedia* **2015**, *69*, 289–298. [CrossRef]
2. Vinuesa, R.; de Arevalo, L.F.; Luna, M.; Cachafeiro, H. Simulations and experiments of heat loss from a parabolic trough absorber tube over a range of pressures and gas compositions in the vacuum chamber. *J. Renew. Sustain. Energy* **2016**, *8*, 023701. [CrossRef]
3. Feng, C.C. *The Measurement of Vortex-Induced Effects in Flow Past Stationary and Oscillating Circular and D-Section Cylinders*; University of British Columbia: Vancouver, BC, Canada, 1968.
4. Griffin, O.M.; Koopmann, G.H. The vortex-exited lift and reaction forces on resonantly vibrating cylinders. *J. Sound Vib.* **1977**, *54*, 435–448. [CrossRef]
5. Griffin, O.M. Vortex-excited cross-flow vibrations of a single cylindrical tube. *J. Press. Vessel Technol.* **1980**, *102*, 158–166. [CrossRef]
6. Griffin, O.M.; Ramberg, S.E. Some recent studies of vortex shedding with application to marine tubulars and risers. *Eur. J. Mech. B Fluids* **1993**, *250*, 481–508. [CrossRef]
7. Brika, D.; Laneville, A. Vortex-induced vibration of a long flexible circular cylinder. *ASME J. Energy Res. Technol.* **1982**, *104*, 2–13. [CrossRef]
8. Hover, F.S.; Miller, S.N.; Triantafyllou, M.S. Vortex-induced vibration of marine cables: Experiments using force feedback. *J. Fluids Struct.* **2011**, *27*, 354–366. [CrossRef]
9. Gharib, M.R. *Vortex-Induced Vibration, Absence of Lock-in and Fluid Force Deduction*; California Institute of Technology: Pasadena, CA, USA, 1999.
10. Khalak, A.; Williamson, C.H.K. Fluid forces and dynamics of a hydroelastic structure with very low mass and damping. *J. Fluids Struct.* **1997**, *11*, 973–982. [CrossRef]
11. Slaouti, A.; Stansby, P.K. Forced Oscillation and Dynamics Response of a Cylinder in a Current Investigation by the Vortex Method. In Proceedings of the BOSS '94 Conference, Cambridge, MA, USA, 12–15 July 1994; pp. 645–654.
12. Zhou, C.Y.; So, R.M.C.; Lam, K. Vortex-induced vibrations of an elastic circular cylinder. *J. Fluids Struct.* **1999**, *13*, 165–189. [CrossRef]
13. Shiels, D.; Leonard, A.; Roshko, A. Flow-induced vibration of a circular cylinder at limiting structural parameters. *J. Fluid Mech.* **2001**, *15*, 3–21. [CrossRef]
14. Leonard, A.; Roshko, A. Aspect of flow-induced vibration. *J. Fluids Struct.* **2001**, *15*, 415–425. [CrossRef]
15. Williamson, C.H.K.; Govardhan, R. Vortex-induced vibrations. *Annu. Rev. Fluid Mech.* **2004**, *36*, 413–455. [CrossRef]
16. Morse, T.L.; Williamson, C.H.K. Prediction of vortex-induced vibration response by employing controlled motion. *J. Fluid Mech.* **2009**, *634*, 5–39. [CrossRef]
17. Franzini, G.R.; Fujiarra, A.L.C.; Meneghini, J.R.; Korkischko, I.; Franciss, R. Experimental investigation of Vortex-Induced Vibration on rigid, smooth and inclined cylinders. *J. Fluids Struct.* **2009**, *25*, 742–750. [CrossRef]
18. Lam, K.; Zou, L. Three-dimensional numerical simulation of cross-flow around four cylinders in an in-line square configuration. *J. Fluids Struct.* **2010**, *26*, 482–502. [CrossRef]
19. Trim, A.D.; Braaten, H.; Lie, H.; Tognarelli, M.A. Experimental investigation of vortex-induced vibration of long marine risers. *J. Fluids Struct.* **2005**, *21*, 335–361. [CrossRef]
20. Yeung, R.W. Fluid dynamics of finned bodies—From VIV to FPSO. In Proceedings of the 12th International Offshore and Polar Engineering Conference, Kitakyushu, Japan, 26–31 May 2002.
21. Bearman, P.W.; Brankovic, M. Experimental studies of passive control of vortex-induced vibration. *Eur. J. Mech. B Fluids* **2004**, *23*, 9–15. [CrossRef]
22. Allen, D.W.; Henning, D.L. Comparison of Various Fairing Geometries for Vortex Suppression at High Reynolds Numbers. In Proceedings of the Offshore Technology Conference, Houston, TX, USA, 5–8 May 2008.
23. Assi, G.R.S.; Bearman, P.W.; Kitney, N. Low drag solutions for suppressing vortex-induced vibration of circular cylinders. *J. Fluids Struct.* **2009**, *25*, 666–675. [CrossRef]

24. Galvao, R.; Lee, E.; Farrell, D.; Hover, F.; Triantafyllou, M.; Kitney, N.; Beynet, P. Flow control in flow–structure interaction. *J. Fluids Struct.* **2008**, *24*, 1216–1226. [CrossRef]

25. Modi, V.J. Moving surface boundary-layer control: A review. *J. Fluids Struct.* **1997**, *33*, 229–242. [CrossRef]

26. Munshi, S.R.; Modi, V.J.; Yokomizo, T. Aerodynamics and dynamics of rectangular prisms with momentum injection. *J. Fluids Struct.* **1997**, *11*, 873–892. [CrossRef]

27. Korkischko, I.; Meneghini, J.R. Suppression of vortex-induced vibration using moving surface boundary-layer control. *J. Fluids Struct.* **2012**, *34*, 259–270. [CrossRef]

28. Wu, C.J.; Wang, L.; Wu, J.Z. Suppression of the von Karman vortex street behind a circular cylinder by a traveling wave generated by a flexible surface. *J. Fluid Mech.* **2007**, *574*, 365–391. [CrossRef]

29. Tchieu, A.A.; Leonard, A. Experimental investigation on the suppression of vortex-induced vibration of long flexible riser by multiple control rods. *J. Fluids Struct.* **2012**, *30*, 115–132.

30. Gbadebo, S.A.; Cumpsty, N.A.; Hynes, T.P. Control of three-dimensional separations in axial compressors by tailored boundary layer suction. *J. Turbomach.* **2008**, *130*, 011004. [CrossRef]

31. Chng, T.L.; Rachman, A.; Tsai, H.M.; Zha, G.C. Flow control of an airfoil via injection and suction. *J. Aircr.* **2009**, *46*, 291–300. [CrossRef]

32. Arcas, D.; Redekopp, L. Aspects of wake vortex control through base blowing/suction. *Phys. Fluids* **2004**, *16*, 452–456. [CrossRef]

33. Fransson, J.H.M.; Konieczny, P.; Alfredsson, P.H. Flow around a porous cylinder subject to continuous suction or blowing. *J. Fluids Struct.* **2004**, *19*, 1031–1048. [CrossRef]

34. Patil, S.K.R.; Ng, T.T. Control of separation using span wise periodic porosity. *AIAA J.* **2010**, *48*, 174–187. [CrossRef]

35. Crawford, C.H.; Karniadakis, G.E. Control of External Flows via Electro-Magnetic Fields. In Proceedings of the AIAA Fluid Dynamics Conference, San Diego, CA, USA, 19–22 June 1995.

36. Weier, T.; Gerbeth, G.; Mutschke, G.; Platacis, E.; Lielausis, O. Experiments on cylinder wake stabilization in an electrolyte solution by means of electromagnetic forces localized on the cylinder surface. *Exp. Therm. Fluid Sci.* **1998**, *16*, 84–91. [CrossRef]

37. Kim, S.; Lee, C.M. Control of flows around a circular cylinder: Suppression of oscillatory lift force. *Fluid Dyn. Res.* **2001**, *29*, 47–63. [CrossRef]

38. Posdziech, O.; Grundmann, R. Electromagnetic control of seawater flow around circular cylinders. *Eur. J. Mech. B Fluids* **2001**, *20*, 255–274. [CrossRef]

39. Zhang, H.; Fan, B.C.; Chen, Z.H. Computations of optimal cylinder flow control in weakly conductive fluids. *Comput. Fluids* **2010**, *39*, 1261–1266. [CrossRef]

40. Zhang, H.; Fan, B.C.; Chen, Z.H. Optimal control of cylinder wake by electromagnetic force based on the adjoint flow field. *Eur. J. Mech. B Fluids* **2010**, *29*, 53–60. [CrossRef]

41. Zhang, H.; Fan, B.C.; Chen, Z.H.; Li, Y.L. Effect of the Lorentz force on cylinder drag reduction and its optimal location. *Fluid Dyn. Res.* **2011**, *43*, 015506. [CrossRef]

42. Zhang, H.; Fan, B.C.; Chen, Z.H.; Li, H.Z. An in-depth study on vortex-induced vibration of a circular cylinder with shear flow. *Comput. Fluids* **2014**, *100*, 30–44. [CrossRef]

43. Lei, C.; Cheng, L.; Kavanagh, K. A finite difference solution of the shear flow over a circular cylinder. *Ocean Eng.* **2000**, *27*, 271–290. [CrossRef]

44. Sarpkaya, T. A critical review of the intrinsic nature of vortex-induced vibrations. *J. Fluids Struct.* **2004**, *19*, 389–447. [CrossRef]

Article

An Experimental Validated Control Strategy of Maglev Vehicle-Bridge Self-Excited Vibration

Lianchun Wang [1], Jinhui Li [1,2], Danfeng Zhou [1] and Jie Li [1,*

[1]   College of Mechatronics Engineering and Automation, National University of Defense Technology,
     Changsha 410073, China; spring_512@163.com (L.W.); li_jinhui@126.com (J.L.); zdfnudt@163.com (D.Z.)
[2]   Scinence and Technology on Near-Surface Detection Laboratory, Wuxi 214035, China
*   Correspondence: jieli@nudt.edu.cn; Tel.: +86-731-845-73388 (ext. 8103)

Academic Editors: Gangbing Song, Steve C.S. Cai and Hong-Nan Li
Received: 12 October 2016; Accepted: 23 December 2016; Published: 4 January 2017

Abstract: This study discusses an experimentally validated control strategy of maglev vehicle-bridge vibration, which degrades the stability of the suspension control, deteriorates the ride comfort, and limits the cost of the magnetic levitation system. First, a comparison between the current-loop and magnetic flux feedback is carried out and a minimum model including flexible bridge and electromagnetic levitation system is proposed. Then, advantages and disadvantages of the traditional feedback architecture with the displacement feedback of electromagnet y_E and bridge y_B in pairs are explored. The results indicate that removing the feedback of the bridge's displacement y_B from the pairs ($y_E - y_B$) measured by the eddy-current sensor is beneficial for the passivity of the levitation system and the control of the self-excited vibration. In this situation, the signal acquisition of the electromagnet's displacement y_E is discussed for the engineering application. Finally, to validate the effectiveness of the aforementioned control strategy, numerical validations are carried out and the experimental data are provided and analyzed.

Keywords: maglev; levitation system; bridge; self-excited vibration; energy

1. Introduction

Compared with the traditional railway train system, the electromagnetic maglev system has advantages of less exhaust fumes emission, lower noise and the ability to climb steeper slopes, which has attracted wide attention in recent years [1–3].

Maglev's rapid development and its potential commercial applications depict a bright future. However, there are still some issues to be solved urgently, such as the vehicle-bridge stationary self-excited vibration [4,5]. When the maglev train is suspended on some special bridges with minor weight per meter, the maglev train and bridges may vibrate vertically and continuously, which degrade the stability of the levitation control and the ride comfort.

In magnetic engineering, due to its clear physical meaning and excellent performance, the cascaded-loop control architecture with displacement-loop and current-loop to control the electromagnetic force is widely adopted [6,7]. According to the formulation of the electromagnetic force, the electromagnetic force is related to the levitation gap and current. To some extent, the electromagnetic force, the levitation gap and the current are interactive and complex. As we all know, from the perspective of the magnetic flux, the electromagnetic force is solely determined by the magnetic flux [8]. The relationship between the electromagnetic force and magnetic flux is simple and clear, if the magnetic flux-loop instead of the traditional current-loop is adopted to control the electromagnetic force, it may be favorable for the stabilization of interaction system. Hence, the magnetic flux will be discussed and adopted in this paper.

To master the underlying principles of the self-excited vibration, plenty of studies have been carried out from different perspectives. Albert et al. [9,10] pointed out that the American magnetic levitation system was successfully suspended in Florida, but it was incapable of achieving stable suspension when the vehicle was moved to the old Dominican university campus. They believed that the over-flexibility of the latter bridge was the main reason for the difficulties of stable suspension.

Wang et al. believed that the self-excited vibration is caused by the inappropriate frequency relationship between the various components of the Maglev vehicle-bridge interaction system [11]. Therefore, the proper frequency distribution is an effective strategy to avoid the resonance. The center manifold method was carried out to discuss the underlying principles of the self-excited vibration in the literature [12,13]. It is believed that the self-excited vibration results from the bifurcations and chaos.

The aforementioned studies about the roots of the self-excited vibration provide us inspiration to avoid the vibration. In this paper, based on the proposed minimum model, the underlying principles of the self-excited vibration will be explored from the real parts of its characteristic roots.

To eliminate the self-excited vibration, the methods, including optimization of the parameters and minimization of the time-delay of feedback channels [14], virtual tuned mass damper algorithm [15] and the virtual energy harvester algorithm [16] are explored by different scholars. They believed that these control strategies can avoid the self-excited vibration for the given bridge. Yau intends to develop a neuro-PI (proportional-integral) controller to control the dynamic response of the maglev vehicles around an allowable prescribed acceleration, numerical simulations demonstrate that a trained neuro-PI controller has the ability to control the acceleration amplitude for running maglev vehicles [17]. However, due to the complexity, the robustness to bridges with different modal frequencies awaits further research.

In this paper, by the analysis of the block diagrams in depth, we find that removing the displacement feedback of bridge instead of the feedback in pairs is an effective technique to enable levitation subsystem passivity and to solve the problem of self-excited vibration theoretically. Furthermore, several implementation issues, including the estimation of bridge's displacement are addressed.

The purpose of the research reported here is to develop a vibration control method that can eliminate the self-excited vibration, and is suitable for the magnetic levitation system.

2. Modeling of Vehicle-Bridge Interaction

Generally, the choice of model's complexity of the maglev vehicle-bridge system depends on its usage. For the validation by the numerical simulation, the model should be precise enough to improve the creditability and precision. For the exploration of the principle and the design of control strategy, the minimum interaction model containing the quintessential parts, a flexible bridge and a levitation unit, may be more practical and effective.

In this section, the nonlinear part of the bridge is ignored because the magnitude of the self-vibration is small enough when compared to the span of the bridge. Furthermore, the bridge is simplified as a Bernoulli–Euler beam because the other dimensions of bridges are much smaller than its length. In addition, the electromagnetic force acting on the bridge and electromagnet is regarded as an equivalent concentrated force. Furthermore, the kinetics coupling between adjacent levitation units and the influence of air springs are neglected [16].

2.1. Modeling of Bridge

Based on the above assumptions, the minimum interaction model is shown in Figure 1. The variables y_B and y_E are the vertical displacements of bridge and electromagnet, respectively. The variable δ is the levitation gap measured by the gap sensor, and m_E is the equivalent mass of electromagnet.

The motion of bridge is described by the following differential equation [4]:

$$EI_B \frac{\partial^4 y_B(x,t)}{\partial x^4} + \rho_B \frac{\partial^2 y_B(x,t)}{\partial t^2} = f(x,t) \tag{1}$$

where the variable x is the axial coordinate of the bridge, t is the time, EI_B is the bending rigidity, ρ_B is the mass per unit length, and $f(x, t)$ is the electromagnetic force acting on the bridge.

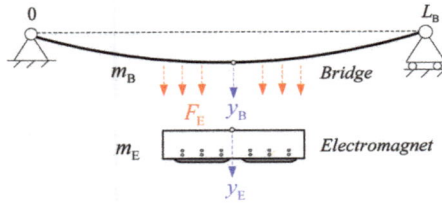

Figure 1. The minimum model of Maglev vehicle-bridge system. Red arrows: F_E is electromagnetic force; blue arrows: y_B and y_E are the vertical displacements of bridge and electromagnet.

For a simply supported bridge, the first modal frequency ω_B and modal shape functions $\phi_B(x)$ are

$$\omega_B = \lambda_B^2 \sqrt{EI_B / \rho_B} \tag{2}$$

$$\phi_B(x) = \sin(\lambda_B x) \tag{3}$$

where λ_B is the space wavelength of the bridge's first modal, and $\lambda_B = \pi / L_B$. The fact observed from the maglev base of china shows that the self-excited vibration is mainly aroused by the first modal of bridges. Hence, in this section, the first modal of bridge is considered solely. Thus, the solutions of Equation (1) may be expressed as

$$y_B(x, t) = \phi_B(x) q_B(t) \tag{4}$$

where $q_B(t)$ is the time-varying amplitude of the modal displacement. Substituting the Equation (4) into Equation (1), multiplying both sides of the aforementioned resultant equation by $\phi_k(x)$, then integrating both sides from 0 to L_B gives

$$\ddot{q}_B(t) + 2\xi_B \omega_B \dot{q}_B(t) + \omega_B^2 q_B(t) = 2\rho_B^{-1} L_B^{-1} \sum_{i=1:n} \phi_B(x_i) F_{Ei}(t) \tag{5}$$

where n is the number of levitation units suspended on a single bridge. Multiplying both sides of the resultant equation by $\phi_B(x)$ gives

$$\ddot{y}_B(x, t) + 2\xi_B \omega_B \dot{y}_B(x, t) + \omega_B^2 y_B(x, t) = 2m_B^{-1} \phi_B(x) \cdot \sum_{i=1:n} \phi_B(x_i) F_{Ei}(t) \tag{6}$$

where $m_B = \rho_B L_B$ is the total mass of bridge. The phases of electromagnetic forces $F_{Ei}(t)(i = 1 : n)$ are exactly in concert and the amplitude of $F_{Ei}(t)$ is proportional to the value of $\phi_B(x_i)$ when the self-excited vibration occurs with the first-order modal frequency [18]. As for Equation (6), with regard to the special case $x = 0.5L_B$, it gives that

$$\ddot{y}_B(t) + 2\xi_B \omega_B \dot{y}_B(t) + \omega_B^2 y_B(t) = \sigma m_B^{-1} F_E(t) \tag{7}$$

where $\sigma = 2\sum_{i=1:n} \phi_B^2(x_i)$, the variable $y_B(t)$ is the modal displacement and the variable $F_E(t)$ is the electromagnetic force of the levitation unit at the location of $x = 0.5L_B$. Equation (7) may be considered as the response of bridge roused by the electromagnetic force $F_E(t)$.

2.2. Modeling of Levitation System

Suppose the turns of a single electromagnet is N, the pole area is A, and the magnetic permeability of vacuum is μ_0. Then, for a single electromagnet [3], the balance equations of electromagnetic force $F_E(t)$ and voltage $u(t)$ are

$$\begin{cases} F_E(t) = \frac{\mu_0 A N^2}{2} \left(\frac{i(t)}{\delta(t)} \right)^2 \\ 2Ri(t) + \frac{\mu_0 A N^2}{\delta(t)} \dot{i}(t) - \frac{\mu_0 A N^2 i(t)}{\delta^2(t)} \dot{\delta}(t) = u(t) \end{cases} \qquad (8)$$

where R is the resistance, $i(t)$ is the current of electromagnet. In light of Equation (8), the balance equation of voltage is related with the four variables, control voltage $u(t)$, current $i(t)$, levitation gap $\delta(t)$ and its derivative $\dot{\delta}(t)$. Besides, the value of $F_E(t)$ refers to the two variables, current $i(t)$ and levitation gap $\delta(t)$.

To simplify the above relationship, the magnetic flux $B(t)$ instead of the current $i(t)$ may be adopted when developing the dynamic equations. In this way, Equation (8) is updated as

$$\begin{cases} F_E(t) = \frac{2A}{\mu_0} B^2(t) \\ 2NA\dot{B}(t) + \frac{4R}{\mu_0 N} \delta(t) B(t) = u(t) \end{cases} \qquad (9)$$

In light of Equation (9), it can be seen that the balance equation of voltage is related with three variables and the electromagnetic force is solely determined by the magnetic flux. Compared with Equation (8), the dynamic equation is much simpler and clearer, which may be favorable adopted for the synthesis of the vehicle-bridge interaction system.

Generally, the natural frequency of air springs is far less than the bandwidth of the levitation control system and the modal frequencies of bridges, so the dynamics of sprung mass is neglected. Combining Equation (7), the movement of electromagnet is

$$m_E \ddot{y}_E(t) = -F_E(t) + (m_C + m_E)g \qquad (10)$$

where the variable $y_E(t)$ is the vertical displacement of electromagnet, g is the acceleration of gravity, m_C is the sprung mass, and m_E is the mass of electromagnet. According to Equations (9) and (10), it can be seen that the steady voltage of electromagnet is

$$u_0 = 2R\delta_0 \sqrt{2(m_C + m_E)g/(\mu_0 N^2 A)} \qquad (11)$$

where the variable δ_0 is the desired clearance between the upper surface of the electromagnet and the lower surface of the guideway.

Traditionally, the cascade control, the inner-loop with current feedback, the outer-loop with states feedback (proportion, damping and acceleration feedback) is widely adopted in maglev engineering. It gives that

$$u(t) = k_c \left[i_{exp}(t) - i(t) \right] + u_0 \qquad (12)$$

$$i_{exp}(t) = k_p e(t) + k_d \dot{y}_E(t) + k_a \ddot{y}_E(t) \qquad (13)$$

where the variable $i_{exp}(t)$ is the desired current of electromagnet and the error $e(t)$ of levitation gap is set as

$$e(t) = y_E(t) - y_B(t) - \delta_{set} \qquad (14)$$

where δ_{set} is the expected levitation gap. To stabilize the maglev vehicle-bridge interaction system with magnetic flux feedback, a similar cascade control scheme, the inner-loop with magnetic flux feedback, the outer-loop with states feedback, is adapted, which gives

$$u(t) = k_B \left[B_{exp}(t) - B(t) \right] + u_0 \qquad (15)$$

$$B_{exp}(t) = k_p e(t) + k_d \dot{y}_E(t) + k_a \ddot{y}_E(t) \qquad (16)$$

where $B_{exp}(t)$ is the desired magnetic flux of levitation gap. To draw the main innovation of this work, combining with Equation (14), Equations (13) and (16) may be rewritten as Equation (17) when the parameter $\bar{k}_p$ is equal to parameter k_p.

$$\begin{cases} i_{\exp}(t) = k_p y_E(t) + k_d \dot{y}_E(t) + k_a \ddot{y}_E(t) - \bar{k}_p y_B(t) - k_p \delta_{set} \\ B_{\exp}(t) = k_p y_E(t) + k_d \dot{y}_E(t) + k_a \ddot{y}_E(t) - \bar{k}_p y_B(t) - k_p \delta_{set} \end{cases} \tag{17}$$

Thus, the minimum vehicle-bridge interaction model with active control is developed.

3. Principle of Self-Excited Vibration

It has been observed that when the maglev vehicle is suspended on the bridge staying still or moving slowly, the self-excited vibration occurs. When the vibration amplitude of bridge is sufficiently small, the interaction system is quasi-static. Hence, the linearized model is practical to simplify the analysis process without introducing noticeable errors.

Combining Equations (7) and (10) in time domains [18], the vertical dynamics of electromagnet and bridge in frequency domains can be converted to

$$\begin{cases} -m_E s^2 y_E(s) = F_E(s) \\ \sigma^{-1} m_B (s^2 + 2\xi_B \omega_B + \omega_B^2) y_B(s) = F_E(s) \end{cases} \tag{18}$$

Similarly, when the magnetic flux inner-loop is adopted, Combining Equations (9), (15) and (16) in time domains, the electromagnetic force and balance equation of voltage in frequency domains can be converted to

$$\begin{cases} F_E(s) = k_F B(s) \\ B(s) = \left((k_p k_B + k_d k_B s + k_a k_B s^2) y_E(s) - \bar{k}_p k_B y_B(s) \right) / (2NAs + k_B) \end{cases} \tag{19}$$

Combining Equations (18) and (19), the maglev vehicle-bridge interaction system may be represented by the following block diagram in Figure 2, where the electromagnetic module (*EM*) block is the voltage equation shown by Equation (9).

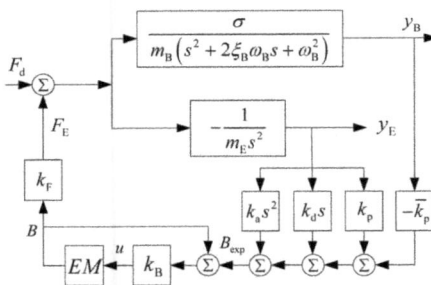

Figure 2. The equivalent block diagram of maglev vehicle-bridge system.

3.1. Stability of Levitation System

The stability of the levitation subsystem itself is a precondition for the avoidance of the maglev vehicle-bridge self-excited vibration. When studying the stability of the levitation system itself, the vertical displacement of bridge y_B is set as zero. In this case, the transfer function from the disturbance F_d to the displacement y_E of electromagnet is

$$T_1(s) = \frac{y_E(s)}{F_d(s)} = \frac{G_E(s)}{1 - G_E(s) H_E(s)} \tag{20}$$

where

$$\begin{cases} G_E(s) = -\frac{1}{m_E s^2} \\ H_E(s) = \frac{k_B k_F}{2NAs + k_B} (k_a s^2 + k_d s + k_p) \end{cases} \tag{21}$$

The characteristic equation of the transfer function $T_1(s)$ is

$$\Delta_1 = 2NAm_E s^3 + (m_E k_B + k_B k_F k_a)s^2 + k_B k_F k_d s + k_B k_F k_p \tag{22}$$

For the levitation subsystem, the parameters k_p, k_d, k_a, k_B are positive and adjustable, and $k_F = 4AB_0/\mu_0$. According to the Routh–Hurwitz stability criterion, the levitation subsystem is stable when Equation (23) is satisfied.

$$(m_E k_B + k_B k_F k_a)k_d > 2NAm_E k_p \tag{23}$$

3.2. Stability of Vehicle-Bridge Interaction System

To study the stability of the maglev vehicle-bridge interaction system, the transfer function from disturbance F_d to the displacement y_B of bridge should be calculated again when the displacement of bridge y_B is considered. In light of Figure 2, it gives that

$$T_2(s) = \frac{G_B(s)}{1 - G_E(s)H_E(s) - G_B(s)H_B(s)} \tag{24}$$

where

$$\begin{cases} G_B(s) = \frac{\sigma}{m_B\left(s^2 + 2\xi_B \omega_B s + \omega_B^2\right)} \\ H_B(s) = -\frac{k_B k_F}{2NAs + k_B} \cdot \bar{k}_p \end{cases} \tag{25}$$

Combining Equations (21), (24) and (25), the characteristic equation of the interaction system is

$$a_5 s^5 + a_4 s^4 + a_3 s^3 + a_2 s^2 + a_1 s + a_0 = 0 \tag{26}$$

where $a_5 = 2NAm_B m_E$, $a_4 = m_B m_E(k_B + 4NA\xi_B \omega_B) + k_B k_F m_B k_a$, $a_3 = m_B m_E\left(2k_B \xi_B \omega_B + 2NA\omega_B^2\right)$ $+ k_B k_F m_B(2k_a \xi_B \omega_B + k_d)$, $a_2 = k_B \omega_B^2 m_B m_E + \sigma m_E k_B k_F \bar{k}_p + k_B k_F m_B(k_p + 2k_d \xi_B \omega_B + k_a \omega_B^2)$, $a_1 = k_B k_F m_B\left(k_d \omega_B^2 + 2k_p \xi_B \omega_B\right)$, and $a_0 = k_B k_F m_B k_p \omega_B^2$. Generally, the roots of Equation (26) are calculable provided that the parameters are definite, which are denoted as $x_{1,2} = R_1 \pm jI_1$, $x_{3,4} = R_2 \pm jI_2$ and $x_5 = R_3$. If the three real parts R_1, R_2 and R_3 all are negative, the interaction system is stable and the self-excited vibration is avoided.

However, the stability of the interaction system is closely related with the bridge's modal frequency ω_B. To illustrate this, the parameters of interaction system are set as $k_p = 1000$, $k_d = 30$, $k_a = 0.4$, $k_B = 30$, $N = 360$, $A = 0.01848$, $\xi_B = 0.005$, and $B_0 = 0.6193$.

According to Equations (15) and (17), the traditional cascade control is adopted when the parameters $\bar{k}_p$ is set as the same with k_p. This is to say, $\bar{k}_p = 1000$. In this case, the real parts R_1, R_2 and R_3 associated with the modal frequency ω_B are shown in Figure 3.

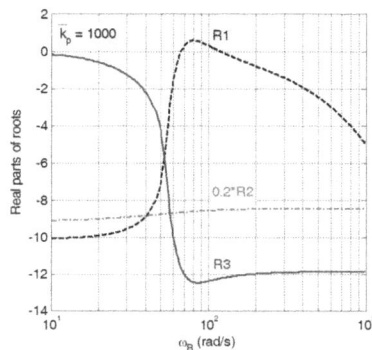

Figure 3. The real parts of characteristic roots for the standard cascade control.

According to Figure 3, it can be seen that the real parts of R_2 and R_3 are negative from 0 to 1000 rad/s. However, the value of R_1 is positive when the modal frequency ω_B belongs to the interval 67.3–118.7 rad/s. This is to say, the interaction system is unstable when the modal frequency ω_B belongs to the interval 67.3–118.7 rad/s. The self-excited vibration will appear.

To avoid the self-excited vibration, tuning the control parameters k_p, k_d, k_a and k_B is an effective technique. However, the ranges of parameters are limited by the performance specification of the levitation subsystem and the noise level of corresponding signals. Hence, a more feasible and robust method should be developed.

3.3. Principle of Self-Excited Vibration from the Perspective of Energy Interchange

From the quiescent state to the vibration state, the bridge needs to absorb energy. However, the interaction system only consists of the electromagnetic levitation system and bridge, so the absorbed energy by bridge is from the exportation of levitation system. Therefore, the characteristic of the energy exportation of the levitation system may be decisive for the occurrence of the self-excited vibration. In this section, we try to discuss the principle of self-excited vibration from the perspective of energy interchange between the bridge subsystem and levitation subsystem.

It has been found that when the vehicle is suspended on some special bridges, standing still or moving under 10 km/h, the self-excited vibration may appear and grow up continuously. Even so, at the beginning of the self-excited vibration, the amplitude of the vibration is tiny enough. In this case, the interaction system may be seen as quasi-static.

When studying the stability of the interaction system around the equilibrium point at the quasi-static states, the analysis process can be simplified by a linearization model without introducing significant errors. Linearizing Equations (8), (10) (13) and (14), the linearized system of frequency domain is given by

$$\begin{cases} u(s) = 2L_0 si(s) + 2Ri(s) - 2F_i s y_E(s) + 2F_i s y_B(s) \\ F_E(s) = 2F_i i(s) - 2F_z y_E(s) + 2F_z y_B(s) \\ F_E(s) = -m_E s^2 y_E(s) \\ u(s) = k_c(k_p + k_d s + k_a s^2) y_E(s) - k_p k_c y_B(s) - (k_c - 2R)i(s) \end{cases} \tag{27}$$

where $L_0 = 0.5\mu_0 A N^2 z_0^{-1}$, $F_i = 0.5\mu_0 A N^2 i_0 z_0^{-2}$, $F_z = 0.5\mu_0 A N^2 i_0^2 z_0^{-3}$. Eliminating the variables $u(s)$, $i(s)$ and $y_E(s)$, the transfer function between $F_E(s)$ and $\dot{y}_B(s)$ is

$$H(s) = \frac{F_E(s)}{\dot{y}_B(s)} = -\frac{\eta_0 m_E s}{\eta_3 s^3 + \eta_2 s^2 + \eta_1 s + \eta_0} \tag{28}$$

where $\eta_0 = F_i k_c \left(k_p - i_0 z_0^{-1}\right)$, $\eta_1 = F_i k_c k_d$, $\eta_2 = 0.5 m_E k_c + F_i k_c k_a$, and $\eta_3 = m_E L_0$. The vibration frequency is assumed as ω_{Vib} and the velocity of the bridge is defined as $\dot{y}_B(t) = 0.1\cos(\omega_{Vib}t + \phi)$. According to Equation (28), the electromagnetic force working on the bridge is

$$F_E(t) = 0.1|H(j\omega_{Vib})|\cos(\omega_{Vib}t + \angle H(j\omega_{Vib})) \tag{29}$$

Furthermore, the averaged power of the electromagnetic force acting on the bridge is

$$P_E(\omega_{Vib}) = \frac{1}{T}\int_0^T F_E(\tau)\dot{y}_B(\tau)d\tau = \frac{1}{200}\text{Re}[H(j\omega_{Vib})] \tag{30}$$

Supposing the damping of bridge is viscous and linear, the damping force is

$$F_D(\omega_{Vib}, t) = 2\xi_k \omega_{Vib} m_B \dot{y}_B(\tau) \tag{31}$$

Herein the averaged power consumed by the damping is

$$P_D(\omega_{Vib}) = \frac{1}{T}\int_0^T F_D(\omega_{Vib}, \tau)\dot{y}_B(\tau)d\tau = 0.01\xi_k\omega_{Vib}m_B \tag{32}$$

The average power accumulated is

$$P_B(\omega_{Vib}) = P_E(\omega_{Vib}) - P_D(\omega_{Vib}) \tag{33}$$

In the normal case of equivalent parameters, the relationships between the vibration frequency and averaged powers are shown in Figure 4.

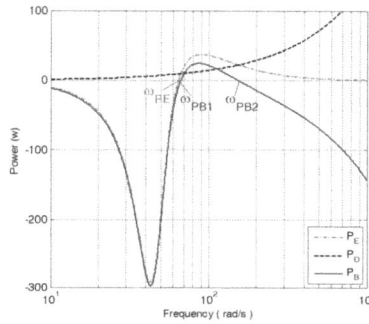

Figure 4. The averaged powers.

According Figure 4, for the crossover frequency ω_{P_E}, $P_E(\omega_{P_E}) = 0$. For any $\omega_{Vib} < \omega_{P_E}$, the averaged power $P_E(\omega_{Vib})$ working on the bridge is negative. This means that the levitation subsystem will absorb the vibration energy of the bridge subsystem when self-excited vibration occurs. On the contrary, for any $\omega_{Vib} > \omega_{P_E}$, the averaged power $P_E(\omega_{Vib})$ working on the bridge is positive. This means that when the self-excited vibration occurs, the levitation subsystem will export energy to the bridge subsystem.

For the bridge subsystem, the energy consumed by its modal damping should be considered. The modal damping attenuates the energy accumulation of the bridge subsystem. This is to say, the larger the modal damping ratio of bridge is, the more energy of bridge will be consumed, and the better the stability of the interaction system will be. However, the modal damping ratio of bridge is determined by its material, and the range is limited.

For any $\omega_{Vib} \in \left(\omega_{PB1} \quad \omega_{PB2} \right)$, the bridge accumulates the averaged power $P_B(\omega_{Vib})$ that is positive. That is, the power consumed by the damping of the bridge is less than the power provided by the levitation subsystem. In this situation, the vibration energy of bridge accumulates and the amplitude of vibration increase continuously until the failure of suspension control.

For any $\omega_{Vib} \notin \left(\omega_{PB1} \quad \omega_{PB2} \right)$, the averaged power $P_B(\omega_{Vib})$ is negative, which indicates that the vibration energy of bridge will decay to zero with the passage of time, and the self-excited vibration is avoided.

4. Suppression Strategy of Self-Excited Vibration

4.1. Influence on Stability with Regard to $\bar{k}_p$

According to Equation (17), the expected magnetic flux B_{exp} consists of the states (displacement y_E, velocity $\dot{y}_E$ and acceleration $\ddot{y}_E$) of levitation subsystem and the state (displacement y_B) of bridge subsystem.

In maglev engineering, the levitation eddy-current sensor can detect the relative displacement $(y_E - y_B)$ between the upper surface of the electromagnet and the lower surface of the guideway. However, the electromagnet's displacement y_E and bridge's displacement y_B can be measured independently. Traditionally, there is no choice but to feed back the displacement of electromagnet and bridge is in pairs $(y_E - y_B)$.

Up to now, no literature indicates that the displacement feedback in pairs is optimal for the levitation stability and the suppression of the self-excited vibration. Taking the stability condition of the levitation subsystem for example, according to the Equation (21), the stability of the levitation subsystem is uncorrelated with the feedback gain of the bridge's displacement $\bar{k}_p$.

Similarly, according to Figure 2, to some extent, the feedback of the bridge's displacement increases the complexity of the block diagram. Furthermore, its influence on the occurrence of the self-excited vibration is unclear and should be explored.

To study the influence on the stability of the interaction system with the feedback of the bridge's displacement $\bar{k}_p$, for an extreme case, we suppose that $\bar{k}_p = 0$. This is to say, the feedback path of the bridge's displacement is removed from traditional control framework. In this case, the block diagram of maglev vehicle-bridge system is updated as Figure 5.

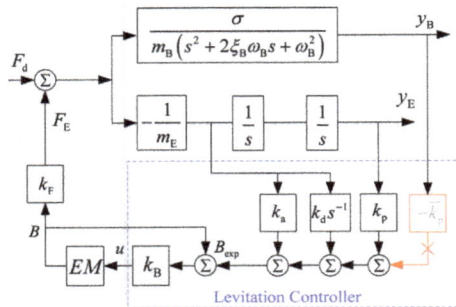

Figure 5. The block diagram of maglev vehicle-bridge system.

In light of Figure 5, the transfer function from disturbance F_d to the displacement y_B of bridge is degraded to

$$T_3(s) = \frac{G_B(s)}{1 - G_E(s)H_E(s)} \tag{34}$$

In this case, the characteristic equation of the transfer function $T_3(s)$ is

$$\Delta_3 = \Delta_1 \cdot \left(s^2 + 2\xi_B \omega_B s + \omega_B^2\right) \tag{35}$$

In light of Equation (35), the characteristic equation of the interaction system is the product of the characteristic polynomial of levitation system itself Δ_1 and $s^2 + 2\xi_B \omega_B s + \omega_B^2$.

Obviously, the characteristic polynomial $s^2 + 2\xi_B \omega_B s + \omega_B^2$ is stable. Hence, we can conclude that the interaction system is stable provided that the levitation subsystem is stable. This is to say, the stability of the interaction system is degenerated into the stability of the levitation subsystem. In this case, the self-excited vibration will be avoided if Equation (23) is satisfied.

To illustrate the conclusion quantitatively, the gain $\bar{k}_p$ is set as zero, and the other parameters are kept the same as the above section. The real parts of characteristic roots of the maglev vehicle-bridge interaction system are shown in Figure 6 when the modal frequency ω_B is varying.

In light of Figure 6, it can be seen that the real parts R_1, R_2 and R_3 are all negative when $\bar{k}_p = 0$. Hence, from the perspective of the characteristic roots, removing the bridge's displacement feedback is beneficial for the stability of the interaction system.

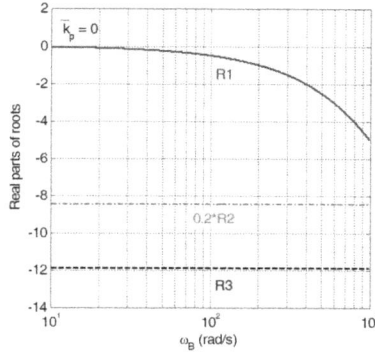

Figure 6. The real parts of characteristic roots when $\bar{k}_p = 0$.

4.2. Energy Variation with Regard to $\bar{k}_p$

In this section, we try to illustrate the validity of the control strategy from the perspective of the energy variation. When the feedback gain of the bridge's displacement $\bar{k}_p$ is set as zero, Equation (27) may be rewritten as

$$\begin{cases} u(s) = 2L_0 s i(s) + 2Ri(s) - 2F_i s y_E(s) \\ F_E(s) = 2F_i i(s) - 2F_z y_E(s) + 2F_z y_B(s) \\ F_E(s) = -m_E s^2 y_E(s) \\ u(s) = k_c(k_p + k_d s + k_a s^2) y_E(s) - (k_c - 2R)i(s) \end{cases} \tag{36}$$

Based on Equation (36), the transfer function between $F_E(s)$ and $\dot{y}_B(s)$ is updated as

$$H(s) = \frac{F_E(s)}{\dot{y}_B(s)} = \frac{2F_z m_E s(2L_0 s + k_c)}{\eta_3 s^3 + \eta_2 s^2 + \eta_1 s + \eta_0} \tag{37}$$

where $\eta_0 = 2F_i k_c k_p - 2F_z k_c$, $\eta_1 = 2F_i k_c k_d - 4L_0 F_z + 4F_i^2$, $\eta_2 = m_E k_c + 2F_i k_c k_a$, and $\eta_3 = 2m_E L_0$. In the case of $\bar{k}_p = 0$, the relationships between the vibration frequency and averaged powers are shown in Figure 7.

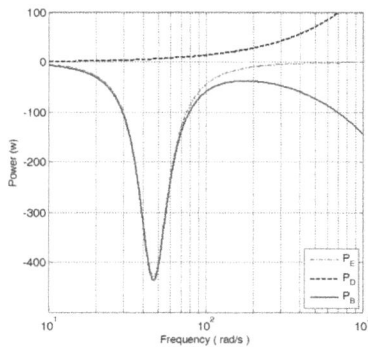

Figure 7. The averaged powers when $\bar{k}_p = 0$.

According to Figure 7, it can be seen that the power P_E is negative over the full frequency range when the displacement feedback of bridge is removed. This is to say, the levitation subsystem is always passive if the feedback of the bridge's displacement y_B is removed. Furthermore, considering the passivity and the dissipation of bridge due to its modal damping, the vibration energy of bridge is delay to zero no matter how large the initial states are. In this case, the self-excited vibration is avoided.

4.3. The Estimation of Electromagnet's Displacement y_E

In an actual magnetic levitation system, there are two real-time signals available. The first signal is the levitation clearance $\delta(t) = y_E(t) - y_B(t)$, measured by the eddy-current sensor, and the other is the acceleration signal of the electromagnet $a_E(t) = \ddot{y}_E(t)$, detected by the accelerometer. Traditionally, the signal of levitation gap is adopted for the outer-loop of the levitation controller. Separately speaking, the feedback gain of the electromagnet's displacement y_E is k_p, and the feedback gain of the bridge's displacement are in pairs is $\overline{k}_p = -k_p$.

When the amplitude of the feedback gains of the electromagnet and bridge's displacement is different, i.e., $\overline{k}_p \neq -k_p$, we should measure the signal of electromagnet's displacement y_E and the bridge's displacement y_B separately. For a special case, $\overline{k}_p = 0$. We still should measure the signal of the electromagnet's displacement y_E.

Considering the signal of the electromagnet's displacement y_E is immeasurable directly, we should develop a method to estimate it. Theoretically, the displacement of electromagnet may be obtained by the double integration of acceleration of the electromagnet:

$$y_E(s) = \frac{1}{s^2} a_E(s) \tag{38}$$

However, in a real maglev system, measurements of acceleration $a_E(t)$ are polluted by its inexact direct bias and ultra low frequency disturbance, which will lead to integral saturation.

To prevent the integral saturation, the estimator $1/(s+\tau)^2$ is adopted to instead of the double-integrator $1/s^2$. This is to say, the estimated value $\hat{y}_E$ of the electromagnet's displacement is

$$\hat{y}_E(s) = \frac{1}{(s+\tau)^2} a_E(s) \tag{39}$$

where τ is the time constant of the estimator. The comparison between the double-integrator and the estimator is shown in Figure 8.

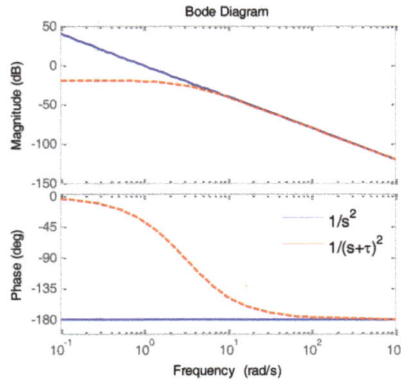

Figure 8. The comparison between the double-integrator and estimator.

Considering Figure 8, when the frequency is less than 10 rad/s, the amplitude of double-integrator is oversized, which may result in excessive transient response when the ultra-low frequency component of $a_E(t)$ is shifted. Hence, the double-integrator is unsuitable for the engineering application.

Luckily, compared with the double-integrator, the amplitude of the estimator is much smaller. Furthermore, when the frequency is larger than 10 rad/s, the amplitude and phase of estimator are highly consistent with the double-integrator, which provides enough phase advances over the upper frequency range. Therefore, compared with the double-integrator, the estimator is more suitable for the engineering applications. When the electromagnet's displacement y_E is replaced by the estimation one, the control Equation (17) is updated to Equation (40).

$$B_{\exp}(t) = k_p \bar{y}_E(t) + k_d \dot{y}_E(t) + k_a \ddot{y}_E(t) \tag{40}$$

However, the phase distortion of the estimator at the low frequency range, which may degrade the stability of the levitation system, should be considered. Generally, the larger the time constant τ is, the less the amplitude of the estimator will be at the ultra low frequency range, and the more seriously its phase distortion will be. Hence, the time constant τ should be selected comprehensively according to the amplitude elimination and phase distortion. When the estimator $1/(s+\tau)^2$ is adopted to replace the double-integrator $1/s^2$, the block diagram of maglev vehicle-bridge system is updated as Figure 9.

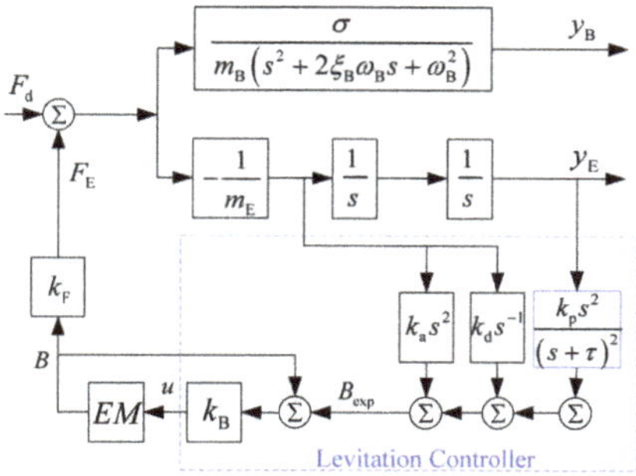

Figure 9. The block diagram of maglev vehicle-bridge system with estimator.

5. Numerical and Experimental Validation

Theoretically, the proposed control strategy can solve the problem of self-excited vibration effectively. Furthermore, it should be checked numerically and experimentally prior to applying to commercial service.

5.1. Numerical Validation

To obtain a reliable conclusion, the conditions of the magnetic levitation project should be simulated as precise as possible. In maglev engineering, the direct component of the acceleration signal is not absolute zero. Hence, its direct component is set as 0.2 m/s². The noise is set as 0.2% when compared to the maximum amplitude of the signal. Besides, the overall nonlinear dynamic model with details, which is shown in Figure 10, is adopted to carry out the numerical simulation. As for this model, ten modules (Due to the limit of Figure's size, only three modules is shown as follow.) are included and distributed along the length direction of the vehicle symmetrically. Each module consists of two levitation units.

During the process of simulation, the nonlinear character of the levitation system, the saturation of the control voltage of the electromagnets, the dynamic responses of air-springs, the real-time estimation of the signals of the electromagnet's displacement and the coupling between the adjacent levitation units are all considered.

In this subsection, the parameters of controller are set as $k_p = 1000$, $k_d = 30$, $k_a = 0.4$, and $k_B = 30$. The parameters of bridge are set as $\xi_B = 0.005$, and $\omega_B = 2\pi \times 13\text{Hz} = 81.68\text{ rad/s}$.

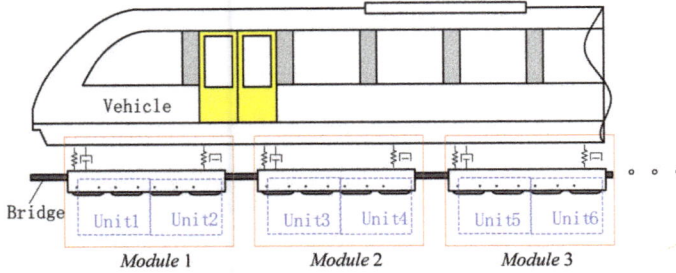

Figure 10. The overall nonlinear dynamic model with ten modules.

The modal frequency ω_B = 81.68 rad/s selected in this case belongs to the unstable interval 67.3–118.7 rad/s. As we expected, the self-excited vibration starts to grow up at t = 2 s. According to Figure 11b, it can be seen that the amplitude of the electromagnet's acceleration is up to 2 m/s², which indicates the electromagnet vibrates violently. The vibration of electromagnet will be transferred to the vehicle, which degrades its ride comfort. The fluctuation of electromagnet displacement is about 0.5 mm. The fluctuation impacts the durability and safety of the bridge.

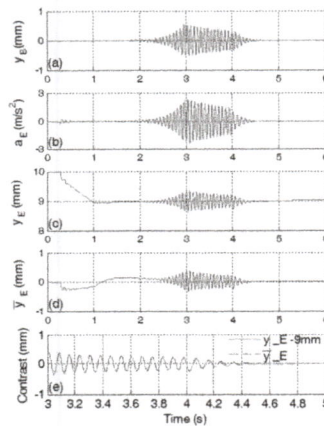

Figure 11. The numerical verification for vibration suppression method, which is activated at t = 4 s: (**a**) the displacement of bridge; (**b**) the displacement of electromagnet; (**c**) the estimated displacement of electromagnet, and (**d**) the contrast signals of the estimated and the real displacement of electromagnet.

According to Figure 11d, it can be seen that the estimated displacement of electromagnet is unfaithful when t < 2 s, which is due to the displacement estimator's transient response. In light of Figure 11e, when the transient response disappears, the estimated signal is converged to the real signal of electromagnet's displacement. To show its validity, the improved control scheme is activated at t = 4 s. It can be seen that the vibration amplitudes of all states rapidly decay to zero. When t > 4.5 s, the self-excited vibration disappears absolutely.

5.2. Experimental Validation

The experiments were carried out on the maintenance platform of the Tangshan maglev test line, as shown in Figure 12. The levitation control system (Beijing Enterprises Holdings Maglev Technology Development Co., Ltd, Beijing, China) consists of a Pulse Width Modulation(PWM) chopper, suspension modules and Power PC-based digital processor. The digital process is capable of performing the proposed vibration control algorithm.

Figure 12. Field experiments on a full-scale maglev train at Tangshan maglev engineering experiment base.

All experimental data were obtained from the monitoring terminal of the notebook computer and the Ethernet-based suspension monitoring network. The data sampling rate was 200 samples per second. The whole weight of the vehicle is 8 ton during experiment.

The full-scale maglev train consists of ten modules, ten air-springs and one cabin. The ten modules are distributed along the length direction of the vehicle symmetrically. For the bridge subsystem, the length of bridge is 18 m, and its mass per meter is about 2.4 ton. The field measurement indicates that its modal damping ratio is about 0.02. Considering the accuracy and reliability of the numerical model in Section 5.1 is close to the real maglev system, we expected the conclusion obtained by the numerical simulation to be validated by the field experiment.

Figure 13 shows the results of self-excited vibration when performing field tests on a maintenance platform. The self-excited vibration appeared when $t < 1$ s. It can be found that the signals of acceleration, levitation gap and current fluctuate violently. The electromagnet vibration degraded the stability of the suspension control and ride comfort of vehicle.

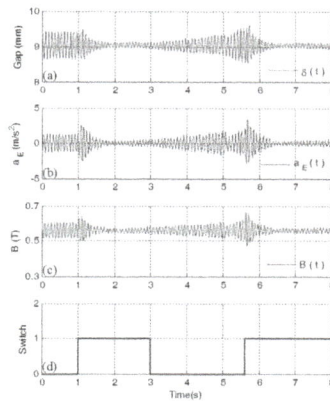

Figure 13. The experimental verification for vibration suppression method, which is activated during [1 3] and [5.5 8]: (**a**) the levitation gap; (**b**) the acceleration of electromagnet; (**c**) the magnetic flux; and (**d**) the switch signal.

At $t = 1$ s, the proposed control strategy was activated. After a short regulation, the signal's fluctuation of the levitation gap and the acceleration of electromagnet were attenuated greatly.

At $t = 3$ s, the control scheme was switched to the traditional cascaded-controller, the self-excited vibration aroused gradually. At $t = 5.5$ s, the control scheme was switched to proposed control strategy again, and the resonance disappear once more.

According to Figures 11 and 13, both the results of the numerical simulation and field experiment indicate that the proposed control scheme is capable of weakening the amplitude of the self-excited vibration to zero in two seconds.

At the same time, we also find that the experimental signals are much more irregular when compared with the signals obtained by the numerical simulation, which mainly results from the high-frequency noise from the eddy-current sensors and the inconsistent between the adjacent levitation units.

6. Conclusions

Firstly, the maglev vehicle-bridge interaction model, including a flexible bridge and several electromagnetic levitation units is proposed, and the comparison between the current-loop and magnetic flux feedback is carried out. The analysis indicated that the performance could be improved by substituting the current-loop with the magnetic flux-loop.

Secondly, the advantages and disadvantages of the traditional control architecture with the displacement feedback of electromagnet y_E and bridge y_B in pairs are explored. The results indicate that removing the feedback of the bridge's displacement y_B from the pairs ($y_E - y_B$) measured by the eddy-current sensor is beneficial for the suppression of the self-excited vibration.

Thirdly, the signal acquisition of the electromagnet's displacement y_E is discussed for the engineering application. The analysis shows that the proposed estimation can avoid the problem of the integral saturation.

At last, the numerical research and the experimental validations on a full-scale maglev train at Tangshan maglev engineering experiment base have been carried out. The data indicated that the proposed control strategy is capable of eliminating the self-excited vibration.

Acknowledgments: This work was financially supported in part by the National Natural Science Foundation of China under Grant 112002230 and Grant 11302252. The authors would like to thank all reviewers for their valuable comments.

Author Contributions: Lianchun Wang and Jie Li proposed the new ideas and designed the experiments; Lianchun Wang performed the experiments; Lianchun Wang, Jinhui Li and Danfeng Zhou analyzed the data; and Lianchun Wang and Jinhui Li wrote the paper.

Conflicts of Interest: The authors declare no conflict of interest.

References

1. Zhou, H.B.; Duan, J.A. A novel levitation control strategy for a class of redundant actuation maglev system. *Control Eng. Pract.* **2011**, *19*, 1468–1478. [CrossRef]
2. Huang, C.M.; Yen, J.Y.; Chen, M.S. Adaptive nonlinear control of repulsive maglev suspension systems. *Control Eng. Pract.* **2000**, *8*, 1357–1367. [CrossRef]
3. Li, J.H.; Li, J. A practical robust nonlinear controller for maglev levitation system. *J. Cent. South Univ. Technol.* **2013**, *20*, 2991–3001. [CrossRef]
4. Li, J.H.; Li, J.; Zhou, D.F.; Yu, P.C. Self-excited vibration problems of maglev vehicle-bridge interaction system. *J. Cent. South Univ. Technol.* **2014**, *21*, 4184–4192. [CrossRef]
5. Zhou, D.F.; Li, J.; Hansen, C.H. Suppression of the stationary maglev vehicle-bridge coupled resonance using a tuned mass damper. *J. Vib. Control* **2013**, *19*, 191–203. [CrossRef]
6. Li, J.H.; Li, J.; Yu, P.C.; Wang, L.C. Adaptive backstepping control for levitation system with load uncertainties and external disturbances. *J. Cent. South Univ. Technol.* **2014**, *21*, 4478–4488. [CrossRef]
7. Glück, T.; Kemmetmüller, W.; Tump, C.; Kugi, A. A novel robust position estimator for self-sensing magnetic levitation systems based on least squares identification. *Control Eng. Pract.* **2011**, *19*, 146–157. [CrossRef]

8. Zhang, W.Q.; Li, J.; Zhang, K.; Cui, P. Stability and bifurcation in magnetic flux feedback maglev control system. *Math. Probl. Eng.* **2013**, *2013*, 537359. [CrossRef]

9. Alberts, T.E.; Hanasoge, A.; Oleszczuk, G. Stable levitation control of magnetically suspended vehicles with structural flexibility. In Proceedings of the 2008 American Control Conference, Seattle, WA, USA, 11–13 June 2008.

10. Alberts, T.E.; Oleszczuk, G. On the influence of structural flexibility on feedback control system stability for EMS Maglev vehicles. In Proceedings of the 19th International Conference on Magnetically Levitated Systems and Linear Drives, Dresden, Germany, 13–15 September 2006.

11. Wang, H.P.; Li, J. Vibration analysis of the maglev guideway with the moving load. *J. Sound Vib.* **2007**, *305*, 621–640. [CrossRef]

12. Zhang, Z.; Zhang, L. Hopf bifurcation of time-delayed feedback control for maglev system with flexible guideway. *Appl. Math. Comput.* **2013**, *219*, 6106–6112. [CrossRef]

13. Wang, H.P.; Li, J.; Zhang, K. Sup-resonant response of a non-autonomous maglev system with delayed acceleration feedback control. *IEEE Trans. Magn.* **2008**, *44*, 2338–2350. [CrossRef]

14. Zhou, D.F.; Hansen, C.H.; Li, J. Suppression of maglev vehicle-girder self-excited vibration using a virtual tuned mass damper. *J. Sound Vib.* **2011**, *330*, 883–901. [CrossRef]

15. Li, J.H.; Li, J.; Zhou, D.F.; Wang, L.C. The modeling and analysis for the self-excited vibration of the maglev vehicle-bridge interaction system. *Math. Probl. Eng.* **2015**, *2015*, 709583. [CrossRef]

16. Li, J.H.; Li, J.; Zhou, D.F.; Cui, P.; Wang, L.C.; Yu, P.C. The active control of maglev stationary self-excited vibration with a virtual energy harvester. *IEEE Trans. Ind. Electron.* **2015**, *62*, 2942–2951. [CrossRef]

17. Yau, J.D. Vibration control of maglev vehicles traveling over a flexible guideway. *J. Sound Vib.* **2009**, *321*, 184–200. [CrossRef]

18. Li, J.H.; Fang, D.; Zhang, D.; Cai, Y.; Ni, Q.; Li, J. A practical control strategy for the maglev self-excited resonance suppression. *Math. Probl. Eng.* **2016**, *2016*, 8071938. [CrossRef]

![applied sciences logo]

applied
sciences

MDPI

Article

A Novel Hybrid Semi-Active Mass Damper Configuration for Structural Applications

Demetris Demetriou * and Nikolaos Nikitas *

School of Civil Engineering, University of Leeds, Leeds, LS2 9JT, UK
* Correspondence: cn09dd@leeds.ac.uk (D.D.); n.nikitas@leeds.ac.uk (N.N.);
  Tel.: +44-011-3343-0901 (D.D. & N.N.)

Academic Editors: Gangbing Song, Steve C.S. Cai and Hong-Nan Li
Received: 19 October 2016; Accepted: 25 November 2016; Published: 30 November 2016

Abstract: In this paper, a novel energy- and cost-efficient hybrid semi-active mass damper configuration for use in structural applications has been developed. For this task, an arrangement of both active and semi-active control components coupled with appropriate control algorithms are constructed and their performance is evaluated on both single and multi-degree of freedom structures for which practical constraints such as stroke and force saturation limits are taken into account. It is shown that under both free and forced vibrations, the novel device configuration outperforms its more conventional passive and semi-active counterparts, while at the same time achieving performance gains similar to the active configuration at considerably less energy and actuation demands, satisfying both strict serviceability and sustainability requirements often found to govern most modern structural applications.

Keywords: vibration control; semi-active; hybrid vibration mitigation systems; high-rise structures

---

1. Introduction

Alleviating the vibration response of tall and slender structures under wind action becomes an increasingly challenging task. Generally speaking, the response of such structures subjected to wind excitation can be thought of as the summation of three components, namely static, background aerodynamic, and resonant dynamic, in the relevant modes of vibration. Mitigating the static and background aerodynamic response can be achieved through supplemental structural stiffness and/or reduction of the mean excitation forces through manipulation of the structural aerodynamics (shape). Still, as structures become taller and more slender, resonant contributions become more and more significant and eventually dominate [1].

One method of successfully and conveniently mitigating the resonant response of structures is by modifying their dynamic properties (frequencies and damping). Amongst the most popular devices used for resonant response reduction are the dynamic vibration absorbers (DVAs), which can be found in passive, active, hybrid and semi-active forms. The passive form of the DVA, the tuned mass damper (TMD), has been studied for more than a century and is shown to be effective and reliable at alleviating structural response under generic dynamic loading; however, this device being tuned to a single vibration mode of the structure thus has performance limited to a narrow band of operating frequencies that in turn compromise the system's attenuation capacity when excited beyond the targeted mode. Overcoming the limitations of the passive TMD, the active version of the DVA, the active mass damper (AMD), achieves resonant response reduction by generating control forces via acceleration and deceleration of auxiliary masses using actuators in a way that at any given time and independent of the excitation and system's characteristics, maximum energy is absorbed. Clearly, while the flexibility and adaptability of active devices allows for better vibration response reduction, this performance enhancement is achieved at the expense of considerable power-force demands and reliability. Adding to the limitations of the purely active AMD configuration, the performance of such devices on high-rise structures is typically limited by the capacity of the installed actuator and the auxiliary mass strokes [2–4]. Despite the attempts made to overcome these limitations, either by using different, more efficient and novel-at-the-time AMD configurations such as the swing-style AMD presented in [5], or the electromagnetic device with semi-active control properties presented in [6], amongst many other configurations [4,7], the crucial absence of a fail-safe mechanism limits the options to structural engineers to an approach that is based on the hybridisation of the AMD device with a component able to prevent instability upon active component failure. To this extent, most practical structural control configurations comprising a form of active DVA are found in an active-passive hybrid state [8], with an inspiring recent application on the 101-storey Shanghai World Financial Center, highlighting the prospects of hybrid control.

The conventional hybrid configuration of a DVA, entitled as active-tuned mass damper (ATMD), is one that requires a passive TMD device to work in conjunction with active control elements such as hydraulic, motor, ball-screw actuators, etc. Such devices are shown to achieve a compromise between performance and reliability at the expense of lower strokes and actuator demands. Studies such as those found in [9–12] are a few amongst the many illustrating the performance gains arising from the use of ATMDs on structural systems under both earthquake and wind excitations. Pushing the boundaries of hybrid control and innovation, Tan and Liu [13] proposed a hybrid mass damper configuration for Canton Tower in Guangzhou China. This configuration requires an AMD to work in parallel with a two-stage TMD, demonstrating significant vibration attenuation under strong wind and earthquake excitations. Following the same path, Li and Cao [14] later proposed a hybrid configuration that uses two interconnected ATMDs for the supply of the control action. More recently, Tso and Yuan [15] proposed an alternative approach to the design of the hybrid vibration absorber that incorporates detached passive and active parts, resulting in a non-collocated setup that was shown to achieve better performance than the traditionally bundled ATMD. In the field of mechanical engineering and away from DVA applications, but following the same logic, Khan and Wagg [16] proposed a hybrid configuration that requires a magnetorheological semi-active damper to work in conjunction with an active actuator placed at the base of the structural system, claiming the first-ever hybrid semi-active/active configuration. The aim of the study was to show how an active actuator can assist the semi-active device in a non-collocated configuration, in an attempt to achieve performance as close to that of a fully active system as possible. Prior to the publication of [16], a different configuration that still makes use of semi-active and active elements has been proposed by the authors of this paper in [17]. The fundamental novelty of the configuration and the main difference to any prior hybrid configuration is the use of collocated semi-active and active elements for the supply of control power directly to the DVA that in turn controls the structural system. In this paper, boundaries of innovation and the limitations of the TMD, AMD and ATMD are surpassed and the idea proposed by Demetriou and Nikitas [17] is extended through the use of a novel semi-active hybrid mass damper (SHMD) configuration proposed in this paper. This device extends the conventional ATMD logic, by employing semi-active dampers working in conjunction with actively controlled elements in a way that, by combining the two components using appropriately designed control algorithms, the potential of timed maximum energy extraction is exploited. To this end, the operating principle of the novel SHMD configuration requires the semi-active elements to be designed such that maximum kinetic energy is extracted from the system at the expense of low energy demands required to control: power operated valves, the fluid discharge through orifices, the magnetic field around a ferrous fluid piston, etc., and then allowing for energy addition to the system using active (hydraulic) actuators, that in turn enhance the system's adaptability to ever-changing loading conditions. In other words, the active control components of the hybrid device are restricted to add energy while the semi-active components perform as usual by extracting energy. Critically, the control algorithm needs to be designed such that when energy is added to the system (and DVA's mass is accelerated), the semi-active component drops to its minimum value such that it does not counteract the action of the active component.

In order to demonstrate the performance gains from the use of this novel SHMD device, comparative studies on a low frequency single-degree-of-freedom (SDOF) system subjected to free and forced vibrations are carried out. The selection of the input conditions was performed in an attempt to quantify the performance gains of the novel configuration over a wider band of operating conditions, through capturing a broader band of excitation frequencies. The study naturally extends the application of the novel configuration on a more realistic 76-storey benchmark structure on which realistic wind loading, actuation and damper stroke limits are applied.

2. Modeling Principles

2.1. General Dynamic Vibration Absorbers (DVA) Modeling Approach

Modeling the novel SHMD device requires a thorough understanding of the modeling principles and procedures followed in the design of passive, active and semi-active systems. In order to do so, the dynamic behaviour of an n-DOF system coupled with a DVA (as depicted in Figure 1) under a random dynamic loading needs to be considered through its equation of motion:

$$M\ddot{x}(t) + C\dot{x}(t) + Kx(t) = Bu(t) + Dd(t) \tag{1}$$

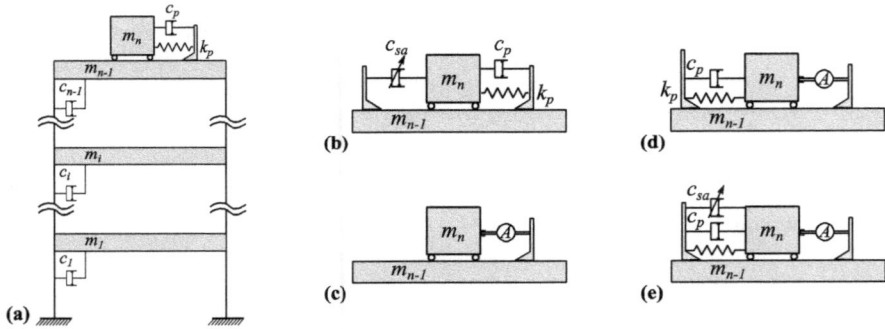

Figure 1. Structural configuration and mathematical models for (**a**) tuned mass damper (TMD); (**b**) semi-active tuned mass damper (STMD); (**c**) active mass damper (AMD); (**d**) active-tuned mass damper (ATMD); and (**e**) semi-active hybrid mass damper (SHMD) systems.

In Equation (1), each overdot declares single differentiation with respect to time, M, C and K are the $n \times n$ mass, damping, and stiffness matrices, respectively; $x(t)$ and $d(t)$ are in order of the displacement, and external force $n \times 1$ column vectors; $u(t)$ is the single scalar control force and $B(n \times 1)$ and $D(n \times n)$ are the influence matrices assigning the control and external force contributions, respectively, to the individual degree of freedoms (DOFs). For each DOF in $x(t)$ being the lateral displacement of the ith ($i = 1, ..., n$) mass, M becomes diagonal, while for the classical viscous damping considered the damping matrix C attains a form identical to the symmetric stiffness matrix K. Without any loss of generality, the DVA is attached to the $(n-1)$th DOF and its motion constitutes the nth DOF. For control implementation purposes, Equation (1) can be represented in the state space domain using a first order differential equation, such that:

$$\dot{z}(t) = Az(t) + Fu(t) + Ed(t) \tag{2}$$

where, $\dot{z}(t)$ represents the first order time-change of the states $z(t) = \begin{bmatrix} x(t) & \dot{x}(t) \end{bmatrix}^T$ of the system, A is the system block matrix containing the system's mass, damping, stiffness properties, F is the control force locator block matrix, and E is the external perturbation locator block matrix, such that:

$$A = \begin{bmatrix} 0 & I \\ -M^{-1}K & -M^{-1}C \end{bmatrix}, F = \begin{bmatrix} 0 \\ M^{-1}B \end{bmatrix}, E = \begin{bmatrix} 0 \\ M^{-1}D \end{bmatrix} \tag{3}$$

With I being the identity matrix of appropriate dimensions (i.e., $(n \times n)$).

2.1.1. Passive Tuned Mass Damper (TMD) Control

A TMD device produces control actions as a result of the relative motion of its mass against the structural mass such that the control force term, $u(t)$ in Equation (2) is calculated at each time step by:

$$u(t) = k_p x_r(t) + c_p \dot{x}_r(t) \tag{4}$$

In this equation, $\dot{x}_r(t)$ and $x_r(t)$ are respectively the relative velocity and displacement between the nth and $(n-1)$th DOFs and c_p and k_p, are the passive damping and stiffness coefficients respectively. To this date, most of the tuning of the mechanical parameters c_p and k_p of a TMD device is achieved via closed-form expressions derived from the minimisation of the rms acceleration response of a single degree of freedom (SDOF) subjected to white noise or harmonic excitation. While this approach is broadly accepted, representing civil engineering structures with an equivalent SDOF system can lead to significant errors in the estimation of their dynamic response. The problem amplifies when one considers the probabilistic nature of the knowledge of the system's properties and the fact that the estimated properties can vary with time (e.g., amplitude dependence, fluid–structure interaction, etc.). Moreover, obtaining TMD mechanical parameters through the use of harmonic or flat spectrum inputs may not always yield optimum values [18]. In this paper, because the motion of long period structures is generally governed by the first modal response, the stiffness coefficient of the auxiliary device is selected just as the mass damper is tuned to the fundamental frequency of the structure, whereas the damping coefficient is obtained using the expressions found in [19,20], and validated and adjusted when necessary through numerical optimisation based on [21,22] for the case of the more complex, wind-excited multi-degree of freedom (MDOF) structural system.

2.1.2. Active Mass Damper (AMD) Control and Hybrid Active-Tuned Mass Damper (ATMD) Control

For a purely active system, the passive control force takes the form of a desired action, $u_a(t)$ determined via a control algorithm such as the Linear-Quadratic-Regulator (LQR), proportional-integral-derivative (PID) controller or similar. With reference to Figure 1c, for the case of AMD control, the force is delivered solely by means of mechanical actuation; thus the actuation force $f_a(t)$ is equal and opposite to the calculated desired action:

$$u_a(t) = -f_a(t) \tag{5}$$

Obviously, for the purpose of limiting the stroke and the requirement of a fail-safe mechanism, an ATMD is found in most practical applications [3,8]. To this end, and with reference to Figure 1d, for an ATMD, the desired force is mathematically expressed as the summation of the passive forces generated by the motion of the mass damper and an additional external force provided by means of mechanical actuation. Because the dynamic characteristics of the mass damper remain unaltered and the desired interaction force has been already calculated by the control algorithm, the required conventional hybrid actuation, $f_{a\_atmd}(t)$ is determined from:

$$u_a(t) = c_p \dot{x}_r(t) + k_p x_r(t) - f_{a\_atmd}(t) \tag{6}$$

In Equation (6), the mechanical properties c_p, k_p of the device can be selected similarly to a purely passive device. Still, typically a higher damping coefficient c_p is used along with the ATMD device for stroke-restraining purposes [3,21].

2.1.3. Semi-Active Tuned Mass Damper (STMD) Control

Similar to an active system, the semi-active counterpart makes use of control algorithms for the selection of appropriate control actions. The first step in the calculation of the semi-active control forces is the calculation of an equivalent active force using active algorithms and Equation (5). Next, the active force is tailored so that it can be physically realised by the semi-active device. In this regard, because of the fact that no energy can be added directly to the system, the semi-active device will produce control forces only when required, i.e., when the damper is requested to "consume" energy. This is achieved by applying semi-active force saturation limits such that the semi-active control force, $u_{sa}(t)$ is calculated by [23]:

$$u_{sa}(t) = f_a(t)\left(\frac{1 - \mathrm{sgn}\left[f_a(t)\dot{x}_r(t)\right]}{2}\right) \qquad (7)$$

$$\mathrm{sgn}\left[f_a(t)\dot{x}_r(t)\right] = \mathrm{sgn}(q_a(t)) \triangleq \{\begin{matrix} 1 & for & q_a \geq 0 \\ -1 & for & q_a < 0 \end{matrix} \qquad (8)$$

The product of $f_a(t)\dot{x}_r(t)$ is the power, $q_a(t)$, of the whole active system device. Similarly, the power of just the semi-active component, $q_{sa}(t)$, is defined as the product of the force that can be physically translated by the device, $u_{sa}(t)$, and its relative velocity, $\dot{x}_r(t)$:

$$q_{sa}(t) = u_{sa}(t)\dot{x}_r(t) < 0 \qquad (9)$$

A schematic representation of the power time histories of both an actively and a semi-actively controlled device is shown in Figure 2. It can be observed that the active device has the advantage of both adding and dissipating energy, as indicated by positive and negative powers, while the semi-active device only consumes power (indicated by only negative power—and its integral energy is also negative).

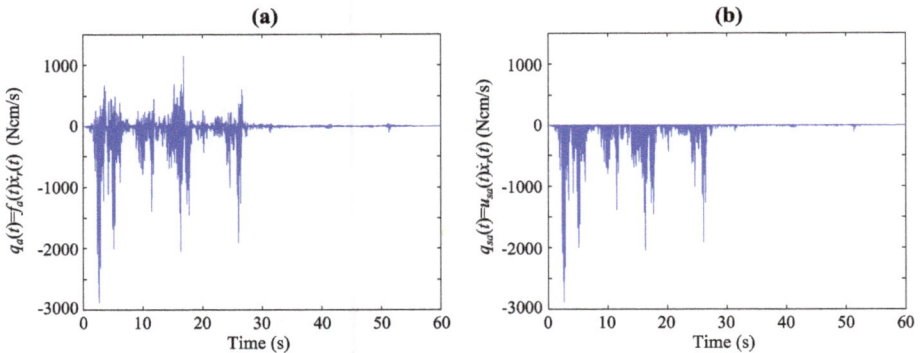

Figure 2. Indicative example of the "power" scheme/demand practised in (a) active; and (b) semi-active control.

When a variable damping (VD) STMD is considered, the method of achieving enhanced performance is by appropriately and in a timely fashion adjusting the damping coefficient of the device within bands, in order for the required control force to be reached. By referring back to the system presented in Figure 1, the semi-active damping force contribution can be expressed as

$c_{sa}(t)\dot{x}_r(t)$. Inspection of Equation (9) easily leads to $c_{sa}(t) < 0$. Updating Equation (6), the resulting overall control force provided at each time instant by a VD-STMD can be expressed mathematically as:

$$u_{sa}(t) = (|c_{sa}(t)| + c_n)\dot{x}_r(t) + k_p x_r(t) \tag{10}$$

In Equation (10), c_n is a small damping coefficient representing the inherent damping of the connection of a semi-active device and the structural system. In this equation, the time-varying semi-active damping coefficient, $c_{sa}(t)$, is the only unknown, making the calculation of the real-time variation of the damping coefficient straightforward.

2.2. Modeling the Semi-Active Hybrid Mass Damper

Through the use of an SHMD, the energy-dissipation capacity of a semi-active device is exploited and energy is added only when required through force actuators. The main difference between an ATMD and the novel SHMD configuration lies in the fact that the actuators of the ATMD both add and dissipate energy, whereas the forcing provision of the SHMD can only add energy. To this end, it can be realised that when the actuators of the ATMD are required to add energy to the system, sufficient power should be provided so that the "braking" force acting on the DVA's mass by the passive damping elements of the ATMD is surpassed for the mass to be accelerated, and sufficient control force can then be applied to the system. On the contrary, the novel SHMD configuration lowers the active actuation demands by lowering the semi-active damping component to its minimum value throughout the period of active actuation. The steps required for the implementation of this configuration and the calculation of the envisaged control action, $u_{shmd}(t)$, are explicitly introduced below and summarised in Figure 3:

(1) Design of a semi-active controller based either on an active controller that is clipped using Equation (7) for semi-active control purposes or using direct output feedback control algorithms such as the ones found in the groundhook control scheme for alleviation of the online computational burden of Equations (7) and (8).

(2) Design an active controller using active control algorithms such as LQR, PID or similar designed to satisfy performance and robustness specifications of the non-linear system (i.e., system including the semi-active controller).

(3) Limit the capacity of the active actuator to only add power to the system:

$$q_a(t) = f_a(t)\dot{x}_r(t) > 0 \tag{11}$$

(4) Incorporation of both active and semi-active forces to the system using:

$$u_{shmd}(t) = u_{sa}(t) + f_a(t) \tag{12}$$

(5) Optimisation of maximum and minimum damping ratios of the semi-active control device for the case of on-off control.

Using the steps described, the resulting control signals relative to the active control counterpart should attain the form shown in Figure 4b,c. Evidently, the active control component of the novel device configuration (Figure 4c) can only supply force in the $q_a \geq 0$ regions, whereas the semi-active control component is able to only supply force in the $q_a \leq 0$ regions. With reference to Figure 4b, the fifth and final step of the SHMD design procedure, the optimisation of the maximum and minimum damping coefficients of the semi-active component determines the slope and the magnitude of the control signal which in turn severely influence the performance of the hybrid system.

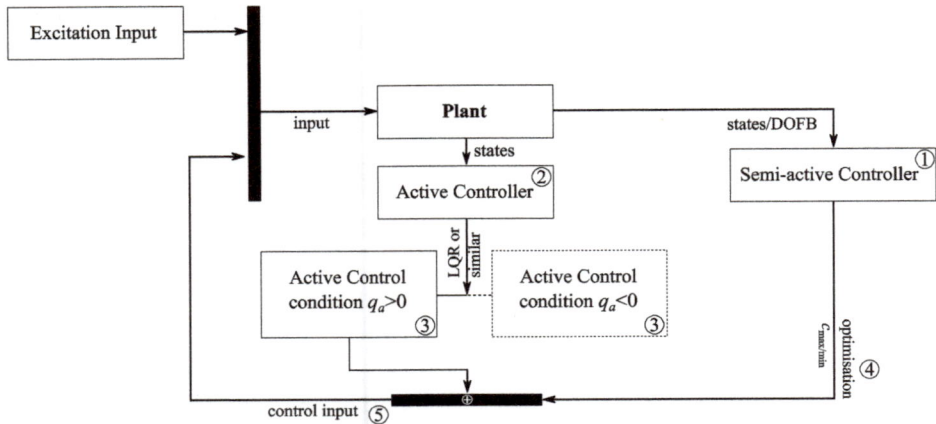

Figure 3. A schematic representation of the procedure followed for modeling the semi-active hybrid mass damper.

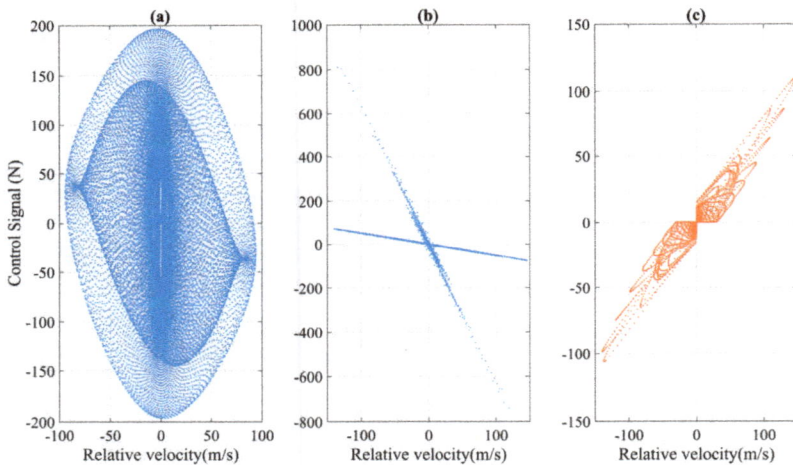

Figure 4. Control signals as a function of relative velocity for the (a) purely active system; (b) semi-active component; and (c) active component of the hybrid configuration subjected to a white noise excitation.

3. Control Methods

Obtaining the semi-active and active forcing components is achieved via the use of control algorithms. In this study, for the purely active control case, the algorithm of choice is the optimal LQR that was proven suitable in a series of studies [3,24–26] for use on flexible structural applications. The design of the controller (i.e., the determination of the weighting matrices which are required in the determination of the control gains) is based on the performance index defined in [25]. For the case of semi-active control, the displacement based groundhook algorithm that belongs to the category of direct output feedback controllers (i.e., the control actions are calculated based on a limited number of measurements) is selected. The choice of this direct output feedback controller for the case of semi-active control is based on the reduction of the computational effort required for the online calculation of Equations (7) and (8), requirement of minimum state measurements as well as its enhanced performance over other conventional direct output feedback controllers as shown in the studies of [21,27]. The mathematical expressions describing the control algorithm used in the derivation of the control actions are found in [28,29].

With reference to Section 2.2, and because of the fact that semi-active control precedes the design of the active controller, the incorporation of semi-active control to the system results in a configuration that is no longer linear but piecewise linear, generating the need for linearisation before a purely active controller is designed. In this study, the linearisation of the piecewise linear system is performed via input/output subspace SSARX identification using MATLAB's (MATLAB2016a, The MathWorks Inc., Natick, MA, USA, 2016) system identification toolbox. To this end, a purely harmonic signal with known frequency and amplitude is used as the external input to the system. The displacement of the structural mass was used as the output. From this, a four-state equivalent linear system is constructed and the state matrices are extracted for use in the active controller design procedure.

4. Numerical Investigation

4.1. Single Degree of Freedom (SDOF) Structural Configuration

In order to quantitatively capture the performance gains of the proposed system on both the transient and steady-state components of the vibration response, four alternatives, namely passive (TMD), semi-active (STMD), active (ATMD) and semi-active hybrid (SHMD)-equipped (low-damped) SDOF structures, are investigated. For the simulations, the mass and stiffness of the SDOF structure is selected such that the resulting mass ratio of the structural mass to the mass of the damper is 1% and the frequency of the system is approximately 1 rad/s, typical for high-rise structural applications. The resulting mass, stiffness and damping matrices used in the simulations are:

$$
M = \begin{pmatrix} 1000 & 0 \\ 0 & 10 \end{pmatrix} \text{ Kg} \quad K = \begin{pmatrix} 1009.9 & -9.9 \\ -9.9 & 9.9 \end{pmatrix} \text{ Ns/m}
$$
$$
C = \begin{pmatrix} 51.22 & -1.22 \\ -1.22 & 1.22 \end{pmatrix} \text{ Ns/m} \quad C_w = \begin{pmatrix} 50.04 & -0.04 \\ -0.04 & 0.04 \end{pmatrix} \text{ Ns/m}
$$

(13)

In Equation (13), C is the damping matrix used for the case of TMD control, and C_w is the damping matrix used for the case of STMD, ATMD and SHMD control. It is evident that for the case of passive control, a damping ratio of 6.1% and stiffness tuning ratio of 0.9 derived using Den Hartog's expressions found in [19] are used for optimal passive behaviour and maximum rms reduction at steady state. For the remaining three control cases, a minimal damping ratio of 0.2% is used in order to capture the inherent damping of the connection of the damper and the structural mass.

4.2. Variable Damping Coefficient Configuration

For the fairness of the comparison and consistency with the optimisation procedure followed for the case of passive TMD control, the selection of the variable damping coefficients for the case of the semi-active and hybrid controlled SDOF systems is performed via examination of the rms acceleration response of the system at steady state. To this end, an investigation of the acceleration response for maximum damping ratios ($\zeta_{max} = c_{max}/2m_d\omega_n$) ranging from 1% to 100% of the critical damping is carried out, the results of which are presented in Figures 5 and 6. With reference to Figure 5a, for the STMD-equipped system, at higher damping ratios, the acceleration response of the structural mass reduces and the distance between the side lobes increases. On the contrary, for the SHMD-equipped system (Figure 5b), it can be observed that at low damping ratios, it has a performance inferior to its STMD-equipped counterpart. Nevertheless, as the damping ratios increase beyond the value of 0.3, the performance of the SHMD-equipped system drastically improves, reducing the acceleration response while at the same time pushing the side-lobes of the response further apart. The comparison of the two systems as a function of the damping ratio is shown in Figure 5c which presents the difference of the acceleration responses of the two systems (i.e., $\ddot{x}_{shmd}(\omega) - \ddot{x}_{stmd}(\omega)$). Owing to the selected sign convention, negative values in Figure 5c indicate performance gains of the SHMD over the STMD system, while positive values indicate performance loss. For clarity, the two-dimensional acceleration response of the STMD and SHMD controlled systems for maximum damping ratios of 0.3, 0.5, 0.75 and 1 is presented in Figure 6. The average response of the systems over the wider range of frequencies is captured by the area under the response curves as illustrated in the same figure.

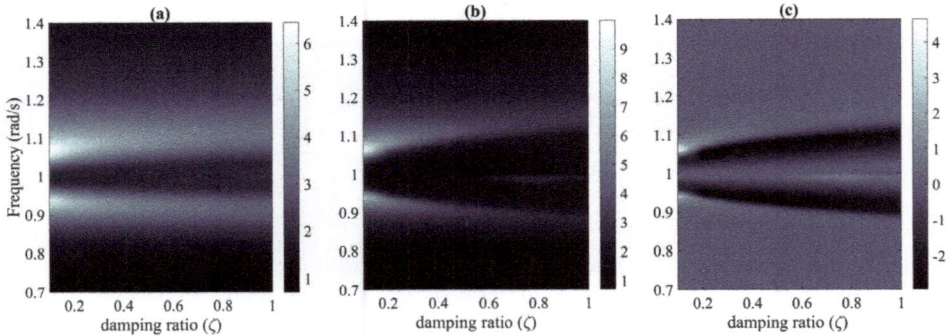

Figure 5. Acceleration response of (**a**) STMD; and (**b**) SHMD; and their (**c**) difference at different damping ratios. (Units of acceleration response in $m^2/s^3/rad$).

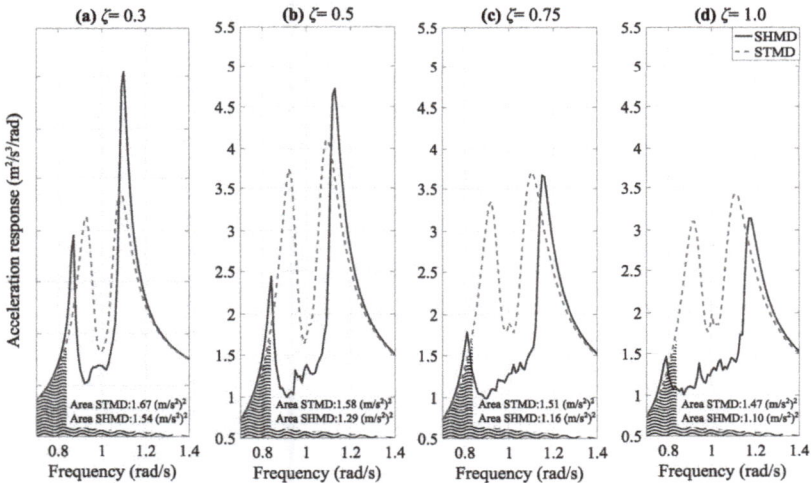

Figure 6. Acceleration response at steady state for damping ratios of (**a**) 0.3; (**b**) 0.5; (**c**) 0.75; and (**d**) 1. Shaded area illustrates average system response.

For the case of the SDOF, the performance of the novel hybrid configuration is investigated at the average maximum damping ratio of 0.5. Similar to the passive (TMD) optimisation procedure, when practical constraints are applied such as force and stroke saturation limits (for the case of the MDOF system), further numerical optimisation is carried out and appropriate values of maximum damping coefficients are selected. For the fairness of the comparison, the SDOF the STMD configuration is also designed with a maximum damping ratio of 0.5.

4.3. Free Vibration Analysis

For the first set of simulations, the SDOF is given an arbitrary initial displacement of 10 cm and is allowed to vibrate freely. Figure 7a,b illustrate the system's displacement along with the active and semi-active forces required by the SHMD system.

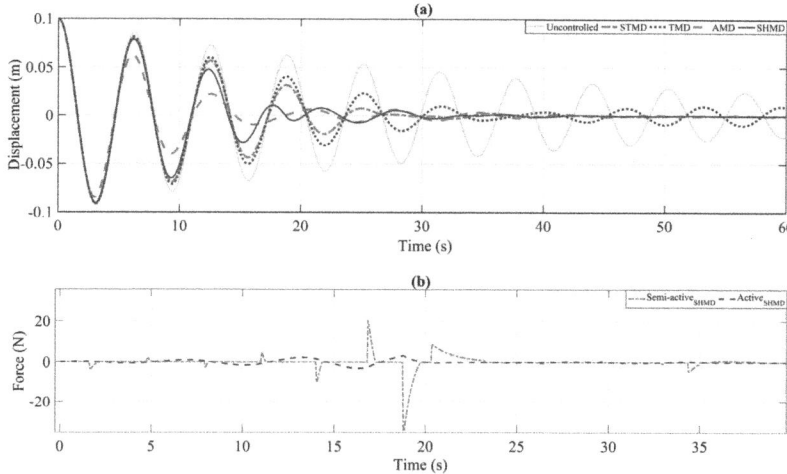

Figure 7. (a) Displacement response time history of different control configurations; (b) control signal of active component and semi-active component of the hybrid configuration.

Clearly, the rate of decay of the system's response is a good primary indication of its effective damping. In this regard, it is shown that at the absence of a DVA, the low damped structure requires a much longer settling time. On the other hand, once a DVA is employed in the form of TMD, STMD, ATMD and SHMD the settling times drastically decrease, thereby demonstrating the effective damping of each of the five structural configurations. More specifically, out of the four DVA configurations, the SHMD and ATMD seem to be superior to their purely passive and semi-active counterparts. As a matter of fact, it is evident that the system coupled with an SHMD device follows closely the trajectory of the AMD-equipped, one particularly at the late part of the vibration response.

4.4. Forced Vibration

Systems equipped with devices such as STMD (and also SHMD) are no longer linear but piecewise linear. For many non-linear systems, the response magnification factor may depend on the type and magnitude of the excitation and the resulting structural response might be of random non-periodic nature. Yet, following the proof of Hac and Youn [30,31], the response of piecewise linear second-order systems to periodic excitation is also periodic, and the amplitude ratio is independent of the excitation amplitude. In other words, exciting the structure using a periodic wave of notional amplitude allows for meaningful performance information in the frequency domain. Figure 8 exhibits the time history response of the structural configurations under harmonic excitation with frequency equal to the structural frequency. For clarity, only the response time histories for the cases of STMD, ATMD and SHMD and TMD are presented. Complementing these results, Figure 9 illustrates the continuous (running) displacement rms for each of the different structural configurations.

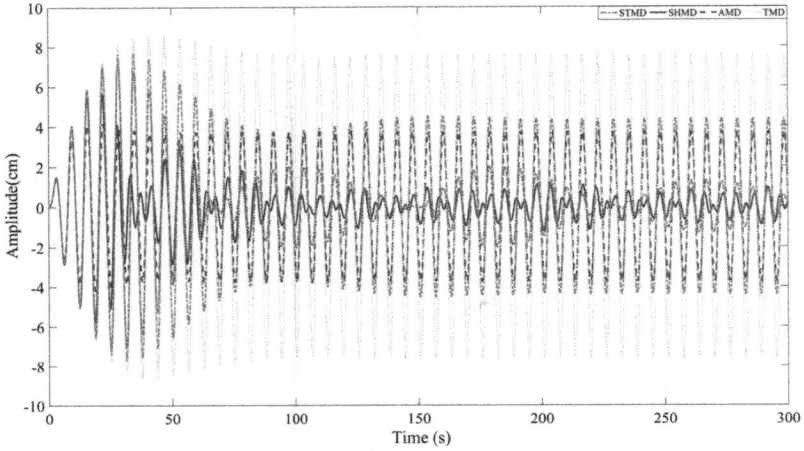

Figure 8. Transient and steady-state response of the different control device configuration under harmonic loading at tuning frequency (excitation frequency 1 rad/s).

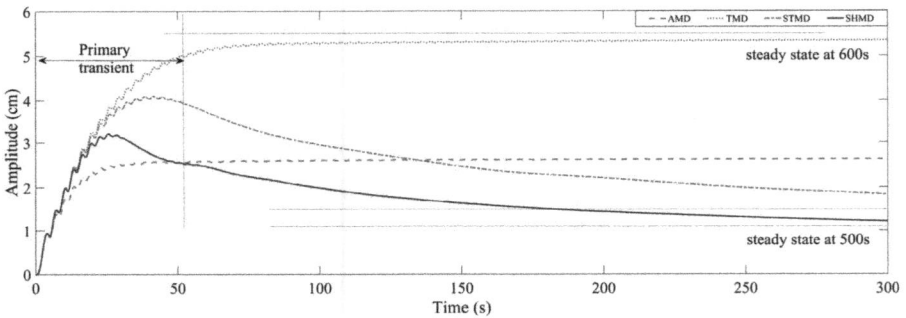

Figure 9. Transient and steady-state Crms response of the different control device configuration under harmonic loading at tuning frequency (excitation frequency 1 rad/s).

It is evident that under resonant forced vibration, the ATMD, STMD and SHMD clearly outperform the more conventional passive TMD under both the transient and steady-state parts of the vibration. Additionally, under the transient component of the vibration, the ATMD and SHMD devices are superior to the STMD. On the other hand, under steady state, the STMD is shown to be significantly better than the ATMD configuration, achieving steady-state response closer to the system equipped with the novel SHMD configuration. Similar remarks can be made after investigating the steady-state and peak frequency response functions (the response of the system at different frequencies, shown here as the ratio of the frequency of the external perturbation, F_e and the natural frequency, F_n of the structure) of the systems. Figure 10a,b illustrate that the novel device configuration achieves the best compromise between steady-state and transient performance.

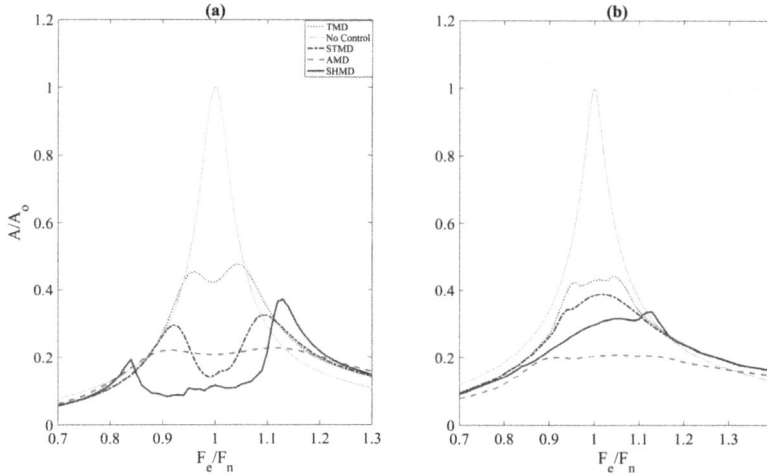

Figure 10. (a) Steady-state; and (b) Peak frequency acceleration response of the different structural configurations.

4.5. High-Rise Structural Configuration

In order to establish the robustness of the novel device and its ability to reduce wind vibration response, it is important to evaluate its performance on realistic high-order systems for which constraints such as actuator force-power demands and damper strokes can be taken into account. To achieve this, the 76-storey benchmark wind-sensitive sway structure proposed by Yang et al. [3] is used in this study. The building has a square 42 m × 42 m cross-section, with a height to width aspect ratio of 7.3 and a low natural frequency that lends it the wind sensitivity attribute. A simplified planar finite element model of the structure is constructed by considering the portion of the building between two adjacent rigid floors as a classical beam element of uniform thickness, leading to 76 rotational and 76 translational degrees of freedom. From these, all the rotational degrees of freedom have been removed using static condensation, leading to a lumped mass sway model with degrees of freedom, representing the displacement of each floor in the lateral direction. The resulting simulated structure has a total mass of 153,000 tons, with the first five frequencies at 0.16, 0.765, 1.992, 3.790 and 6.395 Hz, and corresponding modal structural damping ratios of 1% calculated using Rayleigh's approach. In this study, four alternatives, namely: passive (TMD), semi-active (STMD), active (ATMD) and semi-active hybrid (SHMD) controlled structures are used for the establishment of the comparison metrics. The assemblage of the different control configurations is depicted in Figure 11.

In every control configuration, the dynamic absorber comprises an inertial mass of 500 tons that corresponds to 0.356% of the total structural mass, limited by realistic structural design constraints. For DVA configurations that require tuning of the device (i.e., TMD, SHMD and STMD), appropriate spring stiffness, k_p, is chosen such that the device is tuned to the fundamental frequency of the structure (i.e., ≈ 0.16 Hz).

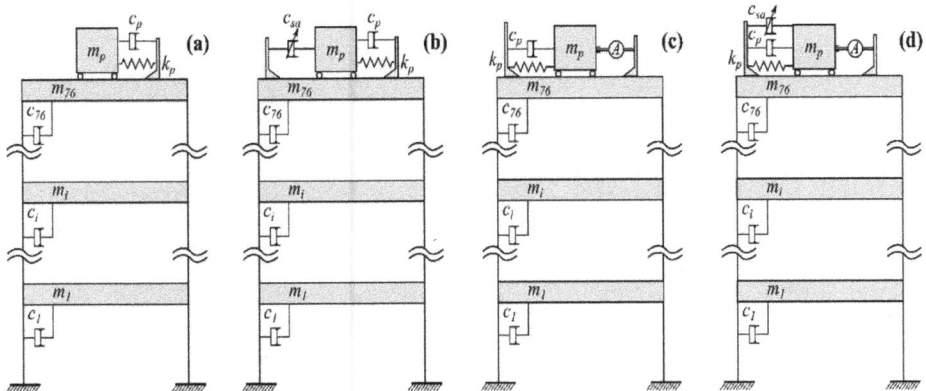

Figure 11. Ensemble of all the different control options (**a**) TMD; (**b**) STMD; (**c**) ATMD; (**d**) SHMD studied herein for the model 76-storey structure of Yang et al. [3].

In order to ensure the fairness of the comparison, it was deemed necessary to restrain the maximum damper stroke of each of the alternatives by increasing the damping coefficient of the device appropriately so as to limit strokes to a maximum of 95 cm. Because control configurations that damper strokes are not a cause of concern, such as the case of the TMD, the damping ratio is numerically optimised (and kept low, approximately to the value calculated using Den Hartog's equations [19]) for maximum rms acceleration response reduction. The resulting damping coefficients that equalise the maximum strokes at maximum rms acceleration response reduction are outlined in Table 1 below:

Table 1. Damping coefficients. For clarity i) TMD, ii) STMD, iii) ATMD, iv) SHMD stand for i) tuned, ii) semi-active tuned, iii) active-tuned and iv) semi-active hybrid mass damper.

Control Strategy	Max Damping Coefficient (kNs/m)	Min Damping Coefficient (kNs/m)	Equivalent Damping Ratio
TMD	47	47	4.7%
STMD	163.4	2.61	16%–2.6%
ATMD	100	100	10%
SHMD	168	39	16.8%–4%

4.6. Evaluation Criteria

The comparison of the different control strategies is based on the stationary response properties of the different control structures. From the response time histories, the rms and peak accelerations and displacements at different storeys were obtained. From the obtained values, 12 performance criteria were identified. The first criterion, J_1, appraises the ability of the control strategy to reduce rms accelerations:

$$J_1 = \max(\sigma_{\ddot{x}1}, \sigma_{\ddot{x}30}, \sigma_{\ddot{x}50}, \sigma_{\ddot{x}55}, \sigma_{\ddot{x}60}, \sigma_{\ddot{x}65}, \sigma_{\ddot{x}70}, \sigma_{\ddot{x}75}) / \sigma_{\ddot{x}75o} \qquad (14)$$

where $\sigma_{\ddot{x}i}$ is the rms acceleration of the ith storey and $\sigma_{\ddot{x}75o}$ is the rms acceleration of the 75th floor (last occupied floor) without control. The second performance criterion evaluates the average performance of six floors above the 49th floor:

$$J_2 = \frac{1}{6}\sum (\sigma_{\ddot{x}i} / \sigma_{\ddot{x}io}) \qquad (15)$$

For $i = 50, 55, 60, 65, 70, 75$; where, $\sigma_{\ddot{x}io}$ is the rms of the ith floor without control. The third and fourth performance indices assess the ability of the control system to reduce top floor displacements:

$$J_3 = \sigma_{x76}/\sigma_{x76o} \tag{16}$$

$$J_4 = \frac{1}{7}\sum_i (\sigma_{xi}/\sigma_{xio}) \tag{17}$$

For $i = 50, 55, 60, 65, 70, 75, 76$; where, σ_{xi} is the rms displacement of the ith floor, σ_{xio} is the rms displacement of the ith storey without control and σ_{x76o} is 10.136 cm. The fifth and sixth indices take into account the rms stroke of the damper (i.e., $i = 77$) and the average power respectively:

$$J_5 = \sigma_{x77}/\sigma_{x76o} \tag{18}$$

$$J_6 = \left\{ \frac{1}{T}\int_0^T [\dot{x}_{77}(t)u(t)]^2 dt \right\}^{1/2} \tag{19}$$

In which, σ_{x77} is the rms stroke of the damper, $\dot{x}_{77}(t)$ is the damper velocity and T is the total time of integration. Similarly to the first performance indices, the next four criteria (i.e., J_7 to J_{10}) evaluate the performance in terms of peak response quantities:

$$J_7 = \max(\ddot{x}_{p1}, \ddot{x}_{p30}, \ddot{x}_{p50}, \ddot{x}_{p55}, \ddot{x}_{p60}, \ddot{x}_{p65}, \ddot{x}_{p70}, \ddot{x}_{p75})/\ddot{x}_{p75o} \tag{20}$$

$$J_8 = \frac{1}{6}\sum_i (\ddot{x}_{pi}/\ddot{x}_{pio}) \tag{21}$$

For $i = 50, 55, 60, 65, 70, 75$;

$$J_9 = x_{p76}/x_{p76o} \tag{22}$$

$$J_{10} = \frac{1}{7}\sum_i (x_{pi}/x_{pio}) \tag{23}$$

For $i = 50, 55, 60, 65, 70, 75, 76$; where, $\ddot{x}_{pi}$ is the peak acceleration of the ith floor with control and $\ddot{x}_{pio}$ is the peak acceleration of the ith floor without control. Similarly, x_{pi} is the peak displacement of the ith floor and x_{pio} is the peak displacement of the ith floor without control and $x_{p76o} = 32.3$ cm. The 11th criterion assesses the ability of the control strategy to minimise the stroke of the damper:

$$J_{11} = x_{p77}/x_{p76o} \tag{24}$$

In which x_{p77} is the peak stroke of the actuator. The last criterion examines the control effort by calculating the maximum required power by:

$$J_{12} = \max |\dot{x}_{77}(t)u(t)| \tag{25}$$

From the above-defined criteria, it can be observed that the better the performance, the smaller the performance indices $J_1, J_2, .., J_{12}$ [3].

5. Simulation Results and Discussion

Four structural configurations consisting of passive, semi-active, hybrid active and semi-active hybrid control devices were considered for investigating the efficacy of the SHMD device for the vibration control of high-rise structures. Figure 12 summarises the peak and rms (displacement and acceleration) responses at every floor. The results of the evaluation for the different performance criteria $J_1, J_2, .., J_{12}$ are presented in Figure 13.

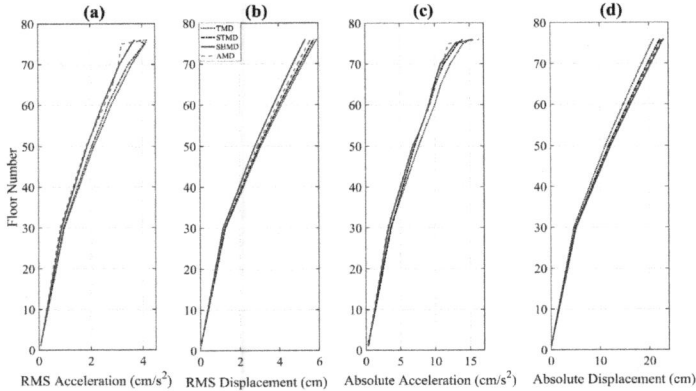

Figure 12. Illustration of the performance of different control measures in terms of (**a**) RMS acceleration; (**b**) RMS displacement; (**c**) absolute acceleration; and (**d**) absolute displacement at different floor levels.

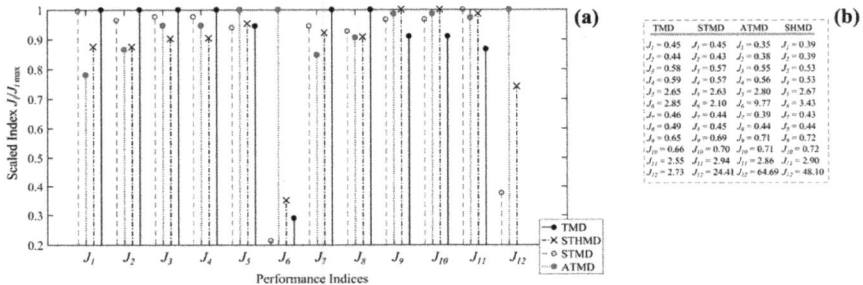

Figure 13. (**a**) Normalised; and (**b**) non-normalised performance indices (lower index indicates better performance).

The results indicate that, for approximately the same damper strokes, the SHMD-equipped structure is able to achieve similar performance as the ATMD-equipped one, while clearly outperforming the passive and semi-actively controlled alternative. With reference to Figures 13 and 14, it is evident that the SHMD device requires much less energy and actuation demands for achieving the aforementioned performance increase. As a matter of fact, the SHMD device requires approximately 26% of the total energy required by the ATMD device (1245 kJ compared to 4863 kJ). This is due to the large control effort and consequently the large amount of energy required to be added by the active actuators (approximately 4125 kJ or 82% of the total required active energy) in order to effectively accelerate the mass so that sufficient control force is provided in order to overcome the "braking" force acted by the passive component of the ATMD. Conversely, in the SHMD configuration, while the actuators are accelerating the mass, the semi-active damping component attains its minimum value, minimising the "braking" force needed to be counteracted by the actuators, thus requiring a lower control power (Figure 14b top). The energy required to be added in the SHMD configured structure is only 1245 kJ compared to 4125 kJ (which accounts for the 82% of the total energy required) (Figure 14). On the other hand, for energy dissipation purposes (Figure 14a,b bottom), the ATMD configuration is required to supply only a fraction (737 kJ and the remaining 18% of the total energy) of energy, while the SHMD-equipped structure requires consumption of a staggering 4600 kJ. However, since energy dissipation in the SHMD configuration occurs exclusively in the semi-active elements, the required energy depends solely on the selected semi-active device. Still, regardless of the device, the energy required for semi-active control is not expected to exceed the order of a few watts [32].

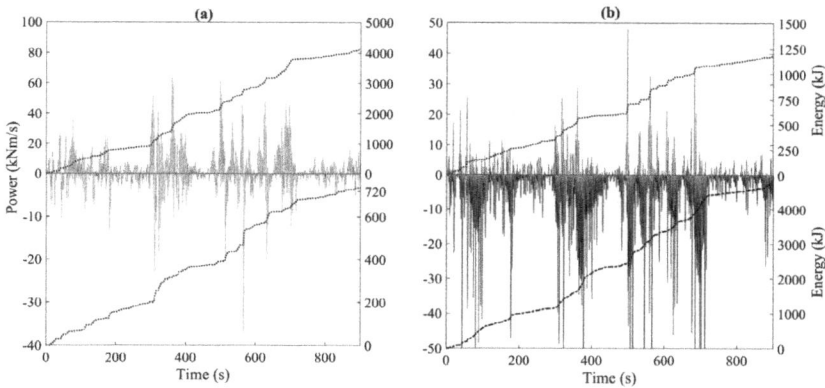

Figure 14. Power and its time integral (dotted line) energy for (**a**) ATMD; and (**b**) SHMD configuration. Positive stands for energy addition and negative for energy dissipation.

For more tolerant damper stroke limits, a lower passive damping ratio can be chosen for the ATMD which will reasonably lower the actuation demands for energy addition. On the contrary, lower damping ratios of the damping device will require the actuators to work harder in dissipating energy by decelerating the mass (and essentially work as an energy-expensive passive damper). The aforementioned arguments are illustrated in Figure 15, in which the power required by a purely active AMD device (i.e., absence of passive damping component) is investigated. As can be observed, the AMD is required to expend most of its energy for dissipation (4500 kJ as opposed to the 720 kJ required by the ATMD counterpart), while only a small fraction of that energy is required for energy addition (approximately 1100 kJ). It should be clarified that no further comparisons can be made with the purely active AMD system, as its performance is theoretically uncapped (the larger the control effort, the lower the response).

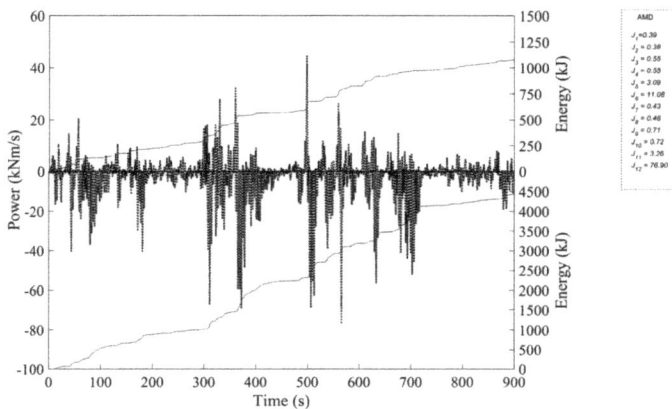

Figure 15. Power and its time integral (dotted line) energy of a purely active mass damper (AMD) system (no passive damping component) along with the corresponding performance indices. Positive stands for energy addition and negative for energy dissipation.

6. Conclusions

In this study, a novel hybrid control device configuration termed semi-hybrid mass damper (SHMD) has been proposed as an alternative design to the traditional hybrid active-tuned mass damper (ATMD) for vibration suppression of dynamic structural systems. The fundamental novelty of this configuration is that it enables modulation of the instantaneous effective system damping via the successive and appropriate action of active and semi-active elements. For this case, the active components of the SHMD device are regulated by an optimal Linear-Quadratic-Regulator (LQR) controller, while the semi-active components are controlled via a direct output feedback displacement based groundhook (DBG) controller. A numerical step-by-step procedure for the calculation of the control actions and the coupling of the devices has been proposed in this paper. Under vibration analyses run on both single degree of freedom (SDOF) and multi-degree of freedom (MDOF) SHMD configured structures, it is shown that the device is effective in reducing both the steady-state, as well as the peak frequency responses of the structural system, achieving similar performance gains to that of an ATMD-equipped structure. However, its achievement is not only the use of this novel hybrid mass damper configuration as a vehicle for enhancing vibration attenuation performance or providing a fail-safe mechanism, it is also shown that the successive action of active and semi-active elements allows an improvement in efficiency both in terms of power and actuation demands. By providing a feasible, reliable, effective and efficient alternative structural control approach, this novel hybrid configuration allows the concept of active control of structures to be extended to one of "active" structures for which both active and semi-active components are integrated and simultaneously optimised to produce a new breed of slenderer, longer and taller structures and structural forms.

Acknowledgments: The authors gratefully acknowledge EPSRC UK (Grant No. EP/L504993/1) and the University of Leeds for the financial support to this study.

Author Contributions: This work constitutes part of the doctoral dissertation of the first author who prepared the manuscript and practiced all the analytical work. The second author supervised, reviewed and edited the work where appropriate.

Conflicts of Interest: The authors declare no conflict of interest.

References

1. Holmes, D.J. *Wind Loading of Structures*; CRC Press: Sound Parkway NW, FL, USA, 2007.
2. Ricciardelli, F.; Pizzimenti, A.D.; Mattei, M. Passive and active mass damper control of the response of tall buildings to wind gustiness. *Eng. Struct.* **2003**, *25*, 1199–1209. [CrossRef]
3. Yang, N.Y.; Agrawal, A.K.; Samali, B.; Wu, J.C. Benchmark Problem for Response Control of Wind-Excited Tall Buildings. *J. Eng. Mech.* **2004**, *130*, 437–446. [CrossRef]
4. Zhang, Y.; Li, L.; Cheng, B.; Zhang, X. An active mass damper using rotating actuator for structural vibration control. *Adv. Mech. Eng.* **2016**, *8*. [CrossRef]
5. Haertling, G. Rainbow ceramics: A new type of ultra-high displacement actuator. *Am. Ceram. Soc. Bull.* **1994**, *73*, 93–96.
6. Scruggs, J.; Iwan, W. Control of a civil structure using an electric machine with semiactive capability. *J. Struct. Eng.* **2003**, *129*, 951–959. [CrossRef]
7. Zhang, C.; Ou, J. Modeling and dynamical performance of the electromagnetic mass driver system for structural vibration. *Eng. Struct.* **2015**, *82*, 93–103. [CrossRef]
8. Ikeda, Y. Active and semi-active vibration control of buildings in Japan—Practical applications and verification. *Struct. Control Health Monit.* **2009**, *16*, 703–723. [CrossRef]
9. Fujinami, T.; Saito, Y.; Masayuki, M.; Koike, Y.; Tanida, K. A hybrid mass damper system controlled by Hinfity control theory for reducing bending-torsion vibration of an actual building. *Earthq. Eng. Struct. Dyn.* **2001**, *30*, 1639–1643. [CrossRef]
10. Nakamura, Y.; Tanaka, K.; Nakayama, M.; Fujita, T. Hybrid mass dampers using two types of electric servomotors:AC servomotors and linear-induction servomotors. *Earthq. Eng. Struct. Dyn.* **2001**, *30*, 1719–1743. [CrossRef]

11. Watakabe, M.; Tohdp, M.; Chiba, O.; Izumi, N.; Ebisawa, H.; Fujita, T. Response control performance of a hybrid mass damper applied to a tall building. *Earthq. Eng. Struct. Dyn.* **2001**, *30*, 1655–1676. [CrossRef]

12. Mitchel, R.; Kim, Y.; El-Khorchi, T. Wavelet neuro-fuzzy control of hybrid building-active tuned mass damper system under seismic excitations. *J. Vib. Control* **2012**. [CrossRef]

13. Tan, P.; Liu, Y.; Zhou, F.; Teng, J. Hybrid Mass Dampers for Canton Tower. *CTBUH J.* **2012**, 24–29.

14. Li, C.; Cao, B. Hybrid active tuned mass dampers for structures under the ground acceleration. *Struct. Control Health Monit.* **2015**, *22*, 757–773. [CrossRef]

15. Tso, M.H.; Yuan, J.; Wong, W.O. Hybrid vibration absorber with detached design for global vibration control. *J. Vib. Control.* **2016**. [CrossRef]

16. Khan, I.U.; Wagg, D.; Sims, N.D. Improving the vibration suppression capabilities of a magneto-rheological damper using hybrid active and semi-active control. *Smart Mater. Struct.* **2016**, *25*, 085045. [CrossRef]

17. Demetriou, D.; Nikitas, N.; Tsavdaridis, K.D. A Novel Hybrid Semi-active Tuned Mass Damper for Lightweight Steel Structural Applications. In Proceedings of the IJSSD Symposium on Progress in Structural Stability and Dynamics, Lisbon, Portugal, 21–24 July 2015.

18. Ricciardelli, F.; Occhiuzzi, A.; Clemente, P. Semi-active tuned mass damper control strategy for wind-excited structures. *J. Wind Eng. Ind. Aerodyn.* **2000**, *88*, 57–74. [CrossRef]

19. Hartog, D. *Mechanical Vibrations*; McGraw-Hill Book Company: New York, NY, USA, 1956.

20. Ghosh, A.; Basu, B. A closed-form optimal tuning criterion for TMD in damped structures. *Struct. Control Health Monit.* **2007**, *14*, 681–692. [CrossRef]

21. Demetriou, D.; Nikitas, N.; Tsavdaridis, K.D. Performance of fixed-parameter control algorithms on high-rise structures equipped with semi-active tuned mass dampers. *Struct. Des. Tall Spec. Build.* **2016**, *25*, 340–354. [CrossRef]

22. Nelder, J.; Mead, R. A simplex method for function minimization. *Comput. J.* **1965**, *7*, 308–313. [CrossRef]

23. Hrovat, D.; Barak, P.; Rabins, M. Semi-Active Versus Passive or Active Tuned Mass Dampers for Structural Control. *J. Eng. Mech.* **1983**, *109*, 691–705. [CrossRef]

24. Yang, N.J.; Akbarpour, A.; Ghaemmaghami, P. New Optimal Control Algorithms for Structural Control. *J. Eng. Mech.* **1987**, *113*, 1369–1386. [CrossRef]

25. Soong, T.T. *Active Structural Control: Theory and Practice*; John Wiley & Sons, Inc.: New York, NY, USA, 1990.

26. Cheng, F.Y.; Jiang, H.; Lou, K. *Smart Structures: Innovative Systems for Seismic Response Control*; Taylor & Francis Group: New York, NY, USA, 2008.

27. Viet, L.D.; Nghi, N.B.; Hieu, N.N.; Hung, D.T.; Linh, N.N.; Hung, L.X. On a combination of ground-hook controllers for semi-active tuned mass dampers. *J. Mech. Sci. Technol.* **2014**, *28*, 2059–2064. [CrossRef]

28. Koo, J.H. *Using Magneto-Rheological Dampers in Semiactive Tuned Vibration Absorbers to Control Structural Vibrations*; Virginia Polytechnic Institute and State University: Blacksburg, VA, USA, 2003.

29. Koo, J.H.; Murray, T.M.; Setareh, M. In search of suitable control methods for semi-active tuned vibration absorbers. *J. Vib. Control* **2004**, *10*, 163–174. [CrossRef]

30. Hac, A.; Youn, I. Optimal semi-active suspension with preview based on a quarter car model. *J. Acoust. Vib.* **1992**, *114*, 84–92. [CrossRef]

31. Pinkaew, T.; Fujino, Y. Effectiveness of semi-active tuned mass dampers under harmonic excitation. *Eng. Struct.* **2001**, *23*, 850–856. [CrossRef]

32. Nagarajaiah, S.; Varadarajan, N. Short time Fourier transform algorithm for wind response control of buildings with variable stiffness TMD. *Eng. Struct.* **2005**, *27*, 431–441. [CrossRef]

MDPI AG

St. Alban-Anlage 66

4052 Basel, Switzerland

Tel. +41 61 683 77 34

Fax +41 61 302 89 18

http://www.mdpi.com

Applied Sciences Editorial Office

E-mail: applisci@mdpi.com

http://www.mdpi.com/journal/applisci

www.ingramcontent.com/pod-product-compliance
Lightning Source LLC
Chambersburg PA
CBHW051725210326
41597CB00032B/5607